DEVELOPMENTS IN AGRICULTURAL AND MANAGED-FOREST ECOLOGY, 7

management of semi-arid ecosystems

OTHER TITLES IN THIS SERIES

1. Plant Modification for More Efficient Water Use
   by J.F. Stone (Editor)
   1975 xii + 320 pp.

2. Tree Ecology and Preservation
   by A. Bernatzky
   1978 viii + 358 pp.

3. Cycling of Mineral Nutrients in Agricultural Ecosystems
   by M.J. Frissel (Editor)
   1978 viii + 356 pp.

4. Interactions Between Non-Pathogenic Soil Microorganisms and Plants
   by Y.R. Dommergues and S.V. Krupa (Editors)
   1978 xii + 476 pp.

5. Ecology of Root Pathogens
   by S.V. Krupa and Y.R. Dommergues (Editors)
   1979 x + 282 pp.

6. Soil Disinfestation
   by D. Mulder (Editor)
   1979 xiv + 368 pp.

DEVELOPMENTS IN AGRICULTURAL
AND MANAGED-FOREST ECOLOGY 7

management of semi-arid ecosystems

Edited by

B.H. WALKER

Department of Botany, University of The Witwatersrand, Johannesburg, South Africa

ELSEVIER SCIENTIFIC PUBLISHING COMPANY
Amsterdam – Oxford – New York 1979

ELSEVIER SCIENTIFIC PUBLISHING COMPANY
335 Jan van Galenstraat
P.O. Box 211, 1000 AE Amsterdam, The Netherlands

Distributors for the United States and Canada:

ELSEVIER NORTH-HOLLAND INC.
52, Vanderbilt Avenue
New York, N.Y., 10017

Library of Congress Cataloging in Publication Data

Main entry under title:

Management of semi-arid ecosystems.

(Developments in agricultural and managed-forest ecology ; 7)
Bibliography: p.
Includes index.
1. Land use--Case studies. 2. Arid regions--Case studies. I. Walker, B. H. II. Series.
HD156.M328 333.7'3 79-4395
ISBN 0-444-41759-1

ISBN 0-444-41759-1 (Vol. 7)
ISBN 0-444-41515-7 (Series)

Printed in The Netherlands

CONTENTS

ACKNOWLEDGEMENTS

This book was conceived during the First International Congress of Ecology at The Hague in 1974. Its development has been gradual, and sometimes fitful, but this has been necessary to ensure some degree of cohesion between the different chapters.

I have been most fortunate in the choice of my co-authors, as all of them have co-operated willingly in the sometimes extensive changes that we eventually made. Editing the book has proved for me to be a tremendous education in the ecological processes and management of the semi-arid world, and for this I am extremely grateful to all of them.

Many people helped me in the preparation of the book, but I mention especially my assistant, Mrs. Joyce Clark, who typed and re-typed almost uncomplainingly, and did so much to put it all in order.

B.H. WALKER
August 1978

Section I

INTRODUCTION

B.H. WALKER

INTRODUCTION

B.H. WALKER

Much of the earth's surface experiences a climate which allows for the development of a more or less continuous vegetative cover, but which is too dry and variable to permit regular, successful, annual dry-land cultivation of cereal or other crops. These are the semi-arid regions of the world where land-use is primarily based on utilizing the natural vegetation, although supplementary, irrigated cultivation may be important. Subsistence-type crop production is often attempted, but frequently fails.

The exact extent of these regions differs according to the criteria used to define "semi-arid". In this book they are not defined by particular levels of rainfall or evapotranspiration (such as given by Meigs, 1953). The definition used is that given in the first sentence.

It is nevertheless useful to have some idea of the extent and distribution of the semi-arid regions of the world, and to provide such a general picture, Fig. 1 has been compiled from various maps and descriptions of the arid and semi-arid zones. It does not conform to any particular climatic classification, but it represents fairly well those parts of the world where land-use is necessarily based on the natural vegetation. Each of the major semi-aird areas depicted on the map is represented in the chapters which follow.

Characteristically, land-use in the semi-arid regions is "backward", in that little modern technology exists and the management practices are generally simple and old-fashioned. It is clear why this should be so. The production potential, and therefore economic potential, of these ecosystems is very low and it is difficult to justify their development in terms of transport, communications and other services. The potential rewards from research into better land-use is far higher in the higher rainfall regions, and these are consequently the areas where advances have been made.

The problem is compounded by the fact that those people with initiative and high capabilities have been attracted away from the semi-arid regions to the more rewarding employment areas in the high-production regions. The semi-arid ecosystems have, therefore, often been managed by a segment of the population which constitutes the least capable, least innovative group, often disinterested in what they are doing, but not capable of changing their circumstances. This, of course, is not always true, particularly where land-use

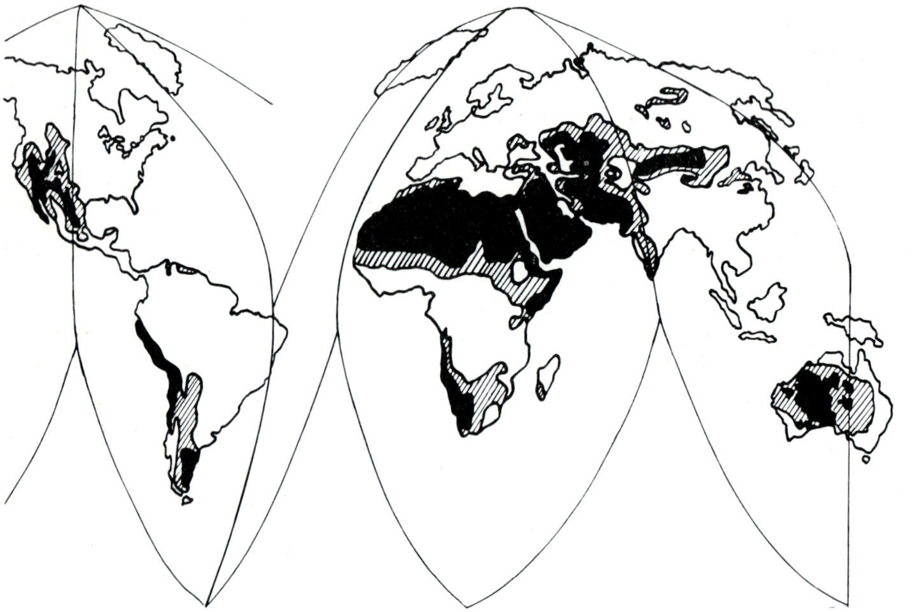

Fig. 1. Arid (■) and semi-arid (▨) regions of the world as defined in the text.

is in the form of large holdings, run on a cash-economy basis. Nevertheless, it is a general anomaly that these fragile ecosystems, which require higher-than-average management, are subjected to indifferent, or even mis-management.

The seriousness of ecological conditions in the semi-arid regions has increased proportionately with the world-wide population increase. More and more people are being forced to make a living in them, and the pace of degradation has increased. Desertification as described by UNESCO and the United Nations Environment Programme (Warren and Maizels, 1976) is increasing. The semi-arid ecosystems constitute an ecotone between desert and the cropping regions of the world. In the latter, bad management certainly occurs and spectactular soil erosion and vegetation damage have resulted. But because of their high potential for plant growth and the much faster rate of plant succession they usually manage to recover. In the semi-arid areas, the pattern is more one of a shift in the position of the ecotone towards the higher rainfall areas.

In the arid and semi-arid ecosystems, should the vegetative cover be damaged, the percentage of bare soil surface increases, plant growth is reduced and a downward spiral begins, which is very difficult to halt. The details will be described in the forthcoming chapters. Owing to the increased pressure on the semi-arid lands, the demand for better (i.e. more efficient, sustainable) management has led to a number of recent developments. They do not necessarily involve large inputs of technology or capital, but could constitute important advances.

It is the aim of this book to examine various semi-arid ecosystems from around the world. It seeks to identify common features which characterize them as semi-arid in terms of both structure and function, then to examine the different types of land-use to which they are currently being subjected. Each will be analyzed with respect to its problems and the management principles which are considered to promote sustained, optimum use. Finally, these principles will be compared in the hope that we can identify common principles for land-use in semi-arid ecosystems.

It is not possible to include an account of each of the semi-arid areas in the world. Those included in the following chapters are representative of the range of types which occur. There will undoubtedly be criticism of the choice and, more particularly, of some of the omissions, but it is suggested that the areas examined allow for an appraisal of semi-arid ecosystems in terms of the objectives of the book, without rendering it too long or repetitive.

REFERENCES

Meigs, P., 1953. World distribution of arid and semi-arid homoclimates. In: Review of Research on Arid Zone Hydrology. Arid Zone Programme, Vol. I. UNESCO, Paris.

Warren, A. and Maizels, I.K., 1976. Ecological Change and Desertification. University College London. Report to Desertification Secretariat, UNEP.

Section II

ECOLOGY AND MANAGEMENT OF THE WORLD'S SEMI-ARID REGIONS

Chapter 1

CATTLE RANCHING IN THE SEMI-ARID SAVANNAS OF EAST AND SOUTHERN AFRICA

D.L. BARNES

Division of Livestock and Pastures, Department of Research and Specialist Services, Salisbury (Rhodesia)

INTRODUCTION

The term "semi-arid savanna" is imprecise and subject to many interpretations, especially when applied over a wide area. For the present purpose it is taken to include a range of physiognomic vegetation types of the tropical and sub-tropical summer rainfall zone of Africa, which, when not degraded by over-use, have a closed herbaceous cover, mainly of grasses, and a scattered to fairly dense population of deciduous trees or shrubs, or a mixture of both. In the terminology of Rattray (1961) it is taken to include woodland savanna, tree savanna, bush savanna, tree—bush savanna and bush—clump savanna, or, as defined by Pratt et al. (1966), woodland, wooded grassland, bushed grassland and wooded and bushed grassland.

Semi-arid savanna vegetation is primarily a reflection of moisture supply, which in turn is influenced by rainfall and local edaphic and topographic conditions, but may reflect also the intensity and frequency of fires. In broad terms it occurs in Africa in areas with a mean annual rainfall of between 350 and 700 mm, which falls during the warmer part of the year, and in which there is a distinct dry season. Because of the low and unreliable rainfall, stable crop production is not possible.

The estimated extent of the semi-arid savanna in Africa, compiled primarily from reference to vegetation and rainfall maps, is shown in Fig. 1.1.

THE ENVIRONMENT

Climate

Using data selected from those presented by Jackson (1961) the pattern of distribution of the mean annual rainfall, by months, is shown in Fig. 1.1, for various sites within the semi-arid savanna.

In the central part of the East African region near the equator, the rainfall pattern is bimodal, with the peaks in the periods November to Decem-

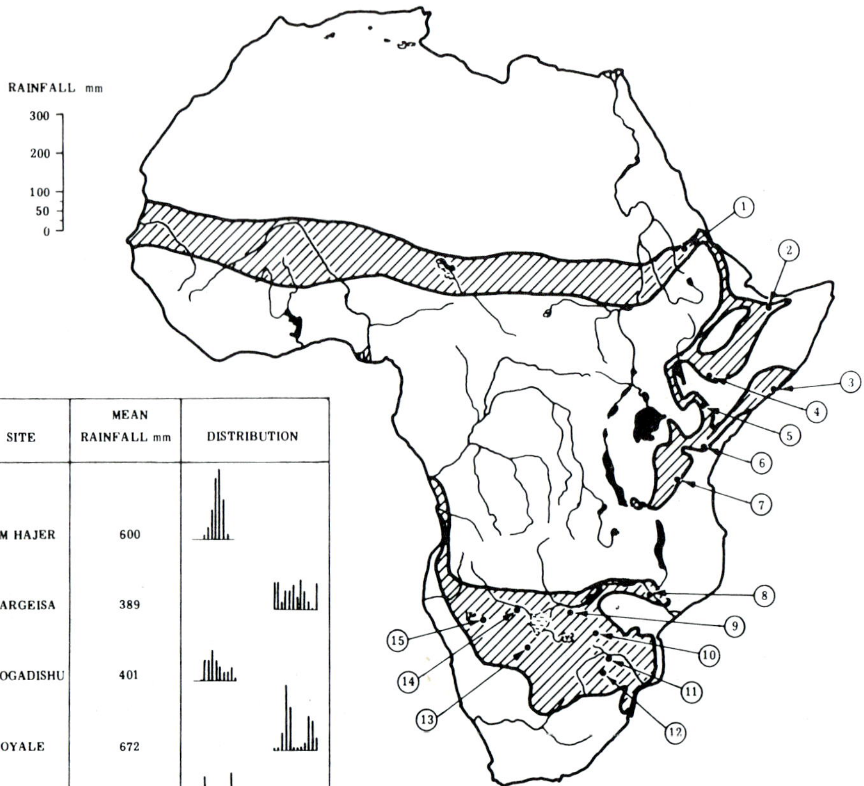

No	SITE	MEAN RAINFALL mm	DISTRIBUTION
1	OM HAJER	600	
2	HARGEISA	389	
3	MOGADISHU	401	
4	MOYALE	672	
5	ISIOLO	564	
6	VOI	536	
7	DODOMA	563	
8	TETE	530	
9	WANKIE	592	
10	BULAWAYO	594	
11	MESSINA	340	
12	PIETERSBURG	508	
13	GHANZI	472	
14	ANDARA	571	
15	TSUMEB	524	

Fig. 1.1. The estimated extent of semi-arid savanna in Africa and monthly distribution of mean annual rainfall from January to December for selected sites in East and Southern Africa.

ber, and March to May, inclusive. Little or no rain falls during the months June to October. Towards the north in East Africa the peaks occur slightly later, and near the coast the rainfall is more evenly distributed throughout the year. To the north and west in the Sudan the rainfall pattern is unimodal with rain falling between June and September; the remainder of the year is dry. To the south in Tanzania the pattern is also unimodal with the distinct wet and dry seasons from December to April and May to November, respectively. A consequence of these differences in pattern is that a given average rainfall does not have the same effectiveness throughout the region and, other things being equal, does not give rise to the same type of vegetation (Heady, 1960).

In the south-central and south-western regions of Africa the rainfall pattern is sharply unimodal. Over most of the area a short rainy season, extending from about December to March, alternates with a virtually rainless period of some 7 to 8 months duration. However, in the southern extent of the region in the Republic of South Africa the wet and dry seasons are less distinct and the dry season is shorter. Accordingly, with a given mean annual rainfall, the vegetation tends to be more mesophytic than that which has developed under the same rainfall further north.

Throughout the semi-arid savanna the distribution of rainfall both within and between years is exceedingly variable. Moreover, as rainfall in the semi-arid regions is frequently induced by convectional movement, rather than convergence of air masses, rainfall is patchy within a given area.

Table 1.1 illustrates the extent of variability in mean annual rainfall at the various sites in East and Southern Africa. Rainfall within seasons is likewise extremely variable.

TABLE 1.1

The variability of mean annual rainfall at various sites in the semi-arid savanna

Location	Years of record	Annual rainfall (mm)			Percentage of years with rainfall below		
		Mean	Least	Most	250 mm	500 mm	750 mm
Isiolo [a]	28	572	375	1006	0	32	89
Tsavo [a]	26	362	62	590	19	88	100
Dodoma [a]	25	580	308	1083	0	32	88
Bulawayo	74	585	199	1202	1	35	80
Messina	49	327	110	640	20	92	100
Tsumeb	64	528	271	1005	0	50	86
Pietersburg	70	486	251	810	0	57	97

[a] From Heady (1960).
The data for East Africa are for the period January to December, those for Southern Africa are for July to June.

Where adequate records are available, the probability that rainfall in any one season or period within seasons will fall below or within selected limits can, of course, be calculated, and Glover and Robinson (1953) have outlined a simple method of doing this. However, prediction of the amount of seasonal rainfall is not possible.

Falling as it does in the equatorial belt, temperature changes throughout the year in East Africa are relatively small and the climate in the semi-arid savanna is generally warm to hot. Differences in mean temperature can largely be attributed to altitude. Over the region as a whole the mean monthly minimum and maximum temperatures vary during the year from about 13 to 40°C. Frost is not a significant feature.

Over most of south-central and south-west Africa the highest temperatures are experienced late in the dry season reaching, at lower altitudes, mean monthly values of about 35°C. Temperatures gradually decrease and are lowest during the period June, July and early August, after which they rise again. At higher altitudes (above 1200 m) frost is common between mid-May and mid-August. Occasional frosts occur at lower altitudes during the same period.

Throughout the semi-arid savanna regions evaporation stress is high, and evaporation from a free water surface is commonly of the order of 2000 to 2500 mm per year and, in the hotter and drier parts of East Africa, may reach 3500 mm or more.

In the semi-arid savanna in general, rainfall and temperature patterns are such that the main limitation to plant growth is water supply, though in Southern Africa temperature may also limit growth in the late growing season. The grasses, which absorb water mainly from the upper layers of soil usually become dormant, or nearly so, early in the dry season, whereas trees and shrubs, which are usually able to draw on subsoil water, remain green for a longer period. Dormancy may also be induced by low temperatures and frost as well as moisture stress. When frost occurs, damage to trees and shrubs is occasionally severe, and part or the whole of the shoot may be killed. The effects are most marked in the low-lying areas and drainage lines. In years when there is a residual supply of water in the subsoil, deep-rooted plants may begin growing in the late dry season before the start of the rains.

A consequence of the variable rainfall in the semi-arid savanna is that productivity, especially of the herbaceous layer, varies radically from year to year and within years. These unpredictable variations in the food supply for livestock are of fundamental importance in relation to stable land-use and will be discussed later.

Vegetation

Floristically the vegetation within the delimitated area of semi-arid savanna in East and Southern Africa is complex, and there are marked differences in vegetation between regions. Published descriptions vary considerably in the detail with which vegetation is described, and in the criteria used to delimit vegetation types.

For convenience three main vegetation complexes are considered, namely, the East African region, the south-central region and the south-western region. Emphasis will be placed on floristic rather than physiognomic aspects.

(1) *The East African region*

Edwards (1951), Phillips (1959), Rattray (1960) and Heady (1960) have published descriptions of East African vegetation and the following account is largely derived from these sources.

In the main the semi-arid savanna in this region is transitional between the mesic highland grassland or forest, or a scattered tree grassland (low tree—high grass) vegetation and a semi-desert vegetation. Where moisture conditions are favourable, as in south-central Kenya and north-central Tanzania, it takes the form of tree savanna characterized by flat-topped *Acacia* trees in an even cover of grass dominated by *Themeda triandra*, but including many other species. *Acacia drepanolobium* is abundant and less common associated species are *A. gerrardii*, *A. seyal*, *A. senegal*, *A. stenocarpa* and *A. nilotica*.

In Tanzania common grasses associated with *Themeda* are *Bothriochloa insculpta*, *Heteropogon contortus*, *Hyperthelia dissoluta*, *Harpachne schimperi*, *Cynodon plectostachyus*, *C. dactylon* and *Pennisetum mezianum*. In Kenya common associated species are *Eragrostis superba*, *Cenchrus ciliaris*, *Cymbopogon pospischilii*, *Bothriochloa insculpta* and *Heteropogon contortus*. *Cynodon plectostachyus* and *C. dactylon* cover extensive areas on old lake sites, and on black clay soils in the plains, *Pennisetum mezianum*, *P. stramineum*, *Setaria phleoides* and *Digitaria* sp. are common.

Where *Themeda* dominant grassland is damaged by over-use one or more of the following species increase and finally assume dominance: *Aristida kenyensis*, *A. adscensionis*, *A. adoensis*, *Eragrostis tenuifolia*, *E. braunii*, *Harpachne schimperi*, *Microchloa kunthii*, *Tragus berteronianus*, *Enneapogon schimperanus*, *Sporobolus* spp., *Chloris pycnothrix* and *Digitaria velutina*. These are relatively unpalatable and unproductive plants and their presence is a sign of degradation.

The drier phases of the semi-arid savanna in East Africa are characterized for the most part by bushes and scattered trees. The grass cover is relatively sparse. *Commiphora* spp. and *Acacia* spp., notably *A. mellifera*, are common bush species while the taller trees commonly comprise *Delonix elata* and *A. tortilis*, with *Adansonia digitata* (the baobab) assuming importance in some areas of southern Kenya and in Tanzania. Common low-growing shrubs are species of *Sericocomopsis*, *Barleria*, *Aerva*, *Disperma* and *Indigofera*. Common bushes and small trees are *Terminalia* spp., *Balanites* spp., *Boscia* spp. and *Salvadora persica*.

In the northern parts of the drier savanna the grass *Chrysopogon aucheri* var. *quinqueplumis* is generally dominant and major associates are *Aristida kelleri*, *A. hordeacea* and *Stipagrostis uniplumis*. In the relatively moist parts *Bothriochloa insculpta*, *B. radicans*, *Sehima nervosum*, *Eragrostis superba* and *Cenchrus ciliaris* may occur. In many areas *Sansevieria* spp., grouped at the base of bushes and occasionally forming dense stands, is a characteristic feature. In the southern area the common grass species are *Chloris roxburghiana*, *Leptothrium senegalense*, *Digitaria* sp., *Enteropogon macrostachyus*, *Tetrapogon* sp., *Cenchrus ciliaris*, *Chrysopogon aucheri* and *Aristida adscensionis*. *Sporobolus helvolus* predominates on heavy black soils with impeded drainage.

(2) *The south-central region*

This region extends over eastern Botswana, the Caprivi strip, and the adjacent area in Zambia, south, west and north-western Rhodesia, the northern Transvaal and the northern Cape regions of the Republic of South Africa and the south-central area of Mocambique adjacent to the Transvaal and Rhodesia. Narrow extensions occur in the Zambesi valley from Lake Kariba to beyond Tete in Mocambique and near the coast through southern Mocambique, eastern Swaziland and northern KwaZulu and Natal in South Africa.

Maps and descriptions of the vegetation have been published in recent years by Rattray (1957, 1960, 1961), Phillips (1959), Wild and Barbosa (1967), Mostert et al. (1971) and Acocks (1975).

Within the region the vegetation varies widely and in a very complex manner. It is impractical here to consider any but major associations.

Colophospermum mopane *associations*

Colophospermum mopane (mopane) is the characteristic tree in tree—shrub savanna or woodland over extensive areas in north-central and north-eastern Botswana, in the southern and south-eastern and parts of western Rhodesia, the north-eastern and extreme northern Transvaal, and the area of Mocambique adjacent to south-eastern Rhodesia and the north-eastern Transvaal. It also occurs in the Zambesi valley extension of the semi-arid savanna in the form of tree savanna and woodland.

In the Makarikari region of Botswana mopane occurs as a tree in fairly pure stands with a sparse understory of shrubs, such as mopane, *Dichrostachys cinerea*, *Maytenus cymosa* and *Grewia flava*. The dominant grasses are *Cenchrus ciliaris*, *Digitaria* sp., *Eragrostis* spp. and *Dicanthium papillosum*. In the east-central area of Botswana it occurs in a tree—bush savanna, mixed with groups of other species.

Mopane is the dominant species in the south of Rhodesia. On the sandy soils derived from gneiss or granite it occurs as a tree but on the heavy clay soils derived from basalt it takes the form of a coppicing shrub, though some scattered trees are present. On the sandy soils it is associated with a wide range of other tree species, prominent among which are *Terminalia sericea*, *Grewia* sp., *Euclea undulata*, *Ormocarpum trichocarpum*, *Acacia nigrescens*, *Combretum apiculatum*, *Grewia* spp., *Sclerocarya caffra* and *Commiphora* spp. Common perennial grasses on the sandy soils include *Eragrostis rigidior*, *E. superba*, *Digitaria pentzii*, *Schmidtia pappophoroides* and many others. Annual and short-lived perennial grasses include *Enneapogon cenchroides*, *Pogonarthria fleckii*, *Brachiaria deflexa*, several *Aristida* species and *Tragus berteronianus*. Rainfall decreases towards the south and with decreasing rainfall, or on the shallower soils, annual grasses assume greater importance. Over-use results in the death of the more palatable perennials and an increase in the proportion of annual grasses and unpalatable dicotyledonous weeds.

On the heavy clays shrub mopane is associated with occasional trees such as *Sclerocarya caffra* and shrubby forms of *Dalbergia melanoxylon* and *Acacia nigrescens*. Common grasses arė *Bothriochloa insculpta*, *B. radicans*, *Schmidtia pappohoroides*, *Cenchrus ciliaris*, *Enneapogon scoparius* and *E. cenchroides*, and several annual *Aristida* spp.,

In the relatively dry environment in the extreme south of Rhodesia and the extreme northern Transvaal mopane occurs as a shrub or small tree in association with *Acacia* and other species. The grass layer is similar to that occurring in the drier parts of Rhodesia, but because of relatively heavy grazing pressure has been reduced over large areas to species of *Aristida* and *Eragrostis*, and *Enneapogon cenchroides* and, in extreme cases, to almost pure stands of the dicotyledonous weed *Tribulus terrestris*.

In the mopane dominant area of Mocambique the physiognomy varies from tree—bush savanna to tree savanna and in the latter, accompanying species are *Ximenia americana*, *Salvadora angustifolia*, *Azima tetracantha*, *Adenium obesum*, *Boscia albitrunca*, *Pachypodium saundersii* and *Euphorbia* spp.

Baikiaea plurijuga *associations*

In north-western Rhodesia, south-western Zambia and extreme north-central Botswana *Baikiaea plurijuga* (Rhodesian teak) dominant woodland occurs on the deep sandy soils derived from Kalahari sand. Common associated trees include *Pterocarpus angolensis*, *Brachystegia spiciformis*, *Guibortia coleosperma*, *Ricinodendron rautanenii* and *Terminalia sericea*. The shrubs *Paropsia brazzeana* and *Baphia massaiensis* are prominent in the lower layer. The grass cover is generally sparse and consists of tall species of *Aristida* such as *A. stipitata* var. *robusta* and *A. pilgeri* with *Triraphis schinzii*, *Tristachya rehmannii*, *Digitaria* spp., *Eragrostis* spp. and *Heteropogon melanocarpus*. The poisonous plant *Dichapetalum cymosum* is widespread.

Terminalia sericea *associations*

Over extensive areas of north-central Botswana, on soils derived from granite, on and immediately south of the watershed in south-western Rhodesia, in the border area between Mocambique and south-eastern Rhodesia and in the western Transvaal, *Terminalia sericea* is a characteristic tree but forms complex and variable associations with numerous other species. Prominent among these are *Burkea africana*, *Peltophorum africanum*, *Pseudolachnostylis maprouneifolia*, *Pterocarpus angolensis*, *Dombeya rotundifolia* and *Piliostigma thonningii.*

The grass cover varies widely. In Botswana *Stipagrostis uniplumis*, *Aristida meridionalis*, *A. congesta*, *Eragrostis* spp., *Schmidtia pappohoroides*, *Triraphis schinzii*, *Pogonarthria* sp., *Anthephora pubescens* and *Heteropogon contortus* are prominent. Further east in western Rhodesia the more common associated grasses are *Hyperthelia dissoluta*, *Diheteropogon amplectens*, *Themeda triandra*, *Heteropogon contortus* and *Schizachyrium sanguineum.* In the north-western Transvaal *Eragrostis pallens*, *E. trichopora*, *Loudetia simplex*, *Schmidtia pappophoroides*, *Aristida* spp., *Stipagrostis uniplumis* and *Brachiaria nigropedata* are common.

Acacia *spp. associations*

On the limited area of black, brown or reddish clays derived from basic rocks in south-western Rhodesia *Acacia* spp. are dominant, notably, *A. nilotica*, *A. karroo*, *A. gerrardii* and *A. rehmanniana.* These are often associated with *Combretum apiculatum*, *C. hereroense*, *Ormocarpum trichocarpum* and *Sclerocarya caffra.* Common grasses on the black soils are *Setaria porphyrantha*, *Bothriochloa insculpta*, *Dicanthium papillosum* and *Ischaemum afrum.* On the red and brown soils *Heteropogon contortus*, *Themeda triandra*, *Bothriochloa insculpta*, *Eragrostis* spp. and *Cymbopogon plurinodis* are common.

In southern Botswana and the north-eastern Cape there is an extensive area of tree—shrub savanna consisting of *Acacia erioloba*, *A. leuderitzii*, *A. mellifera* and *Boscia albitrunca* associated with a number of other species. Characteristic grasses are *Stipagrostis uniplumis*, *Schmidtia pappophoroides*, *S. kalahariensis*, *Aristida meridionalis*, *Anthephora pubescens* and *Eragrostis lehmanniana.*

On the heavy soils in the central Transvaal several *Acacia* spp., notably *A. tortilis*, *A. nilotica* and *A. gerrardii* form a tree—shrub savanna with *Dichrostachys cinerea*, *Zizyphus mucronata* and *Grewia flava*, while on similar soils in the western Transvaal *A. karroo* is the characteristic species and forms an open shrub savanna. The grass species vary with soil type. On the self-mulching black clays in the central Transvaal the principal species include *Ischaemum afrum*, *Sehima galpinii*, *Setaria* sp., *Themeda triandra*, *Elionurus argenteus*, *Digitaria* sp. and *Eragrostis chloromelas.* On red clay soils *Themeda triandra* and *Cymbopogon plurinodis* are dominant.

To the east of the Drakensberg range in the central and southern Transvaal, Swaziland and KwaZulu the semi-arid savanna vegetation is characterized by *Acacia nigrescens* and *Sclerocarya caffra* associated with a large number of other species of trees and shrubs including many *Acacia* spp. On the heavier soils *Themeda triandra* is dominant, though it is replaced by creeping grasses such as *Digitaria* sp. and *Aristida bipartita* when degraded by over-grazing. On the sandy soils *T. triandra* is less common and is associated with a wider range of other species, including *Eragrostis* spp., *Aristida* spp., *Heteropogon contorus*, *Tristachya leucothrix*, *Elionurus argenteus* and *Digitaria* sp. With over-grazing *Digitaria eriantha*, *Aristida congesta*, *Eragrostis* spp., *Perotis patens* and *Pogonarthria* sp. assume dominance.

Combretum *spp. associations*

To the north of the *Acacia nigrescens—Sclerocarya caffra* associations in the eastern Transvaal, *Combretum* spp., especially *C. apiculatum*, and *Colophospermum mopane* assume greater importance. In this area common grasses are *Andropogon schinzii*, *Cymbopogon plurinodis*, *Heteropogon contortus*, *Themeda triandra* and *Schmidtia pappophoroides*. Under heavy grazing pressure these are replaced by *Eragrostis* sp. cf. *trichophora*, *Schmidtia pappophoroides* and *Aristida congesta*.

Over large areas of the central and north-western Transvaal the vegetation is extremely mixed. It comprises various associations of trees and shrubs and the grass species present and the quality of the grazing vary widely.

(3) *The south-western region*

In a preliminary vegetation map of South West Africa, Geiss (1971) has differentiated five main types of savanna or woodland within the semi-arid regions of South West Africa and has listed the major tree and shrub species in each. The following description has been derived from these lists and the brief descriptions of the common grass species in the area given by Rattray (1960).

There is a decreasing rainfall gradient from the region comprising south-eastern Angola, north-eastern South West Africa and north-western Botswana towards the west and south-west. Within the semi-arid savanna rainfall decreases from about 650 mm to about 350 mm. In the wetter parts towards the north-east, in which sandy soils are characteristic, the vegetation is a tree savanna or woodland. Common tree and shrub species are *Baikiaea plurijuga*, *Pterocarpus angolensis*, *Dialium englerianum*, *Burkea africana*, *Ricinodendron rautanenii*, *Lonchocarpus capassa* and *Guibortia coleosperma*. *Hyphaene ventricosa* occurs on the drainage lines and on open flats. Characteristic grass species are *Eragrostis* spp., *Urochloa* spp., *Panicum* sp., *Themeda triandra* and *Hyparrhenia* spp. Towards the west this association gives way to one in which mopane is dominant and occurs either as a tree or shrub, depending on local conditions. In the drier parts it is associated with *Commiphora* spp., *Sesamothamnus benguellensis*, on the Kunene River, and *S. guerichii*, which is of wider distribution. Common grass species are *Ento-*

plocamia aristulata, *Stipagrostis hirtigluma* and *Schmidtia kalahariensis.* In the less arid parts, *Eragrostis* spp., *Aristida* spp., *Urochloa* spp. and *Panicum* spp. are common. *Vetiveria nigritana* often occurs in areas with impeded drainage.

Towards the south-west species of *Acacia* predominate, notably *A. reficiens*, *A. hebeclada*, *A. mellifera*, *A. erubescens* and *A. fleckii.* In some areas *A. tortilis* occurs. In the more mountainous areas to the south *A. hereroensis* is common and is associated with *Combretum apiculatum*, *Euclea undulata*, *Dombeya rotundifolia*, *Tarchonanthus camphoratus* and several other species. Abundant grass species are *Eragrostis rigidior*, *Heteropogon contortus*, *Anthephora pubescens*, *Monelytrum luederitzianum*, *Eragrostis trichophora*, *Brachiaria nigropedata* and *Cenchrus ciliaris. Panicum maximum*, as is frequently the case in Africa, is common under tree canopies. Rattray (1960) states that *Themeda triandra*, *Trachypogon spicatus* and *Hyparrhenia* sp., though not important now, might have been so in the past.

On the predominantly aeolian sands in east-central South West Africa and west-central Botswana the camelthorn (*Acacia erioloba*) is the typical tree and is associated with the shrubs *Acacia hebeclada* sub sp. *hebeclada*, *Ziziphus mucronata*, *Tarchonanthus camphoratus* and others. Important associated grasses include *Eragrostis lehmanniana*, *Schmidtia pappophoroides*, *Stipagrostis* spp. and *Aristida meridionalis.*

The significance of fire

Fire in vegetation, resulting from natural phenomena such as lightning or, more usually, the accidental or deliberate actions of man, has undoubtedly been of profound importance in the moulding of vegetation in tropical and sub-tropical Africa. However, the effects of fire vary widely, depending on the time, frequency and intensity of the fires, the vegetation and soil types involved, and the pattern and intensity of use of the vegetation by livestock and game. In the negative sense, the elimination of fire by artificial protection or the reduction of grass fuel by grazing, may also affect both the form and composition of vegetation.

Having evolved in habitats subject to periodic fires, trees and shrubs in the semi-arid savanna, at least when they are well established, are characteristically fire tolerant, in that they either escape severe damage, or, when cut back to ground level by fire, regrow from the stem base. Likewise the herbaceous species are well able to withstand fire. In the case of perennials, the underground parts survive, while annuals re-establish from seed.

In certain respects fire is distinctly harmful. It destroys the grass and litter cover and in consequence the effectiveness of rainfall is reduced, sometimes drastically. This is due in part to increased runoff, which is especially marked on clay and silt-rich soils which seal readily, and in part to increased evaporation from the surface soil. Thus moisture stress may severely limit productivity, especially in the early growing season. In certain environments the increase in aridity consequent on frequent burning causes severe degradation. For example, on ungrazed *Acacia* savanna on clay soils in south-western Rhodesia, burning every year in the late dry season resulted in the death of the dominant perennial grasses *Themeda triandra*, *Heteropogon contortus* and *Cymbopogon plurinodis*. They were replaced during the rainy season by a sparse cover of pioneer grass species such as *Brachiaria eruciformis* and *Aristida scabrivalvis*. However, with the same treatment on *Burkea—Terminalia* savanna on deep sandy soils, changes have been negligible (West, 1965).

In circumstances where large amounts of dead grass and litter accumulate, infrequent fires may be very destructive. Many perennial species lose vigour if the old dead shoots are not removed for several years; in addition, fires following several years of protection are very hot. The overall effect is the death of many perennial grass tufts. *Themeda triandra* is especially sensitive to litter accumulation and in high and medium rainfall areas dies out if the top hamper is not removed by grazing, mowing or burning (Staples, 1930; Edwards, 1942). However, in the drier environment of the semi-arid savanna *T. triandra* may persist as a dominant for many years even under complete protection, and is not apparently harmed by infrequent burning every 3 to 5 years in the absence of grazing (West, 1965). Generally speaking, it seems that any damaging effects of fire on the herbaceous component in semi-arid savanna may, in the main, be ascribed to effects on moisture relations.

The harmful effects of burning in semi-arid environments are aggravated by subsequent heavy use of the vegetation by livestock or game, especially where this takes place in the early growing season. Livestock and game animals are especially attracted to the uncontaminated fresh green grass in burned areas, and when these comprise only a small portion of the total area available, damage may be severe.

On the other hand, fire may have beneficial effects, in that it inhibits invasion by undesirable woody plants. In many vegetation types regular hot fires every few years have the effect of preventing the establishment of seedlings of woody plants, and occasionally some established trees may be killed. Such fires are most effective if they occur late in the dry season and the burned area is protected from grazing in the early growing season. This prevents over-use of the early grass growth with consequent loss of vigour and competitive ability. Experimental results indicate that fire early in the dry season is unlikely to be as effective and, at least in relatively wet environments, may even result in an increase in tree and shrub density (Trapnell, 1959). Thus, especially in *Acacia* savanna subject to low grazing pressure, periodic fires can be regarded historically as the major factor in the development of open savanna. However, heavy stocking with grazing animals, with purposeful or consequential elimination of fire, has all too frequently in Africa, swung the balance in favour of the woody plants. For example,

Donaldson (1969) has described how, on very sandy soils in the Molopo area in the northern Cape Province, over-stocking has within 20 years transformed a vegetation comprising open grassy plains dominated by the palatable and desirable grass species *Anthephora pubescens* with scattered tall trees of *Acacia giraffae* and *Boscia albitrunca*, to a dense tree—shrub savanna dominated by the tree *Acacia reficiens* and shrub-like thicket-forming plants such as *A. mellifera*, *A. hebeclada*, *Dichrostachys cinerea* and *Rhigozum obovatum* and a shrub form of *Terminalia sericea*. The understory is sparse and comprises mainly *Stipagrostis uniplumis*, *Eragrostis lehmanniana* and *Schmidtia bulbosa* associated with *Aristida stipitata* and *A. graciliflora*. Because of the poor grass cover the use of fire to reverse the trend is quite impractical and, at least initially, artificial and costly thinning of the tree and shrub components is necessary.

Even where tree and shrub encroachment is not serious, controlled burning as a means of favouring grass growth at the expense of woody plants is usually only feasible in the wetter phases of the semi-arid savanna. Fire should be used on an opportunist basis, when there has been exceptionally good growth of grass in the previous rainy season. Sufficient herbage should be left unburned so that reserve grazing is available in the event of drought.

In the drier phases of the semi-arid savanna the very large unpredictable fluctuations in the grass production between seasons makes the use of fire very risky. The probability that burning will be followed by drought is high, and apart from the loss of reserve grazing at a time when it might be badly needed, the grass cover is likely to be damaged.

Tsetse fly in the semi-arid savanna

Trypanosomiasis, infection with the disease-causing protozoa of the genus *Trypanosoma*, which are parasitic on blood, virtually precludes cattle production over some 6¼ million square kilometres in tropical Africa (Hagan and Wilmhurst, 1975). The trypanosomes causing disease in cattle are *Trypanosoma congolense*, *T. vivax* and *T. brucei*. However, the disease caused by *T. brucei* is often of little pathological significance. These organisms multiply in the blood of domestic and wild animals and are transmitted by blood-sucking flies. Flies of the genus *Glossina* (tsetse) are by far the most important of these and, apart from transmitting the trypanosomes, serve also as a host for part of the life-cycle of the organisms.

Exposure to repeated infection by *T. congolense* and *T. vivax* usually results in debilitation and death. However, the degree of virulence of the infection with a particular species of trypanosome varies greatly in different circumstances and different areas. For example, infection with *T. vivax* in West Africa is usually fatal, whereas in East Africa the disease is usually mild (Mulligan, 1970). The period of exposure to trypanosome infection may also influence the outcome of the disease. Tolerance to infection has been noted in some strains of Zebu cattle (Ford, 1971).

Thirty-four species, sub-species and races of *Glossina* are recognized, of which 22 probably play some part in the transmission of trypanosomiasis in Africa. In the semi-arid savanna of East and Southern Africa three species of the *morsitans* group, namely *G. morsitans*, *G. pallidipes* and *G. swynnertoni* and one of the *fusca* group *G. longipennis* are of primary importance. *G. morsitans* and *G. pallidipes* occur in East and south-central Africa and *G. morsitans* is present in Botswana in the Okavango region. *G. swynnertoni* and *G. longipennis* are characteristic species in drier habitats in East Africa. The general distribution in Africa of the *morsitans* group which, apart from the species mentioned above, includes *G. austeni* and *G. longipalpis*, and of *G. longipennis* as delimitated by Ford (1971) is shown in relation to the semi-arid savanna in Fig. 1.2. The challenge is greatest in East Africa but the threat in Southern Africa is appreciable and is only contained by constant vigilance.

Control of trypanosomiasis can be effected by preventing transmission of the disease (vector control) or by means of prophylactic or curative drugs.

Vector control. Each species of *Glossina* appears to be limited to particular habitats and the type of vegetation is often an indicator of the species likely to be found. Changing the habitat, especially by drastic clearing of the woody vegetation, has been successfully used to control tsetse fly. However, where two or more species of tsetse fly occupy the same area modification of the vegetation in a way that is unfavourable to one species may make the habitat more favourable for another. For example, if thickets are opened up with the object of controlling *G. pallidipes*, conditions may favour an increase in *G. morsitans*. To be effective, clearing must be permanent and in most circumstances in Africa the expense of clearing precludes its use on a large scale.

The strategic application of insecticides has been shown to provide a feasible method of vector control at a reasonable cost. Because of their behaviour and life-cycle — part of the life-cycle involves underground development — tsetse are exposed to insecticidal attack for only limited periods. Persistent insecticides such as DDT are thus the most effective. However, successful control of *Glossina* has been achieved with repeated aerial application of a non-persistent insecticide, endosulfan, the timing of the second and subsequent applications being estimated at that required to kill all newly emerged flies before they can deposit larvae (Chapman, 1976). Aerial ultra-low-volume applications offer promise, as do ground application of insecticides to known dry season resting sites.

A further measure which has proved to be effective, if not popular, has been the elimination of the favoured food hosts of *Glossina* among wild game. In Rhodesia, warthog (*Phacochoerus aethiopicus*), bushpig (*Potamochoerus koiropotamus*), kudu (*Tragelaphus strepsiceros*) and bushbuck

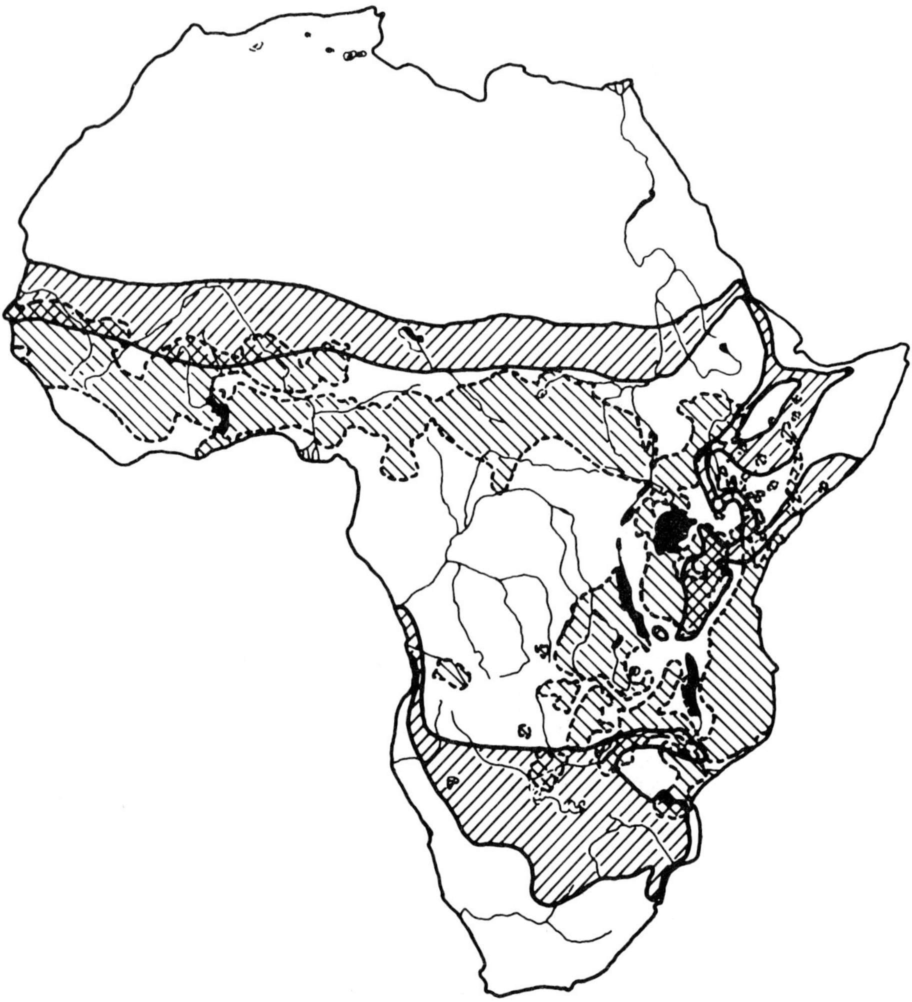

Fig. 1.2. The distribution of the *morsitans* group and *Glossina longipennis* of the *fusca* group of tsetse (▧) in relation to the semi-arid savanna (▨) of Africa.

(*Tragelaphus scriptus*) have been shown to be preferred food hosts and their elimination combined with a programme aimed at driving out or eliminating buffalo (*Syncerus caffer*) and elephant has resulted in effective control, especially where such measures are supplemented with the ground application of residual insecticides to dry season resting and refuge sites (Lovemore, 1969).

A variety of methods of biological control have been tried or considered, as, for example, the use of parasites and predators on tsetse pupae or adults, and the release of males sterilized by chemical or radioactive means. The main problem is to produce sterilized males in sufficient numbers to compete with the wild population. Success thus depends on the development of methods which will allow for the capture or rearing of large numbers of males which can be sterilized and released. This has not yet been achieved, but is by no means infeasible. Vale and Hargrove (1975) have recently shown, for example, that very large numbers of flies can be captured using the odour from cattle.

The use of drugs. Trypanosomicidal drugs have been successfully used to protect cattle in tsetse infested areas and to cure cattle which have been infected with trypanosomes. Although such chemical therapeutic treatment is most valuable in the short-term, it cannot in the long-term be regarded as more than palliative. The most satisfactory solution to the control of trypanosomiasis is the control of the vector.

Immunization and immunity. The possibility of exploiting the antigenic properties of trypanosomes is of considerable interest. However, trypanosomes possess a very complicated antigenic structure and have the faculty of antigenic change in response to production of antibodies by the host. So far, it has not been possible to produce an effective vaccine.

The fact that wild game are resistant to trypanosomiasis has prompted suggestions that, instead of attempting to control tsetse, the land which is infested should be used for game production. Such a suggestion has a certain merit, especially where the elimination of tsetse is likely to result in severe degradation because of over-use by domestic livestock. However, as Lovemore (1969) has pointed out, the problem of dealing with the interface between game areas used for ranching would remain, and control measures would have to be adopted on such interfaces in perpetuity.

Harvester termites

Termites which feed on dead and living herbaceous material are of considerable ecological significance in relation to cattle production in the semi-arid savanna of Africa. The most important species is *Hodotermes mossambicus* (Hagen), though other species such as *Macrotermes* spp. and the snouted harvester termites *Trinervitermes* spp. may also be locally important. The nest system of *H. mossambicus* is well below the soil surface and, apart from the removal of herbaceous vegetation, the only signs of their presence are small irregularly-spaced dumps of loose soil 5—20 cm high, and scattered foraging holes about 1 cm in diameter on the soil surface. When the latter are not in use they are sealed. *Macrotermes* spp. construct mounds and the nest system is both above and below the original soil surface. *Trinervitermes* spp. construct a cellular mound and the nest structure is largely confined to this.

The extensive and complex nest system of the common harvester termite (*H. mossambicus*) consists of sub-spherical breeding hives up to about 60 cm in horizontal diameter, smaller supplementary hives (occupied by foraging parties and used to store material harvested in the vicinity) and small chambers used for temporary storage of food. All are interconnected by numerous passages and galleries. The area occupied by the nest system fluctuates seasonally according to the availability of surface food supplies. In drought years the termites forage over a wider area. According to Coaton (1950) the area occupied may vary seasonally from about 0.4 to 6 ha. However, Nel (1968) made use of mutual hostility shown by workers from different colonies to estimate that the mean foraging area of termite colonies in the central Orange Free State was 92 m^2 and that their density was 110 ha^{-1}. Be this as it may, there is ample evidence that saturated populations of harvester termites often occur, and that their presence may result in extensive and drastic denudation of the grass cover (Coaton, 1950, 1958; Lee and Wood, 1971; Coaton and Sheasby, 1975). Local denudation may occur in the absence of grazing animals, but damage is usually more widespread in their presence. Coaton (1958) has presented evidence that, with heavy infestation, termites are capable of removing amounts of herbage of the order of 1 to 3 t per ha of dry matter. Both living and dead material may be taken (Coaton and Sheasby, 1975).

Provided adequate food is available, harvester termites are favoured by dry soil conditions; conversely, excessive infiltration of rainfall in wet seasons results in a sharp decline in the population (Bissett, 1974, 1975). They are thus generally most important in the drier phases of the semi-arid savanna, and on sites where edaphic conditions are such that infiltration is inhibited. For example, on certain heavy soils in Rhodesia, destruction of the grass cover by localized over-grazing or by artificial means, as in road or firebreak construction, results in a sealing of the soil surface by raindrop action and soil wash. Infiltration is severely restricted and such sites are colonized by harvester termites (Bissett, 1974; Macdonald, 1974). Likewise, patches of sodic soil which have a relatively impervious deflocculated clay subsoil and a sparse vegetative cover are heavily infested.

The maintenance of a satisfactory vegetative cover, especially on heavy soils which seal readily, is thus important in limiting harvester termite populations. Conversely, drastic reduction of the cover will favour termite activity. Termite populations are likely to be greatest in seasons of low rainfall, a time when competition between livestock and termites will be most severe.

Chemical control of harvester termites is feasible using spray applications of persistent insecticides such as Dieldrin or poisoned baits involving Dieldrin, sodium fluosilicate or carbaryl (Coaton, 1958; Nel, 1968; Findlay, 1971; Bissett, 1975). However, with saturated populations on land of low productive potential the cost is likely to be excessive. Moreover, it might be necessary to repeat the control treatment every few years.

Once denudation has occurred, control by means of sprays is not effective, and the use of baits will not result in any immediate return. In the light of present knowledge, it seems that the most likely possibility for economic control would be spot application of baits or sprays to residual colonies following seasons of good rainfall. In this case termite populations will be at a low level and the cost will be minimal. Naturally, such measures should be supplemented by careful grazing control, and stocking rates should be such as to prevent undue pressure on the vegetation in seasons of low rainfall. However, further study of the biology and control of the harvester termites is clearly desirable.

PRESENT LAND-USE AND PRODUCTION METHODS

Livestock producers in the semi-arid savanna of Africa can be divided into three categories, true pastoralists who depend solely on livestock for their living, semi-pastoralists who, while keeping stock, also grow grain crops for their own use, and commercial ranchers or state-run companies.

True pastoralists

The true pastoralists, comprising, according to Brown (1971), some 50 million people in Africa south of the Sahara, lead a nomadic or semi-nomadic life. Their movements are primarily aimed at providing the best possible natural grazing for their livestock and are normally co-ordinated with different rainfall patterns and the seasonal availability of forage and water in the area over which they move. Their diet is derived largely from the products of their stock and comprises mainly milk, meat and, in some cases, blood. Besides cattle, sheep, goats, donkeys and, in the drier areas north of the equator, camels are kept. The small stock may contribute significantly to the food supply and at times may be bartered for grain (Allan, 1965). Brown (1971) has calculated that on average three standard livestock units of 500 kg livemass are required in East Africa to support each adult equivalent of the population, but this may vary from 2.5 to 4.5, depending on ecological conditions. Allan (1965) has suggested a figure of six livestock units per individual but considers that there is much variation, and that there is insufficient information to accurately assess requirement.

Livestock management is minimal and largely confined to close herding during the day and enclosure overnight in a stockade for protection from carnivores and raiding tribes. Except for seasonal movement to natural salt licks no supplementary food is provided (Skovlin, 1971). There is virtually no control of breeding. Bulls remain with the cow herd throughout the year and heifers are allowed to breed before they are well grown. Calves are not weaned, but, because the cows are milked to supply food for the pastoralists, the calves may be herded separately from the cows during the day.

There is no deliberate control of stock numbers, and herds commonly increase over a run of favourable rainfall seasons, only to be reduced, often drastically, as a result of starvation during dry years. In the Kajiado District of Kenya Masailand, for example, cattle numbers were reduced by some 55% between 1957 and 1961, primarily as a result of drought. Significantly, sheep numbers decreased by only 17%, whereas goat numbers did not change (Maloiy and Heady, 1965).

During the dry season cattle may be watered only once every 3 days (Maloiy and Heady, 1965; Pratt 1969), and under these conditions the dry season grazing area extends as far as 19 km from water.

In the absence of management, and under conditions where severe overuse of the vegetation results in semi-starvation, output of meat is very low and in commercial terms probably seldom exceeds 10% of the herd numbers (Skovlin, 1971). Moreover, the use of animals for the production of milk and blood must, in the circumstances, severely limit meat production.

Traditionally, land tenure is communal and individual responsibility for herds is divorced from individual responsibility for the land. As in most of semi-arid Africa, population growth has resulted in severe pressure on the land. Degradation and erosion is characteristic of all semi-arid grazing areas which have a high population, or where only limited water supplies are available for stock. Practices which were logical and non-destructive with a moderate pressure are, under excessive pressure, destroying the very resources on which the people depend.

Semi-pastoralists

In the moister phases of the semi-arid savanna some crop production, albeit sporadic and at a low level, is possible. With increasing population pressure the pastoralist usually attempts to augment subsistence food supplies by growing grain crops. In these circumstances nomadism is not common. As with the true pastoralists, land tenure is commonly communal, with individual ownership of stock.

Inoculation against disease, and dipping for tick control might be enforced by government agencies. Apart from this, livestock management is essentially on the level of that practised by the true nomad. Overnight confinement in stockades, and the drawing of milk from beef cows to the detriment of the calf, is common, and productivity is generally at a low level. Over-stocking is common and many stock die of poverty in drought years.

The levels of production achieved under traditional (cattle post) management at various sites in Botswana as compared with "reasonably acceptable" (ranch) management is shown in Table 1.2 (Animal Production Research Unit, Botswana, 1975). Reasonably acceptable management levels were defined as involving (a) a degree of fencing to allow for the control of breeding herds, the separation of young stock and the retention of standing hay

TABLE 1.2

Productivity under cattle post and ranch production systems

	Systems	
	Cattle post	Ranch
Calving (%)	46.4	74.0
Calf mortality (%)	10.2	8.5
Weaning (%)	41.7	67.7
Weaning weight (at 210 days) (kg)	122.5	177.4
Post-weaning gain (7—18 months) (kg)	84.1	100.5
Weight of weaner calf/cow year (kg)	51.1	120.1
Weight of 18 month-old calf/cow year (kg)	86.1	188.2

for dry season feeding, (b) attention to disease control and mineral supplementation, and (c) the provision of adequate water supplies for livestock.

Commercial ranching

Commercial ranching enterprises, whether owned and run privately, or run by government agencies, represent the most sophisticated level of land-use in the semi-arid savanna. Apart from the investment in stock, capital outlay is primarily for fencing, water supplies, facilities for cattle handling and dipping for the control of ectoparasites. In some cases capital may be invested in bush control, that is, the thinning or clearing of unwanted woody growth.

On many ranches small flocks of goats or sheep are kept, and a deliberate attempt is often made to ensure the presence of suitable species of wild game. However, the primary interest is cattle, and wild herbivores which compete with cattle for food, or are agents of disease transmission to cattle are, as far as possible, eliminated, as are carnivores which prey on cattle.

The degree of development and standard of management on different ranches naturally varies widely. Ideally, however, the following practices are adopted.

Breeding is controlled so that calving takes place during a short period, and at a time which results in the closest coincidence over the year between the pattern of change in nutritive quality of the herbage and the nutritive requirements of the cow and calf. Selected bulls are used and are bred to specific groups of cows or heifers. Unwanted bulls are castrated. Heifers are put to the bull only when adequately grown.

Where necessary, supplementary food, especially protein-rich concentrates are fed during the dry season. In the most advanced enterprises culling and selection is based on records of production.

Diseases are controlled by means of inoculation and the control of ectoparasites which transmit disease. Internal parasite infestation is kept to a minimum by dosing and preventive measures, such as the fencing-off of swampy areas.

Some form of rotational grazing is usually practised and stock numbers are kept at a level which obviates over-use of the vegetation in seasons of low rainfall. Adequate water supplies are provided, and the siting of these, and the layout of the paddocks, is designed to favour uniform use of the available vegetation. Firebreaks are maintained, usually along fence lines, as a precaution against uncontrolled fires and in suitable circumstances in the more humid savanna, to permit controlled burning. Burning is carefully timed and is primarily aimed at the control of woody growth.

Levels of productivity naturally vary from ranch to ranch depending on the quality of vegetation and the level of management. Under efficient management production per cow may considerably exceed those shown for "ranch management" production systems in Table 1.2. However, it is likely that on average, the level of production is of the order of that shown (Skovlin, 1971; Madallali, 1975; Bembridge and Steenkamp, 1976).

THE MANAGEMENT OF SEMI-ARID SAVANNA FOR CATTLE PRODUCTION

Selective use of vegetation

In spite of superficial appearances the natural savanna vegetation of Africa is extremely heterogeneous. It comprises a wide range of plant species and these are distributed in complex mosaics of different communities, reflecting soil variation, the macro- and micro-relief and the effect of these factors on plant soil—water relations and micro-climate. Moreover, different components of the vegetation vary on a different scale. Changes in the composition of the herbaceous layer are likely over relatively short distances and those of the woody component over greater distances.

A consequence of this heterogeneity is marked selective use of the vegetation by domestic livestock and wild herbivores, at both the species and community level. Feeding animals show a preference for certain species of plants, parts of plants or communities of plants. Different species of animals usually differ in their food preferences, though these often overlap, and preferences usually change over the year and in different circumstances (Petrides, 1975; Gammon, 1976).

Table 1.3 shows the differences in the diets of cattle and goats, estimated by observation every 2 min, on the minute, during daylight hours, in mopane savanna on gneiss-derived soil in the south of Rhodesia. The values are means of determinations during alternate months over a period of 2 years, and are a measure of the percentage of the total feeding time which was spent feeding on the species shown. The stocking pressure was moderate with ample grass available.

TABLE 1.3

Diet preferences of steers and goats feeding on tree—shrub savanna in Rhodesia — expressed as percentage of total feeding time which was spent on individual plants

Species	Steers		Goats	
Grasses				
Digitaria pentzii	20.85	74.02	13.96	29.32
Panicum maximum	16.11		9.20	
Urochloa mosambicensis	11.57		3.05	
Pogonarthria fleckii	5.56		—	
Brachiaria deflexa	3.97		1.20	
Other grasses	15.96		1.91	
Herbs				
Crotalaria damarensis	1.72	16.24	5.70	18.41
Other herbs	14.52		12.71	
Woody plants				
Colophospermum mopane	4.05	9.29	6.71	51.77
Combretum apiculatum	2.24		10.67	
Grewia flavescens	—		4.43	
Commiphora mollis	—		4.19	
Sclerocarya caffra	—		3.50	
Terminalia prunioides	—		3.50	
Grewia spp. (hybrid population)	—		2.18	
Other woody plants	3.00		16.59	
Unidentified plants	0.45		0.50	
	100.00		100.00	

Selective feeding is invariable and is in fact essential to the well-being of the feeding animals. Its partial control and the minimizing of its undesirable effects are part of efficient management.

Responses to short-term climatic changes

The erratic nature of the rainfall in the semi-arid savanna regions of Africa has been emphasized previously. The plant communities are in the long term adapted to the fluctuations in water supply, but such adaptation does not preclude short-term changes in composition, especially of the herbaceous component. Changes are greatest in the drier phases of the semi-arid savanna, where the proportion of perennial to annual grassess, for example, may decrease and increase, respectively, during a series of dry or wet years. Overuse of the herbaceous component by the grazing animal, especially during the growing season, commonly induces changes similar to those engendered by continued moisture stress. However, the two factors, drought and overgrazing, must normally be considered together, the effects being additive.

By comparison, the tree and shrub component of savanna is relatively stable, though woody species commonly increase if the herbaceous cover is badly damaged by over-grazing. However, in times of severe and continued drought, trees and shrubs may die. In dry tree—shrub savanna in the extreme south of Rhodesia, for example, three consecutive seasons of very low rainfall, combined with over-use of the herbaceous component, resulted in the death of a large number of trees and shrubs of the species *Adansonia digitata*, *Commiphora* spp., *Combretum apiculatum*, *Terminalia prunioides*, *Grewia* spp., *Colophosphermum mopane* and *Acacia fleckii* (West, 1968).

In the more mesic areas of the semi-arid savanna the vegetation is more stable. The grasses are mainly perennials, and though changes do occur with short-term fluctuations in climate, major changes are undoubtedly man-induced, primarily as a result of over-grazing and trampling by domestic livestock. The weakening of the herbaceous cover favours an increase in the density of the woody plants, which have a continuing suppressive effect.

Differences in the feeding value of vegetation types

Throughout the semi-arid savanna it has been found that the vegetation in the more arid phases provides more palatable and nutritious food for livestock, especially during the dry season, than that in the less arid phases. In Southern Africa such vegetation, comprising widely different plant associations, is commonly termed "sweetveld". Cattle on sweetveld normally maintain or gain condition during the dry season with little or no protein supplementation. Comparative terms "mixed veld" and "sourveld" are also used, the latter referring to vegetation in higher rainfall areas in which the herbaceous layer comprises grass species which are palatable and nutritious for only the early and mid-growing season. Mixed veld is intermediate in nutritive value and can be regarded as characteristic of the less arid areas of semi-arid savanna. Protein is the major nutrient deficiency during the dry season and the maintenance of body-mass in mixed veld usually depends on the provision of protein-rich supplements.

The reasons for the differences in quality have not been adequately investigated, but the comparatively high value of sweetveld has been variously ascribed to the relatively high proportion of annual grasses in the grass cover, the less woody nature of the perennial grasses, and the presence of browse plants, the leaves of which are available during much of the dry season, and which have a comparatively high protein content. The herbaceous component of sweetveld is easily damaged by continued heavy grazing during the growing season. Palatable perennial species are weakened and in extreme cases killed. Conversely, sweetveld usually benefits greatly from light use or resting during the growing season. Moreover, it is common experience that sweetveld grasses can be grazed at a more mature stage of development than mixed and sourveld, without reducing animal performance. Thus, with conservative stocking rates, high animal production without deterioration of the grass cover is quite feasible.

Occurring as it does in arid regions, the carrying capacity of sweetveld is lower than that of mixed veld, being on a year-round basis of the order of one 500 kg bovine livestock unit to 6 to 12 ha or more, depending mainly on rainfall and the tree and shrub density. In the wetter phases of the semi-arid savanna carrying capacities usually vary from about 4 to 6 ha per livestock unit.

The relative feeding value of savanna plants

The most valuable plants for animal production are those which are readily eaten by livestock and which are of high quality, in the sense that they have a high content of digestible nutrients. Usually these two factors are correlated, in that the voluntary intake of plants increases with increasing quality.

Thus the food preferences of grazing animals broadly reflect the feeding value of the species present. Food preferences must of necessity be considered in relation to the kinds and amounts of the different foods on offer and Petrides (1975) has defined a "preferred food species" as one which is proportionately more frequent in the diet of an animal than it is in the available environment. Within any community, relative preferences may be determined by measuring percentage utilization of the different species present.

From a completely different standpoint, a "preferred food" may also be regarded as one which is eaten in the greatest quantity when a range of foods is offered to animals, and each is in excess of the amount of each food eaten by the animals.

Foods which are eaten in the greatest quantities by an animal population are termed "principal foods" (Petrides, 1975). These are foods which form the greatest proportion of the animals' diet, but are not necessarily the most preferred, in either of the senses in which the word is used above.

The accurate measurement of the diet of grazing animals and the proportions of different foods on offer in heterogeneous vegetation is costly and time-consuming. However, percentage utilization may often be estimated relatively easily, using, for example, grazed-plant techniques, such as those recently reviewed by Barnes (1976). The relative percentage utilization of species within communities will provide a good indication of their relative value for animal production.

Attempts have been made to evaluate savanna species by sampling herbage at intervals throughout the year and determining parameters such as crude protein content and in vitro digestibility.

This approach has been adopted by the Animal Production Research Unit (1975) in Botswana, where samples of leaves of the important species of grass are collected at monthly intervals throughout the year from a wide range of sites in grazed vegetation. The crude protein content and in vitro digestibility of these samples are determined. Values are corrected for site effect by least squares analysis and this facilitates the ranking of species. In a similar assessment the important components of the vegetation were sampled in savanna grazed by cattle in Kenya (McKay and Frandsen, 1969). Proximate analyses for crude protein, crude fibre, ether extract, total ash, nitrogen-free extract and dry matter content were done.

Monthly sampling of the leaves and twigs of some common browse plants in dry savanna in the south of Rhodesia has facilitated assessment of their relative value. Crude protein content of the leaves and twigs and the in vitro digestible dry matter content of the leaves was determined. The mean values of these parameters derived from measurements on samples taken at monthly intervals over two growing seasons are shown, together with the estimated digestibility coefficient for crude protein in the leaves, in Table 1.4. The latter values were calculated using the relation $Y = 0.9X - 3$ (Elliott and Topps, 1963), where Y and X denote digestible crude protein and crude protein percentages, respectively. The relatively low values for digestible dry matter are noteworthy. It seems that these plants are primarily useful as a source of protein.

With realistic sampling procedures, especially when considered in conjunction with preference ratings and diet assessments, data obtained in the above manner provide a useful basis for assessing the relative value of savanna plants. However, it should be recognized that the quality of the different species sampled may be differentially affected by the degree of use; palatable grasses, for example, will be more intensively grazed and are likely to be at an earlier stage of development than unpalatable species when sampled.

TABLE 1.4

Crude protein contents of the leaves and twigs and digestibility of leaves of common browse plants in arid savanna in Rhodesia

Species	Crude protein (% dry matter)		Digestion coefficients for leaves (%)	
	Leaves	Twigs	Crude protein	Dry matter
Melhania acuminata	15.9	6.9	71.1	48.6
Grewia flavescens	15.1	6.9	70.1	53.5
Grewia spp. [a]	14.4	7.4	69.2	35.3
Combretum apiculatum	12.8	6.5	66.6	47.4
Colophospermum mopane	12.3	5.0	65.6	41.3
Commiphora mollis	10.9	5.1	62.4	46.1

[a] Hybrid complex between *G. bicolor*, *G. monticola* and *G. subspathulata*.

Hence, inherent differences will be confounded with differences due to stage of development.

Intensification of production

In the sense of producing more and better quality food, as distinct from increasing the efficiency of utlization of food by means of grazing management, the intensification of pasture production normally involves one or more of the following measures: (a) the replacement of undesirable plants with species which are more palatable and productive, (b) the improvement of the supply of plant nutrients, especially nitrogen, (c) the improvement of plant soil—water relations.

The low and uncertain rainfall and the low productive potential of the semi-arid savanna set severe limits to the possibilities for intensification, and the feasible level of capital input. Improvement of plant soil—water relations, for example, is restricted to the maintenance of a satisfactory vegetative cover, with the aim of reducing losses of rainfall from runoff and evaporation. Likewise, although fertilizing with nitrogen, supported by superphosphate, has been shown in semi-arid grassland to result in greatly increased yields (Mills, 1966), and thus greater efficiency of rainfall use, the practice cannot be justified economically in present circumstances.

All things considered, the possibilities for intensification and improvement seem to be largely confined to modification of the vegetation, with the aim of increasing the proportion of palatable and productive plants. In due course it might well be economically feasible to introduce improved pasture grasses and legumes, especially in the more humid phase of the semi-arid savanna. This aspect is beyond the scope of this discussion, which will be confined to the possible manipulation of the natural savanna vegetation with a view to increasing cattle production.

In savanna vegetation with a moderate to high density of trees and shrubs the woody species are invariably dominant, and by successfully competing for water, nutrients and light suppress the herbaceous layer. Complete or partial clearing of the woody plants commonly results in a marked increase in the productivity of the grass understory. In the drier phases of the semi-arid savanna clearing may also favour an increase in the proportion of perennial species.

In some vegetation types the trees and shrubs have little or no value as browse. Hence, they are in effect weeds which compete with the herbaceous component, and their removal is likely to be advantageous. However, where palatable browse plants are present any increase in yields, or change in composition of the herbaceous component, must be set against the loss of browse. In this case it is important to have adequate estimates of (a) the diets and food preferences of the animals concerned, (b) the relative value of the different plants present, and the quantity and quality of the food they provide during the year, (c) the quantitative and qualitative reactions of the vegetation to clearing or thinning, and (d) the safe level of utilization, especially of the browse component.

These facts have by no means been adequately established for even the major vegetation types in the semi-arid savanna of East and Southern Africa. However, in keeping with the results shown in Table 1.3, it is general experience that where cattle have access to adequate quantities of both grass and browse, the latter comprises only a small part of the diet (Whyte, 1947; Heady, 1960; McKay and Frandsen, 1969; McKay, 1970; Animal Production Research Unit, Botswana, 1975). Large amounts of browse are eaten only where grazing pressure is heavy and grass is in short supply. In this case cattle performance is likely to be adversely affected (McKay and Frandsen, 1969; McKay, 1970).

It can be assumed, therefore, that, except where the tree and shrub density is low, it would generally be advantageous to thin out woody plants drastically or perhaps remove them completely, even where these provide palatable browse. In the absence of information on such aspects as the nutritional importance and productivity of the browse and the safe level of utilization for sustained production, the number of plants of any particular species which should be left is largely a matter of guess-work.

Responses to complete clearing

Responses of the grass component of the vegetation to clearing have been estimated at various sites in Rhodesia over a number of years by harvesting replicated plots at the end of the growing season on cleared and uncleared savanna. The mean yields of grass herbage and the range of yields at each of four sites is shown in Table 1.5.

The relative yield increase with clearing varied at the different sites from about 160 to about 400%. It is significant that the large response at the Tuli site is associated with a very low mean yield on uncleared savanna. Similar large increases have been recorded in the higher rainfall areas of Rhodesia, where yields on uncleared land were low.

Safe stocking rates for livestock are usually determined by the amount of grass available in seasons of lowest production. Thus particular importance is attached to the relative increases in production, as determined from the lower values of the ranges. On this basis, increases ranged from 134% at Matopos, on granite-derived sands, to 443% at Tuli.

Clearing has also been found to result in marked changes in grass species composition. The difference occurring in 1974 between cleared and uncleared plots at the sites listed in Table 1.5 are shown in Table 1.6 in terms of the percentage contribution of different species to total herbage yield (I.A.W. Macdonald, personal communication, 1974).

The data apply to a situation in which the grasses were harvested once a

TABLE 1.5

The effects on yields of grass of clearing trees and shrubs at various sites in semi-arid savanna in Rhodesia

	Mean seasonal rainfall prior to harvest (mm)	Mean yields (kg dry matter ha^{-1})		Relative yield increase (%)
		Not cleared	Cleared	
Tuli (on sandy soil derived from acid gneiss—cleared 1963)				
1965—1975	414	515	2085	405
Range	132—753	175—945	775—3195	443 [a]
Matopos (on sandy soil derived from granite—cleared 1962)				
1963—1975	579	1220	1930	158
Range	321—1032	795—1715	1065—3175	134
Matopos (on clay soil derived from basic rock—cleared 1962)				
1963—1967 } 1971—1975 }	566	1545	2635	171
Range	301—949	610—2920	1170—4130	192
Nyamandhlovu (on sandy soil derived from sandstone—cleared 1963)				
1964—1975	565	1275	2225	174
Range	377—854	980—1790	1455—2970	148

[a] Relative yield increase with lower limits of range.

year when dormant at the end of the growing season. It is not axiomatic that similar differences would have occurred under grazing. Nevertheless, it is significant that not all the changes in species composition with clearing are favourable. For example, there was a marked decrease in certain desirable species, notably *Digitaria pentzii* and *Panicum maximum* at Matopos on sandy and clay soils, respectively, and *Urochloa mosambicensis* at Tuli. In addition there was an increase in certain distinctly unpalatable species such as *Trachypogon spicatus* and *Eragrostis rigidior* on sand at Matopos and *Aristida pilgeri* at Nymandhlovu. Clearly, possible changes in species composition of the herbaceous layer are an important consideration in relation to clearing of the woody components of savanna.

Partial clearing

The results given above relate to complete clearing of the woody component. Various studies in different parts of the world have shown that the inverse relation between the productivity of the herbaceous understory and the density of the tree or shrub component is commonly curvilinear (Jameson, 1967; Clary, 1969; Walker et al., 1972). The usual relation is well described mathematically by means of an equation proposed by Grosen-

TABLE 1.6

Differences in grass species composition in 1974 on cleared and uncleared semi-arid savanna at various sites in Rhodesia

Species	Species yield as a percentage of total yield	
	Not cleared	Cleared
Tuli (on sandy soil derived from acid gneiss—cleared 1963)		
Schmidtia pappophoroides	31.1	72.1
Urochloa mosambicensis	12.6	1.8
Eragrostis rigidior	10.3	14.9
Aristida adscensionis	9.4	4.3
Digitaria pentzii	8.2	—
Stipagrostis uniplumis	7.7	0.9
Others	20.7	6.0
	100.0	100.0
Matopos (on sandy soil derived from granite—cleared 1962)		
Digitaria pentzii	45.4	11.5
Heteropogon contortus	15.6	7.0
Trachypogon spicatus	8.7	26.4
Schizachyrium jeffreysii	7.6	12.7
Rhynchelytrum repens	7.4	10.7
Eragrostis rigidior	2.8	16.4
Others	12.5	15.3
	100.0	100.0
Matopos (on clay soil derived from basic rock—cleared 1962)		
Panicum maximum	62.6	8.0
Eragrostis trichophora	18.3	2.3
Heteropogon contortus	3.4	13.1
Urochloa spp.	4.2	50.7
Others	11.5	25.9
	100.0	100.0
Nyamandhlovu (on sandy soil derived from sandstone—cleared 1963)		
Eragrostis rigidior	31.2	14.9
Digitaria pentzii	20.0	15.4
Schmidtia pappophoroides	17.0	26.0
Schizachyrium jeffreysii	8.4	14.4
Aristida pilgeri	7.3	19.8
Others	16.1	9.5
	100.0	100.0

baugh (1965)

$$Y = H + A\,(1 - e^{-BX})^{M+1}$$

where Y = herbage mass, X = density (or similar variable), H and A are the intercept and slope terms of the regression equation, and M and B are constants.

Density has been variously expressed by parameters such as basal area of stems, canopy cover and number of plants per unit area. The implication of such relations is that the response in terms of herbage mass per unit of "clearing input" increases progressively with progressive degrees of clearing.

Little information is available on the effect of various densities of overstory on the herbaceous component of semi-arid savanna. However, Donaldson and Kelk (1970) published data relating the density of blackthorn (*Acacia mellifera* subspecies *detinens*) to mean hay yields over a 3-year period in the Molopo area of the northern Cape Province. Yields of grass increased exponentially as the number of mature plants per unit area decreased; from 119 kg ha^{-1} with a density of 1071 plants per ha^{-1} to 1071 kg ha^{-1} with complete clearing.

In vegetation comprising a mixture of different species of trees and shrubs, it is of course difficult to define the relative competitive effect of different species, and plants of different size. Likewise, extrapolation of results obtained in a particular community to dissimilar communities is unlikely to be valid.

Policy and problems in clearing

Justification for clearing the woody component of semi-arid savanna so as to favour cattle production rests on two main premises. Firstly, it must be economic, and secondly, the economic level of animal production achieved should be greater than that which can be achieved by other possible land-use practices, as for example, the use of two or more species of animal with different diet preferences.

Economic and technological considerations

As shown earlier, the response of different vegetation types to clearing varies widely. Thus the responses in terms of cattle production will also vary; these must be assessed financially and be set against the cost of clearing. Where the potential carrying capacity is high and the responses to clearing large, the practice may be patently profitable. Usually, however, the position is not clear-cut, and the problem can only be resolved on the basis of detailed knowledge of responses, and the cost of effective control of woody plants.

In certain circumstances control of woody vegetation may be achieved at reasonable cost by mattocking or stumping — combined, for example, with charcoal production (Skovlin, 1971), by ring-barking (Ward and Cleghorn, 1964) or by mechanical means (Heady, 1960). Fire and browsing animals may also be used to aid control (West, 1965; Du Toit, 1972; Trollope, 1974). However, generally speaking, the large-scale control of trees and shrubs calls for the use of chemical arboricides, and from appraisal of the literature it is clear that the technological base for chemical control of the important tree and shrub species in East and Southern Africa is weak. It is evident that most species can be killed by the use of arboricides such as picloram and 2, 4, 5-T at relatively high dosage rates. However there have been few systematic attempts to establish minimum effective dosage rates for individual species by means of well-designed experiments on both trees and coppice, covering a wide range of sites and environmental conditions, and including such variables as method and season of application and the size and stage of growth of the plant.

In consequence, the results of experiments have been inconsistent, and it is rarely possible to make unequivocal recommendations for economic chemical control of savanna trees and shrubs. On the other hand, knowledge of the effects of clearing on cattle production is also extremely limited. Harrington (1973), for example, has reported that the reduction in the population of *Acacia hockii* by up to 2400 bushes ha^{-1} with arboricide treatment did not result in increased cattle production. Rational appraisal of the economics of clearing thus depends on further study of both control measures and responses.

Alternatives to cattle production

It is often claimed that the productivity of a multi-species population of wild ungulates in savanna vegetation is greater than that of domestic livestock. These claims are largely based on the supposition that the diet preferences of the various species tend to be complementary, thus making for more efficient use of the vegetation. Other advantages quoted are the ability of many ungulates to go without water for long periods, their mobility, and their relative resistance to endemic diseases (Talbot et al., 1961). Where water supplies are inadequate for cattle, or in country infested with tsetse, the use of land for wildlife is often logical, though in the modern world safari hunting is frequently much more lucrative than cropping for meat production (Johnstone, 1975). However, in spite of the research advances in recent years, the scientific base for the management of such complex ecosystems with a view to sustained meat production is far from adequate. There is little firm information on which to base estimates of potential productivity, and there are difficulties with respect to the slaughter and marketing of wild animals, their control by individual landholders and their significance as carriers of disease.

Under enlightened management and where water supplies are adequate, cattle make efficient use of the herbaceous cover. There is no reason to believe that the use of grazing ungulates, as distinct from those that prefer to browse, would result in appreciable increases in production. However, superimposition of domestic or wild herbivores which prefer to browse might well lead to higher production (see Chap. 2).

There is a need for objective study of this possibility, bearing in mind that the supply of browse feed in the semi-arid savanna varies within and between seasons. Very few estimates have been made of even the standing crop of browse in various vegetation types, and very little is known about the productive potential of browse in semi-arid savanna, or optimum defoliation procedures. There is the possibility also of increasing the browse supply by lopping or felling trees and shrubs. Until such studies are undertaken the subject will remain, as it is now, largely a matter for speculation.

Grazing and fire as manipulative tools

In the previous section it was shown that clearing the woody growth commonly results in changes in the grass species composition in savanna, but that such changes are not necessarily for the better. There remains the question whether composition can be changed by other means in a determinate way, favouring desirable species at the expense of the undesirable.

Discounting in the present discussion the possibility of artificially establishing desirable species from seed, the only practicable means for possible manipulation is the use of specific grazing procedures or, in the wetter phases of the semi-arid savanna, controlled burning.

Except for its effect on the woody component, and thus indirectly on grass composition, there appears to be no evidence that fire can be used effectively to induce favourable changes in the composition of the herbaceous layer in semi-arid savanna (West, 1965). Likewise, except where sound grazing management has enabled degraded veld to recover, there appears to be no evidence that grazing procedures as such bring about any marked beneficial changes in species composition. For example, in recent years a wide range of intensive rotational grazing procedures have been applied experimentally to grassland in semi-arid areas of Rhodesia. No differential effects on grass species composition have been detected. In the same area, repeated mowing for several consecutive seasons, at various times and combinations of times during the growing season, on two vegetation types, has not affected botanical composition differentially (Matopos Research Station, 1972).

Natural savanna vegetation in good condition usually contains some inferior or unpalatable herbaceous species. Treatments which favour the vigour and persistence of the desirable components apparently also favour the less desirable. However, where semi-arid savanna has been over-used or misused for long periods it is not always possible to predict the potential composition under proper use. In such instances carefully sited exclosures may provide a useful indication of the potential for improvement. In appropriate cases, similar changes can then be induced by conservative stocking and sound grazing management.

Grazing management

The usual aim of grazing management is to make efficient use of the grazing resource while conserving and, if possible, improving it. The degree of control which is feasible or desirable will naturally vary in different circumstances, but, at a sophisticated level, efficient grazing management would involve the following considerations.

(1) the provision of sufficient and properly located stock-watering points;
(2) the correct choice of stocking rate;
(3) the provision of paddocks;
(4) the use of suitable grazing procedures;
(5) the feeding, where necessary, of protein-rich supplements and minerals.

(1) Stock water supplies

Lack of water supplies for stock is one of the main limitations to the efficient use of the semi-arid savanna in Africa for livestock production (Heady, 1960; Allan, 1965; McKay, 1968). Frequently, also, poor water distribution leads to over-use and trampling and hence severe degradation of the vegetation, in the vicinity of the more permanent water supplies. Nevertheless, in many parts of Africa the development of stock water supplies without a concomitant control of stock numbers may merely result in the degradation of additional land. Thus, while the provision of water must be regarded as a pre-requisite for the efficient use of land, this aspect must not be considered in isolation.

The availability in recent years of rigid polythene pipes at relatively low cost has greatly facilitated the development of water supplies. On commercial ranches and in development schemes it has become common practice to pipe stock water over distances of 25 km or more. A rule-of-thumb guide based on observation in Rhodesia is that the distance between water points should not exceed 5 km. However, there is a need for critical study of the effect of distance from water, and the siting of water points on herbage utilization, in different circumstances. There is a case also for further study of the effect of water deprivation on animal performance. For example, it has been shown in Botswana that watering every second day as compared with daily watering, did not affect steer performance when walking distance was not associated with watering frequency (McKay, 1968). In areas where water is scarce and herding is feasible such studies could point the way to more efficient use of grazing land without detriment to animal production.

(2) Stocking rates

Stocking rate and production per head and per unit area

The relations between production per head and per unit area and stocking rate, expressed as a number of animals per unit area, are of considerable interest, and have recently been discussed by Jones and Sandland (1974), Sandland and Jones (1975) and Connolly (1976).

The former authors have suggested that with stocking rates in excess of a critical level, beyond which gain per head decreases with increasing stocking rate, a linear model relating gain per animal to stocking rate, expressed as animals per unit area, is empirically justifiable. However, Connolly (1976) has cautioned against the use of a linear model on the basis of limited evidence. He showed that if the relation was in fact quadratic, assuming linearity could lead to large bias.

The discussion has highlighted the importance of establishing the true nature of the relation between stocking rate and animal gain in different circumstances by means of critical experiments with four or more stocking rates. Moreover, in semi-arid savanna the amount of herbage available between years varies widely and there is a corresponding variation in stocking pressure. Accordingly, it is also desirable to establish the variations in animal performance between seasons in relation to stocking rate. Fig. 1.3 serves to illustrate hypothetically the likely magnitude of the extreme changes in the seasonal patterns of gain per animal and gain per unit area in

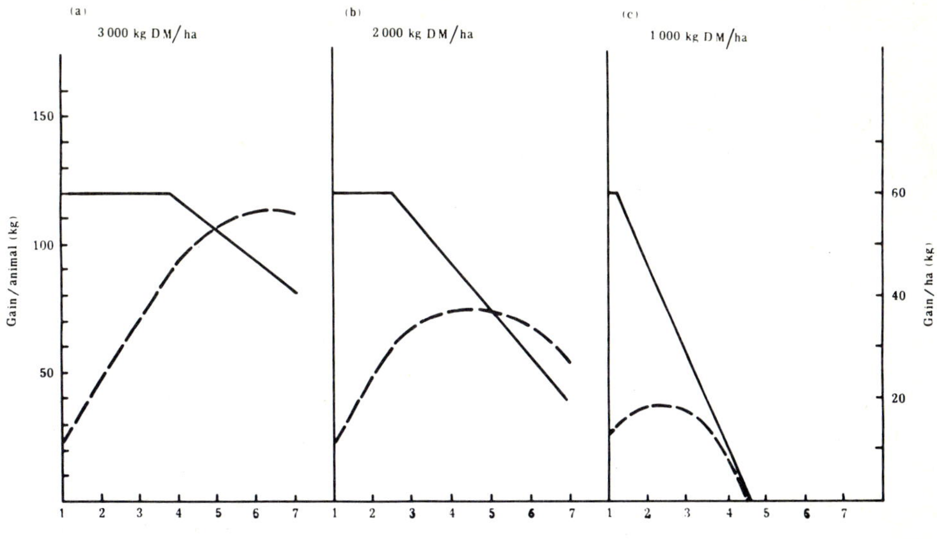

Fig. 1.3. Hypothetical patterns of gain per animal (———) and gain per hectare(— — —) in relation to stocking rate at three levels of herbage production (3000, 2000 and 1000 kg dry matter (DM) per hectare).

relation to stocking rate. In plotting the curves the following assumptions were made:

(1) that the gain per animal in relation to stocking rate, expressed as animals per unit area, follows the pattern shown in Fig. 1.3(b) for a level of herbage production of 2000 kg dry matter ha^{-1} $year^{-1}$;

(2) that at stocking pressures higher than the critical stocking pressure shown in Fig. 1.3(b) (the point of inflexion), gain per animal is directly related to the amount of herbage available per animal in any one season;

(3) herbage produced in any one season, and not used in that season by the grazing animals, has no production potential as food in the next season.

In Fig. 1.3(a) and (c) herbage yield levels of 3000 and 1000 kg dry matter ha^{-1} have been assumed. Fluctuations of this magnitude are by no means uncommon in semi-arid savanna (Table 1.5).

In practice it is usually not feasible to alter stock numbers appreciably from year to year, nor is it possible to predict the level of herbage production in future seasons. Thus, conservative stocking, at a level at which adequate grazing will be available in seasons of low herbage production, is usually essential for the economic and ecological stability of the ranch enterprise. This is especially true with the relatively inflexible systems such as weaner production, where all the progeny of the breeding herd, except the heifers used in replacing culled cows, are sold each year.

The need to stock at a conservative rate also means, of course, that the full potential of favourable seasons cannot be realized. It follows that one of the keys to increased production is to obtain maximum carrying capacity during seasons of low herbage production, while maintaining a satisfactory level of animal performance. If, for example, the adoption of a particular grazing procedure gives an increase in carrying capacity of say 25% in seasons of poor herbage growth, increased productivity of this order can be expected in all seasons. That is, the "safe stocking rate" can now be set at a higher level.

Steer gains at various stocking rates in experiments in a semi-arid region of Rhodesia have been reported by Kennan (1969). Generally there was a negative linear correlation between stocking rate and gain per steer, though in one experiment the relation was curvilinear.

Carew (1976) used these data to assess the influence of stocking rate on profit margins per unit area. He showed that profits were greatest at stocking rates which gave near maximum gains per steer and that the optimum rates in terms of profit were much lower than those which gave maximum gain per unit area.

Stocking rate and the grazing resource

In the drier phases of the semi-arid savanna in particular, the herbaceous cover is characteristically sensitive to over-use (Heady, 1960; West, 1968; Donaldson, 1969; Kennan, 1969; Walter 1971). When the stocking rate exceeds that which is safe in seasons of poor grass growth, stress on the vegetation in dry years is very severe. As mentioned earlier, the position is commonly aggravated by failure of stock water supplies, and an upsurge in the harvester termite population.

Damage to the grazing resources as a result of over-stocking is widespread in the semi-arid savanna of Africa, and the most significant measure which could be adopted to put cattle production in the semi-arid savanna on a sound basis, would be the application of correct stocking rates, based on knowledge of carrying capacity in the poorer seasons.

Generally speaking, the use of drought-reserve fodder banks to supply additional feed to ranch stock in times of stress is unlikely to be practicable or economic. However, on a national or regional basis there is a case for investigating the possibility of setting up intensive feed-lots in which suitable young stock or culled females could be fattened for slaughter as an emergency measure. Though such facilities would only be used intermittently, they would, if they were available on a permanent basis, act as a safety valve and permit more efficient use of the grazing resource in the more favourable years. In addition, the waste resulting from loss of condition, or even death of stock in drought years, could be minimized.

(3) Paddocking

Apart from obvious advantages in terms of grazing procedure, animal management and controlled breeding, paddocking greatly facilitates management of vegetation. In appropriate circumstances it allows for rotational grazing, controlled burning combined with resting, or the exclusion of stock from areas in which, for example, there is a danger of plant poisoning, or which are liable to be damaged if grazed at certain times. Moreover, it is often practicable to separate vegetation types of different acceptability to livestock, thus reducing wasteful selective use of the vegetation. Stock confined to relatively unpalatable areas, without access to more palatable areas, are likely to make better use of the former than where they have access to both types.

Careful alignment of fences may often obviate, or greatly reduce, damage or erosion as a result of the formation of stock trails. Likewise, the provision of corridors on stock routes may sometimes be desirable, although where these are poorly sited they may do more harm than good. Little information is available on the effect of size of paddock as such on the efficiency of utilization of vegetation and on animal production.. It is generally assumed that the efficiency of utilization, and therefore animal production, increases with decreasing paddock size. Nevertheless, optimum paddock size is likely to vary in different circumstances in a complex manner; on the face of it, the most rewarding approach to the problem would be to study the utilization of vegetation in relation to distance from water points, and to use the information to assess the most suitable paddock size as well as the optimum distance between water points. However, it is likely that there are interactive effects between paddock-shape and stocking pressure on what might be termed "utilization-distance" and these must be taken into account. Long, narrow paddocks, for example, radiating from a water point might be utilized in a different way to square paddocks of the same area with the water point in the corner.

A recent development in southern Africa is the development of "cells" in which paddocks radiate from a central core in which water and dipping and handling facilities are placed. In some instances as many as 30 to 40 or more paddocks have been used. The arrangement facilitates cattle management and supervision and is especially suitable in relatively uniform vegetation in country with subdued relief.

(4) Grazing procedure

It is general experience that individual animal performance is invariably at a high level with continuous grazing (set-stocking) at conservative, though not necessarily ultra-conservative rates. However, in natural vegetation especially, set-stocking frequently leads to over-use of certain preferred communities or species and hence to localized degradation and undesirable changes in species composition. Frequently also, there is under-use of potentially valuable constituents. By sub-dividing an area into paddocks and practising rotational grazing it is, to some extent, possible to control the frequency and intensity of defoliation of the preferred plants and to reduce selective grazing.

On the other hand, the supposedly harmful effects of set-stocking, as well as the benefits of rotational grazing, are frequently exaggerated. It is commonly assumed, for example, that set-stocking results in frequent close defoliation of the preferred species, and that intensive rotational grazing allows for controlled defoliation, resulting in optimum yields of herbage of the desired quality.

However, Gammon (1976) recently studied the utilization by cattle of different communities, and defoliation patterns of individual grass tillers by species, under year-long set-stocking and simulated intensive rotational grazing in natural grassland in Rhodesia. The rotational system involved six paddocks and the periods of stay were 6 days in the early growing season and 12 days subsequently. Overall stocking intensity was the same in both systems. Defoliation patterns were found to be very similar under both systems with only a marginal decrease in the frequency and severity of defoliation with rotational grazing as compared with continuous grazing.

Similarly, studies of defoliation patterns under set-stocking with sheep on sown pasture in Britain, have shown that frequency of defoliation of individual tillers is a function of stocking pressure (Hodgson and Ollerenshaw, 1969). Morris (1969) working on similar pasture continuously grazed by sheep, showed that where the stocking pressure was in balance with the net growth of the pasture, neither the frequency nor the severity of defoliation was excessive.

Close control of defoliation patterns of individual species during the growing season in natural savanna, by means of rotational grazing, even where a large number of paddocks are available per herd, seems to be an unattainable ideal. A large number of species are usually involved, diet preferences of livestock change over the year, and the amount of herbage available, and therefore the grazing pressure, varies widely within and between years. Moreover, as has recently been demonstrated by Mufandaedza (1976) in Rhodesia, the response to defoliation in terms of nutrient yields may vary widely between species and ecotypes of grasses; that is, the optimum defoliation treatment is likely to differ between species and ecotypes.

Considering such complexities, and the variability and heterogeneity of natural vegetation, the only wholly reliable basis for defining efficient grazing procedures appears to be to compare different procedures in the field, and to determine effects on animal production and the condition and composition of the vegetation.

The term "rotational grazing" has no precise meaning; innumerable procedures are possible depending on the number of paddocks available per herd and the specific rate of movement of stock at various times in the season or year.

The degree of intensification which is possible or desirable varies widely in different circumstances. Broadly speaking, the less intensive systems involving narrow ratios of paddocks to herds (usually 3.0 or less) are designed primarily with a view to maintaining, or improving, the grazing resource. The more intensive systems, with wider ratios of paddocks to herds, are usually designed with a view to increasing carrying capacity by imposing more efficient defoliation procedures, as well as maintaining or improving the resource.

Systems involving rotational resting

In the less intensive systems the animals are usually kept in each paddock for relatively long periods, varying from several weeks to several months, and provision is made for rotational resting of paddocks for different periods during the growing season, in successive years. In some systems, paddocks are rested in rotation for a full year. Usually as large an area as possible is grazed at the start of the growing season to reduce the grazing pressure in any one paddock to a minimum. Provision may be made for periodic controlled burning of paddocks, the burn being followed by, and sometimes preceded by, rest.

Grazing procedures of this type have been advocated, for example, by Scott (1947), Kennan et al. (1955) and Mostert et al. (1971) for various vegetation types in Southern Africa and by Heady (1960) for East Africa.

They are purportedly based primarily on the provision, in a flexible manner, of rests during three periods of growth which are regarded as critical for the maintenance of vigour and for seed production. These are, the start of growth, when the plants are drawing on stored reserves, the reproductive period, when flowering and seed set occurs, and the period of maturation and senescence when translocation of carbohydrates to the roots and stem bases takes place.

Grazing systems involving rotational resting have been successfully applied in experiments and in ranch practice in East and Southern Africa, but generally speaking do not appear to have been widely accepted. There is limited evidence that they are often beneficial, especially when applied at conservative stocking rates on degraded vegetation (Heady, 1960), or as a means of obviating degradation as a result of continuous grazing (Louw, 1973). However, rotational resting systems have not always been found to be superior to continuous grazing (McKay, 1968). Moreover, unless stress due to excessive stocking pressure on part of the area is avoided, individual animal performance may be depressed as compared with continuous grazing (Kennan, 1962; McKay, 1968).

Intensive rotational grazing

In the more intensive rotational grazing systems four or more paddocks are used per herd, and a rapid rotational grazing procedure with relatively short periods of stay is advocated. Interest in intensive systems was stimulated in recent years by the development of so-called non-selective grazing (NSG) or the Howell-Acocks system (Acocks, 1966) as a means of reclaiming degraded veld. Acocks (1966) suggested that high stocking densities for short periods of 2 weeks or less would result in non-selective use of the herbaceous component and that minimum rest periods of 6 weeks would allow for restoration of reserves between grazings. It was advocated that in the early stages of its application on degraded veld, long rests of 12 months duration should be incorporated in the system. A minimum of twelve paddocks per herd was required.

Circumstantial evidence was presented to support claims that very large increases in carrying capacity could be obtained while at the same time reclaiming the grazing resource. A reduction in the performance of individual animals was expected, and was regarded as part of the cost of reclamation.

In the light of experience, NSG has proved in the main to be impracticable. In the first instance the number of stock which could be carried for a full year was insufficient to achieve the desired stocking pressure on individual paddocks during the growing season. Secondly, where something akin to NSG was achieved, the nutritional stress on the animals and the severity of defoliation of palatable plants was found to be excessive. However, the idea of using a relatively large number of paddocks per herd gave rise to two modifications of NSG, termed in South Africa, controlled selective grazing (CSG) and short-duration grazing (SDG).

(a) Controlled selective grazing. This type of grazing procedure was initially developed on the farm Soutpan about 40 km north of Pretoria (Pienaar, 1968). It was described as an open (flexible) rotational grazing system in which one to three herds grazed 15 paddocks. The periods of stay in the paddocks, the stocking intensity, and the order in which the paddocks were grazed were based on the following precepts. During the main growing season paddocks were not grazed until most of the grasses were flowering. Animals were removed from paddocks before there was a danger of regrowth of previously grazed plants being eaten, and when sufficient leaf material remained on the palatable species to ensure rapid regrowth.

No attempt was made to force animals to eat species unacceptable to them. In practice, periods of stay varied from about 3 to about 21 days. In view of the relatively low quality of the herbage on offer a urea—molasses lick was made available even during the growing season. During the dormant season fairly close grazing of the palatable species, to remove dry herbage, was considered desirable. Stocking rates were relatively low as compared with NSG.

(b) Short-duration grazing. Short-duration grazing as applied to natural vegetation is essentially similar to the type of intensive rotational grazing commonly practised on planted pastures. The period of stay varies over the year, depending mainly on the amount of herbage available and the number of paddocks per herd. In its original form as proposed by Savory (1969), emphasis was placed on avoiding repeated defoliation of grazed tillers while animals remained in a paddock, and the supposed beneficial effects of trampling on the soil and seedling micro-environment. The high stocking density for short periods was regarded as a means of breaking up soil caps, covering seed and laying litter composed of old dead grass. The claim was made that with increasing numbers of paddocks per herd and correspondingly shorter periods of stay and with periods of absence of 45—60 days, depending on circumstances, the stocking rate could be progressively increased as compared with less intensive systems. It was advocated that rotational grazing be continued throughout the dry season.

Unlike the less intensive systems, CSG and SDG, and similar "multi-paddock" grazing procedures, involving the use of four or more, usually six or seven and sometimes 12 or more paddocks per herd, have been widely accepted in Southern Africa. General experience is that, provided the stocking rate is not excessive, degraded veld, especially, often improves markedly as a result of the use of these systems. However, little critical information is available about the effects on livestock production, carrying capacity and veld condition, as compared with less intensive grazing systems or continuous grazing.

Unequivocal information on these points can only be obtained for a given environment using well-designed grazing trials in which different grazing procedures, including continuous grazing, are compared at a suitable range of stocking rates. Moreover, comparatively small differences in grazing procedure may radically influence the results.

The importance of the latter aspect is illustrated by the results of a grazing trial conducted over a 5-year period in Rhodesia. In this trial, which involved 72 paddocks and 12 herds of steers, assessment was made of the effects on production and botanical composition of three periods of stay in paddocks, namely 5, 10 and 20 days in all combinations with four and eight paddock-units to give a range of six widely differing grazing procedures. Each procedure was applied at two stocking rates, one rate being double the other. For a period of about 45 days in the early growing season the paddocks were grazed on a rapid rotation in all treatments. The scheduled treatments were applied for the remainder of the year.

Botanical composition was not differentially affected by the treatments. Relative treatment effects on the pattern of body-mass gains were much the same in each of the 5 years. Animal performance during the dry season was essentially similar with all six grazing procedures, but was slightly higher at the lower stocking rate. However, during the mid and late growing season grazing procedure had marked effects on gains and the relative mean gains for each of the twelve treatment combinations over the 5-year period are shown in Table 1.7.

Critical examination of the results revealed that the primary factor influencing steer performance during the mid and late growing season was grazing intensity. There was a highly significant negative linear relation between grazing intensity in paddocks during each period of stay (in terms of livestock unit days per hectare) and body-mass gains per steer in each of the 5 years of the trial (Denny and Barnes, 1977). Maximum gains per steer were only obtained where the grazing intensity was low, of the order of five to 10

TABLE 1.7

Relative body-mass gain per steer in the mid and late growing season with six rotational grazing procedures applied at two stocking rates (means for the five years 1971—1975)

Period in stay in each paddock (days)	Stocking rate			
	Four paddocks		Eight paddocks	
	High	Low	High	Low
5	97	97	68	94
10	91	100	71	82
20	63	90	45	74

livestock unit days per hectare. It follows that, with a given grazing intensity, individual animal performance will be similar, irrespective of the number of paddocks used per herd.

Differences between the effects of the different grazing procedures were much greater at the higher than at the lower stocking rate. By inference, it seems that in comparisons of intensive rotational grazing with continuous grazing, differential effects will be found only at relatively high stocking rates, or in seasons of relatively low herbage production. Evidence in keeping with this conclusion has recently been presented by Robinson and Simpson (1975). Working in Australia they compared continuous and rotational grazing of sown pastures at three stocking rates, 10, 20 and 30 wethers per ha. Differences in favour of rotational grazing were found at the intermediate stocking rate and only 2 of the 4 years covered by the study. Significantly, in the years in which no differences were found, rainfall was either well below or well above average. They suggest that increases in animal production can only be obtained by the use of rotational grazing at relatively high stocking rates, but that no response can be expected at excessively high rates. In various ways the findings of Conway (1963), McMeekan and Walshe (1963), Walker (1968) and Young and Newton (1973) support this conclusion.

Claims that intensive rotational grazing provides a unique means of favourably modifying veld composition have not been substantiated by the results of multi-paddock grazing trials in Rhodesia (Denny et al., 1977; Denny and Barnes, 1977; Denny and Steyn, 1977). By contrast, substantial benefits have been recorded on certain veld types in South Africa (Howell, D., 1976; Howell, L.N., 1976).

In the context of the semi-arid savanna, the present indications are that rotational grazing of a number of paddocks, using procedures akin to CSG or SDG facilitates the maintenance of a healthy and vigorous grass cover. Provided the stocking rate is not excessive, and the animals are not stressed by being kept too long in each paddock, especially during the growing season, it seems that a high level of individual animal performance can be maintained. If desired, rotational resting of certain paddocks may be incorporated in intensive rotational grazing systems. However, the outstanding issues are whether intensive rotational grazing will result in an increase in carrying capacity and production per unit area, and where, and in what circumstances, it offers a means of favourably modifying the composition and condition of veld.

The experimental evidence available indicates that there are no advantages, and that there may be disadvantages in using a large number (in excess of six or eight) paddocks per herd. Consideration of the relation between periods of stay and absence and increasing numbers of paddocks per herd leads to the same conclusion. However, it is not uncommonly believed that carrying capacity will increase in proportion to the number of paddocks per herd. In some instances 30 or more paddocks per herd are advocated. Intensification of grazing procedure involves an appreciable outlay in feeding and water supplies and the demands on the skill of the grazier increase sharply with increasing number of paddocks per herd. There is a need to resolve this question, as well as those posed above, by means of critical experiments in major plant associations.

(5) The feeding of protein-rich supplements and minerals

The quality of the food in savanna vegetation in terms of energy and protein varies widely during the year, being highest in the early and mid-growing seasons and lowest during the dry season, when it is usually at sub-maintenance level. It is well established that, provided sufficient herbage is available, the feeding of protein-rich supplements, or, in some instances non-protein nitrogen, makes it possible for the animals to consume enough herbage to meet their energy requirements for maintenance or a slow gain in mass. The improvement in nutritional status results in marked improvement in the reproductive performance of breeding stock. Hence, the feeding of protein increases the efficiency of utilization of the herbage on offer.

It has also been shown that the feeding of phosphorus during the growing season, when animals are gaining mass, often enhances gains, and occasionally also the feeding of common salt is beneficial (Van Niekerk, 1975).

Appropriate supplementary feeding is thus part and parcel of efficient grazing management, and the cost of supplementary feeding will only be at a minimum when grazing management is at a high level.

CONCLUSION

The semi-arid savanna of Africa comprises delicately balanced ecosystems which are subjected to erratic fluctuations in moisture supply and in which production potential per unit area is relatively low. Nevertheless, it constitues a valuable resource, the conservation of which is of great importance to the welfare of many millions of people.

Over large areas of the semi-arid savanna population pressure is intense and over-use by domestic livestock has resulted in severe degradation. The problems are compounded by the attitudes of the majority of the inhabitants towards population control, stock ownership and land-use. In the traditional tribal grazing areas individual responsibility for stock is usually divorced from responsibility for the grazing resource. Activities are directed towards survival, and the individual tribesman sees large families and large herds as a source of security. In the face of rapid population growth, administrators have on the whole been singularly unsuccessful in persuading tribesmen that, in the context of modern Africa, the opposite applies.

Without a proper balance between plant production and animal numbers, the rational and stable use of semi-arid savanna for animal production is impossible. The future of large tracts of Africa thus depends, in the first instance, on drastic changes in traditional attitudes towards land-use among relatively unsophisticated and uneducated indigenous peoples. This can only be brought about by concerted and well-planned programmes of rural reform and education, linked with an economic plan for the country as a whole. The need to create an understanding among the people of the broad requirements for stability of the soil—plant—animal complex, and to develop a social and economic infrastructure which will favour a change.

Appreciable areas of semi-arid savanna in Africa are used for commercial cattle ranching. Generally, but by no means invariably, such land is in relatively good condition and production is at a reasonably high level. Moreover, the social and economic framework is such as to provide a satisfactory basis for further improvement. In the main, present knowledge is adequate for stable production, provided stocking rates are conservative. The challenge is to define procedures which will allow for the maximum level of sustained production from the land. In this respect knowledge is deficient, and important gaps have been indicated in earlier discussions. Further progress depends in the first instance on well planned research to remedy these deficiencies.

ACKNOWLEDGEMENTS

I thank Mr. T. Muller and Mrs. K.E. Bennett of the National Herbarium, Salisbury, for kindly checking the botanical nomenclature.

REFERENCES

Acocks, J.P.H., 1966. Non-selective grazing as a means of veld reclamation. Proc. Grassl. Soc. S. Afr., 1: 33—39.

Acocks, J.P.H., 1975. Veld types of South Africa. Memoirs of the Botanical Survey of South Africa, No. 40, 2nd edn. Botanical Research Institute, Pretoria, 128 pp.

Allan, W., 1965. The African Husbandman. Oliver and Boyd, London, 505 pp.

Animal Production Research Unit, Botswana, 1975. A Five Year Programme of Integrated Beef Cattle and Range Research in Botswana 1970—1975. Part I. Technical Results. Ministry of Agriculture, Botswana, 182 pp.

Barnes, D.L., 1976. A review of plant-based methods of estimating food consumption, percentage utilisation, species preferences and feeding patterns of grazing and browsing animals. Proc. Grassl. Soc. S. Afr., 11: 65—71.

Bembridge, T.J. and Steenkamp, J.D.G., 1976. An agro-economic investigation of beef production in the Matabeleland and Midlands Provinces of Rhodesia. Rhod. Agric. J., 73: 29—34.

Bissett, J.L., 1974. In: Annual Report, 1973—74. Division of Livestock and Pastures, Department of Research and Specialist Services, Rhodesia, pp. 120—124.

Bissett, J.L., 1975. In: Annual Report, 1974—1975. Division of Livestock and Pastures, Department of Research and Specialist Services, Rhodesia, pp. 169—174.

Brown, L.H., 1971. The biology of pastoral man as a factor in conservation. Biol. Conserv., 3: 93—100.

Carew, G.W., 1976. Stocking rate as a factor determining profitability of beef production. Rhod. Agric. J., 73: 111—115.

Chapman, N.G., 1976. Aerial spraying of tsetse flies (*Glossina* spp.) in Rhodesia with ultra low volumes of endosulfan. Trans. Rhod. Sci. Assoc., 57: 12—21.
Clary, W.P., 1969. Increasing sampling precision for some herbage variables, through knowledge of timber overstory. J. Range Manage., 22: 200—201.
Coaton, W.G.H., 1950. Termites and their control in cultivated areas in South Africa. Bull. No. 305, Dep. Agric. Government Printer, Pretoria.
Coaton, W.G.H., 1958. The hodotermitid harvester termites of South Africa. Sci. Bull. No. 375. Dep. Agric. Government Printer, Pretoria.
Coaton, W.G.H. and Sheasby, J.L., 1975. National survey of the Isoptera of Southern Africa. 10. The genus *Hodotermes* Hagen (Hodotermitidae). Cimbebasia Ser. A, 3: 105—138.
Connolly, J., 1976. Some comments on the shape of the gain-stocking rate curve. J. Agric. Sci., Camb., 86: 103—109.
Conway, A., 1963. Effect of grazing management on beef production. 11. Comparison of three stocking rates under two systems of grazing. Ir. J. Agric. Res., 2: 243—258.
Denny, R.P. and Barnes, D.L., 1977. Trials of multi-paddock grazing systems on veld. 3. A comparison of six grazing procedures at two stocking rates. Rhod. J. Agric. Res., 15: 129—142.
Denny, R.P. and Steyn, J.S.H., 1977. Trials of multi-paddock grazing systems on veld. 2. A comparison of a 16-paddocks-to-one-herd system with a four-paddocks-to-one-herd system using breeding cows. Rhod. J. Agric. Res., 15: 119—127.
Denny, R.P., Barnes, D.L. and Kennan, T.C.D., 1977. Trials of multi-paddock grazing systems on veld. 1. An exploratory trial of systems involving twelve paddocks and one herd. Rhod. J. Agric. Res., 15: 11—23.
Donaldson, C.H., 1969. Bush Encroachment with Special Reference to the Blackthorn Problem of the Molopo Area. Dep. Agric. Tech. Serv. Government Printer, Pretoria.
Donaldson, C.H. and Kelk, D.M., 1970. An investigation of the veld problems of the Molopo area. 1. Early findings. Proc. Grassl. Soc. S. Afr., 5: 50—57.
Du Toit, P.F., 1972. The goat in a bush—grass community. Proc. Grassl. Soc. S. Afr., 7: 44—50.
Edwards, D.C., 1942. Grass-burning. Emp. J. Exp. Agric., 10: 219—231.
Edwards, D.C., 1951. The vegetation in relation to soil and water conservation in East Africa. Commonw. Bur. Pastures Field Crops Bull. No. 41: 28—43.
Elliott, R.C. and Topps, J.H., 1963. Studies of protein requirements of ruminants. 1. Nitrogen balance trials on two breeds of African cattle given diets adequate in energy and low in protein. Br. J. Nutr., 17: 539—548.
Findlay, J.B.R., 1971. An investigation into the use of insecticidal baits for the control of *Hodotermes mossambicus* (Hagen). Phytophylactica, 3: 97—102.
Ford, J., 1971. The Role of the Trypanosomiases in African Ecology — A Study of the Tsetse Fly Problem. Clarendon Press, Oxford, 573 pp.
Gammon, D.M., 1976. Studies of the patterns on defoliation, herbage characteristics and grazing behaviour during continuous and rotational grazing of Matopos sandveld of Rhodesia. D.Sc. (Agric.) thesis, University of the Orange Free State, Bloemfontein, 300 pp.
Geiss, W., 1971. A preliminary vegetation map of South-West Africa. Dinteria No. 4, November 1971, 69 pp.
Glover, J. and Robinson, P., 1953. A simple method of assessing the reliability of rainfall. J. Agric. Sci., Camb., 43: 275—280.
Grosenbaugh, L.R., 1965. Generalisation and reparametrization of some sigmoid and other non-linear functions. Biometrics, 21: 708—714.
Hagan, D.H. and Wilmhurst, E.C., 1975. Controlled bovine trypanosomiasis, the disease spread by tsetse. World Crops, 27: 248—250.
Harrington, G.N., 1973. Bush control. A note of caution. East Afr. Agric. For. J., 39: 95—96.
Heady, H.F., 1960. Range Management in East Africa. Government Printer, Nairobi, 125 pp.
Hodgson, J. and Ollerenshaw, J.H., 1969. The frequency and severity of defoliation of individual tillers in set-stocked swards. J. Br. Grassl. Soc., 24: 226—234.
Howell, D., 1976. Observations on the role of grazing animals in revegetating problem patches in veld. Proc. Grassl. Soc. S. Afr., 11: 59—63.
Howell, L.N., 1976. The development of multi-camp systems on a farm in the southern Orange Free State. Proc. Grassl. Soc. S. Afr., 11: 53—57.
Jackson, S.P., 1961. Climatological Atlas of Africa. CCTA/CSA. Joint Project No. 1, Lagos—Nairobi. Government Printer, Pretoria, 55 plates.
Jameson, D.A., 1967. The relationship of tree overstory and herbaceous understory vegetation. J. Range Manage., 20: 247—249.
Johnstone, P.A., 1975. Evaluation of a Rhodesian game ranch. J. S. Afr. Wildl. Manage. Assoc., 5: 43—51.
Jones, R.J. and Sandland, R.L., 1974. The relation between animal gain and stocking rate in grazing trials: Derivation of a model from experimental results. J. Agric. Sci., Camb., 83: 335—342.
Kennan, T.C.D., 1962. Veld management in the semi-intensive, semi-extensive and extensive farming areas of Southern Rhodesia with special reference to research findings. Proc. First Interdepart-

mental Conf. Pasture Workers, Bulawayo, 1962. Cyclostyled, pp. 74—91.
Kennan, T.C.D., 1969. A review of research into the cattle—grass relationship in Rhodesia. Proc. Veld Manage. Conf. Bulawayo, May 1969. Government Printer, Salisbury, pp. 5—26.
Kennan, T.C.D., Staples, R.R. and West, O., 1955. Veld management in Southern Rhodesia. Rhod. Agric. J., 52: 4—21. Reprint Bull. No. 1802. Government Printer, Salisbury.
Ledger, H.P., 1964. The role of wildlife in African agriculture. East Afr. Agric. For. J., 30: 137—141.
Lee, K.E. and Wood, T.G., 1971. Termites and Soils. Academic Press, London, 251 pp.
Louw, A.J., 1973. 'n Vergelykende studie tussen ononderbroke beweiding en vasgestelde seisoenwissel-weidings-stelsels op natuurlike veld te Towoomba navorsingstasie. Tech. Bull. No. 114, Dep. Agric. Tech. Serv. Government Printer, Pretoria.
Lovemore, D.F., 1969. Tsetse flies as a limiting factor to the full utilisation of the natural grazing of Rhodesia by domestic stock. Rhod. Sci. News, 3: 342—345.
Macdonald, I.A.W., 1974. In: Annual Report, 1973—1974, Division of Livestock and Pastures, Department of Research and Specialist Services, Rhodesia, pp. 117—120.
Madallali, S.A., 1975. Animal production in Tanzania. Proc. III World Conf. Anim. Prod. Sydney University Press, pp. 351—357.
Maloiy, G.M.O. and Heady, H.F., 1965. Grazing conditions in Kenya Masailand. J. Range Manage., 18: 269—272.
Matopos Research Station, 1972. Annual Report For Year Ending 30th September 1972. Cyclostyled, pp. 17—26.
McKay, A.D., 1968. Rangeland productivity in Botswana. East Afr. Agric. For. J., 34: 178—193.
McKay, A.D., 1970. Range management research at E.A.A.F.R.O. East Afr. Agric. For. J., 35: 346—349.
McKay, A.D. and Frandsen, P.E., 1969., 1969. Chemical and floristic components of the diet of Zebu cattle (*Bos indicus*) in browse—grass range pastures in a semi-arid upland area of Kenya. Trop. Agric. (Trinidad), 46: 279—292.
McMeekan, C.P. and Walshe, M.J., 1963. The interrelationships of grazing method and stocking rate in the efficacy of pasture utilisation by dairy cattle. J. Agric. Sci., Camb., 61: 147—163.
Mills, P.F.L., 1966. Effects of nitrogen on the yields and quality of hay from three types of veld at Matopos. Rhod. Agric. J., 63: 9—12, 18, 21.
Morris, R.M., 1969. The pattern of grazing on continuously stocked swards. J. Br. Grassl. Soc., 24: 65—76.
Mostert, J.W.C., Roberts, B.R., Henslinga, C.F. and Coetzee, P.G.F., 1971. Veld management in the O.F.S. region. Dep. Agric. Tech. Serv. Bull. 391. Government Printer, Pretoria.
Mufandaedza, O.T., 1976. Effects of frequency and height of cutting on some tropical grasses and legumes. 1. *Hyparrhenia filipendula* (Hochst.) Stapf and *Heteropogon contortus* (L) Beauv. ex. Roem. and Schult. Rhod. J. Agric. Res., 14: 21—38.
Mulligan, H.W. (Editor), 1970. The African Trypanosomiases. George Allen and Unwin, London, 950 pp.
Nel, J.J.C., 1968. Agressive behaviour of harvester termites *Hodotermes mossambicus* (Hagen) and *Trinervitermes trinervoides* (Sjösted). Insectes Soc., 15: 145—156.
Petrides, G.A., 1975. Principal foods versus preferred foods and their relation to stocking rate and range condition. Biol. Conserv., 7: 161—169.
Phillips, J., 1959. Agriculture and Ecology in Africa. Faber and Faber, London, 424 pp.
Pienaar, A.J., 1968. Weiveldbeheer op Soutpan: Toepassing in die suur tot suuragtige gemengde Bosveld. Farmers' Day at Soutpan. Lecture Ser. No. 1. Dep. of Agric. Tech. Serv., pp. 9—14.
Pratt, D.J., 1969. Management of arid rangeland in Kenya. J. Br. Grassl. Soc., 24: 151—157.
Pratt, D.J., Greenway, P.J. and Gwynne, M.D., 1966. A classification of East African rangeland. J. Appl. Ecol., 3: 369—379.
Rattray, J.M., 1957. The grasses and grass associations of Southern Rhodesia. Rhod. Agric. J., 54: 197—234.
Rattray, J.M., 1960. The Grass Cover of Africa. Food and Agricultural Organisation of the United Nations, Rome, 168 pp.
Rattray, J.M., 1961. Vegetation types of Southern Rhodesia. Separate from Kirkia, 2: 68—93.
Robinson, G.G. and Simpson, I.H., 1975. The effect of stocking rate on animal production from continuous and rotational grazing systems. J. Br. Grassl. Soc., 30: 327—332.
Sandland, R.L. and Jones, R.J., 1975. The relation between animal gain and stocking rate in grazing trials: and examination of published theoretical models. J. Agric. Sci., Camb., 85: 123—128.
Savory, C.A.R., 1969. Principles of range deterioration, reclamation and management. Proc. Veld Manage. Conf., Bulawayo. Government Printer, Salisbury, Rhodesia, pp. 83—87.
Scott, J.D., 1947. Veld management in South Africa. Bull. No. 278, Union of South Africa, Dep. Agric. Government Printer, Pretoria.
Skovlin, J.M., 1971. Ranching in East Africa: a case study. J. Range Manage., 24: 263—270.
Staples, R.R., 1930. Studies in veld management. A second report on certain veld burning and grazing experiments. Sci. Bull. 91, Union of S. Africa Dep. Agric.

Talbot, L.M., Ledger, H.P. and Payne, W.J.A., 1961. The possibility of using wild animal production on East African rangelands. 8th Int. Congr. Anim. Husb., Hamburg, pp. 205—210.
Trapnell, C.G., 1959. Ecological results of woodland burning experiments in Northern Rhodesia. J. Ecol., 47: 129—168.
Trollope, W.S.W., 1974. Role of fire preventing bush encroachment in the Eastern Cape. Proc. Grassl. Soc. S. Afr., 9: 67—71.
Vale, G.A. and Hargrove, J.W., 1975. Field attraction of tsetse flies (*Diptera: Glossinidae*) to ox odour, the effects of dose. Trans Rhod. Sci. Assoc., 56: 46—50.
Van Niekerk, B.D.H., 1975. Supplementary feeding of range animals in grassveld areas of Southern Africa. Proc. III World Conf. Anim. Prod. Sydney University Press, pp. 273—279.
Walker, B., 1968. Grazing experiments at Ukiguru, Tanzania. 2. Comparisons of rotational and continuous grazing systems on natural pastures. East Afr. Agric. For. J., 34: 235—244.
Walker, J., Moore, R.M. and Robertson, J.A., 1972. Herbage response to tree and shrub thinning in *Eucalyptus populnea* shrub woodlands. Aust. J. Agric. Res., 23: 405—410.
Walter, H., 1971. Ecology of Tropical and Subtropical Vegetation. Oliver and Boyd, Edinburgh, 557 pp.
Ward, H.K. and Cleghorn, W.B., 1964. The effects of ring-barking trees in *Brachystegia* woodland on the yield of veld grasses. Rhod. Agric. J., 61: 98—105, 107.
West, O., 1965. Fire in vegetation and its use in pasture management with special reference to tropical and subtropical Africa. Commonw. Bur. Pastures Field Crops. Mimeographed publication No. 1/1965.
West, O., 1968. The vegetation of Southern Matabeleland: a study of the reasons for its present degradation and possible means of rehabilitation. Proc. Symp. Drought Develop. 1st Rhod. Sci. Congr. The Association of Scientific Societies in Rhodesia, Salisbury, pp. 85—103.
Whyte, R.O., 1947. In: The Use and Mis-Use Of Shrubs and Trees As Fodder. Imperial Agricultural Bureaux Joint Publication No. 10, pp. 94—109.
Wild, H. and Barbosa, L.A. Grandvaux, 1967. Vegetation map of the *Flora Zambesiaca* area. Supplement to Flora Zambesiaca. Collins, Salisbury, 71 pp.
Young, N.E. and Newton, J.E., 1973. A comparison of set stocking v. rotational grazing with sheep at three stocking rates. J. Br. Grassl. Soc., 28: 183.

Chapter 2

GAME RANCHING IN AFRICA

B.H. WALKER

Department of Botany, University of The Witwatersrand, Johannesburg (South Africa)

INTRODUCTION

The semi-arid savanna regions of Southern and East Africa, as defined in the Introduction to this book, occupy between one-third and one-half of the total land area (see Chap. 1, Fig. 1.1). It is not possible to be more precise, since the definition of semi-arid is itself imprecise. It is valid, though, to state that they cover no less than one-third of the region.

Apart from edaphically determined, pure grasslands, and the virtually pure scrub of some induced semi-desert zones (such as the Karoo in South Africa) the vegetation is basically savanna in that it is a mixture of trees, shrubs and a herbaceous layer of grasses and non-graminoid forbs. The distribution of these savannas is discontinuous and is largely determined by rainfall. There is, in fact, no clear cut-off between savanna and other vegetation, since increasing rainfall leads towards a gradual increase in the development of woodland, which nevertheless still has an understory of shrubs and herbs. Our attention though, is here limited to those savanna areas where annual arable dry-land agriculture is not a viable form of land-use.

ECOLOGICAL CHARACTERISTICS OF THE SEMI-ARID SAVANNAS

In Chap. 1 Barnes describes the climate of this region in some detail, and therefore only the dominant features will be discussed here.

The ecology of these savannas is largely determined by climate, and the over-riding feature of the climate is the low and generally very variable rainfall. For example, approximately one-third of East Africa has a rainfall probability of less than 500 mm in four years out of five (Glover et al., 1954). The variability has two dimensions. Temporally, it varies both inter- and intra-seasonally, and spatially it varies considerably over short distances (Anonymous, 1963). Differences of 100% in rainfall can occur in successive years, with pronounced effects on vegetative growth (Kelly and Walker, 1976), as described in the next section.

Periodic droughts are thus a feature of the climate and, within seasons, are associated with peaks in temperature. At such times temperatures are well in excess of the range within which growth takes place, and surface temperatures may exceed 65°C for much of the day.

Vegetation

Details of species composition of the main community types are again given in Chap. 1 and this account concentrates on the ecological aspects of the vegetation.

Soil type, rainfall and fire interact to produce vegetation types which vary in structure from grassland through parkland and scrub to dry woodland. The individual effects of each are extremely difficult to unravel, but their combined effects result in the following characteristics.

Annual primary production is very variable, and closely follows the pattern of rainfall; but although rainfall is demonstrably a dominant control factor for vegetative production, it is in fact the seasonal soil moisture regime which directly influences plant growth. Soil moisture is in turn influenced by the rate of water infiltration and the water-holding capacity of the soil. On heavy-textured soils the development of a soil surface cap assumes great significance. The caps consist basically of clay from dis-aggregated soil, resulting from raindrop action on exposed surfaces. They are frequently compounded by the development of a micro-floral crust, dominated by blue-green algae but also including a variety of other lower plants.

The role of algal crusts is debatable. Some research workers (Shields et al., 1957; Fogg et al., 1973) consider that, owing to their nitrogen-fixing properties, their overall effect is beneficial. Against any such benefits, however, must be weighed their effects on soil water relationships, which are not clear. Loope and Gifford (1972) have shown algal crusts to cause an increase in infiltration in south-eastern Utah. However, their method of measurement may have influenced the results, since they pre-wet the soil surface before the infiltrometer runs. This is important because, at least in central Africa, the dried mucilaginous sheaths of many of these algal crusts are hydrophobic (unpublished results), and intensify the action of the soil cap in preventing water infiltration, particularly where rainfall generally comes in the form of short, hard storms. Kelly and Walker (1976) have shown in south-eastern Rhodesia that infiltration through a soil surface covered by litter (where there is no cap) is nine times faster than that through a bare soil surface.

The action of grass tufts is also of considerable importance. Glover et al. (1962) have demonstrated the effect of bunch grasses in funnelling rainfall through their crowns into their own rhizospheres. The rate of infiltration through the crown of a grass tuft is very high and emphasizes the importance of a high basal cover.

This leads to a consideration of the next important aspect, namely the proportions of perennial and annual grasses. Two factors emphasize the importance of maintaining a high proportion of perennial grasses. Firstly, they have a caespitose structure, and a sward of perennials has a far higher basal cover than one of annual grasses. Infiltration of water in a perennial sward is therefore much higher than in annual swards. Secondly, the failure of annual grasses to grow in drought years gives them a highly variable pattern of inter-seasonal production, as shown in Fig. 2.1, and although annuals play an important role in these semi-arid areas it should be an objective of management to maintain as high a proportion of perennials as possible.

A characteristic of savanna vegetation which is of great significance to game ranching is the variation in phenology of the woody and grass species. The woody species generally break vegetative dormancy well in advance of the grasses. In Southern Africa the flush of woody species begins in August, some 3 months before the rains set in, and Rushworth (1975), for example,

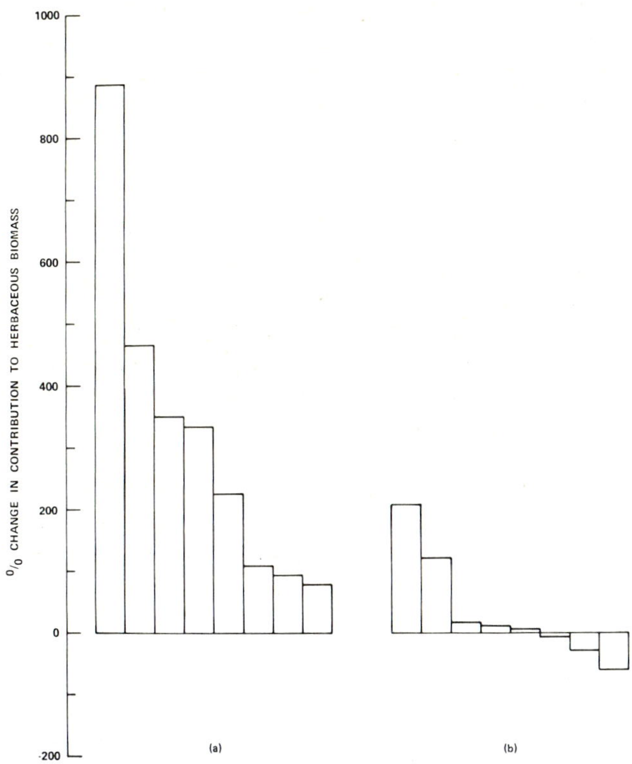

Fig. 2.1. Results from a comparative study of the effects of land-use on the functioning of a semi-arid savanna ecosystem (Kelly and Walker, 1976). Changes in the contribution by (a) annual (plus micro-perennial) and (b) perennial grasses to total herbaceous peak standing crop from the 1970—1971 to the 1971—1972 season.

has shown that this early flush in the Wankie National Park provides a high-quality food supply during the hot, end-of-dry-season period when both the quantity and quality of grass are at a minimum. Sinclair (1975) has established that the amount of green grass (high-quality grazing) during this period in the Serengeti is the limiting factor to animal welfare.

The proportions of grass and woody plants are a function of climate, soil-type and fire. Relative to the life-span of the woody plants the first two are constant, and fire is therefore the important variable with regard to the changes which occur in these proportions. It has been well established (Phillips, 1965; West, 1965; Scott, 1972) as a major determinant of the structure of savanna vegetation, and the frequency, intensity and season of burning all influence the differential effect of fire on trees, shrubs and grasses. In general, an increase in frequency leads to a reduction in woody species and the development of the herbaceous layer (see Fig. 2.2). Under natural conditions fires usually occur as a result of lightning at the end of the dry season (West, 1965), when the woody plants have flushed and are therefore vulnerable, but the grasses are still dormant. Unlike the higher rainfall areas, where regular burning leads to a fire sub-climax grassland, a high frequency of fires in semi-arid ecosystems can be very deleterious, resulting in a

Fig. 2.2. Effects of regular burning on savanna vegetation. The foreground is an annual-burn plot of a long-term burning experiment in the Matopos National Park, Rhodesia. The background is a control plot. The experiment was 22 years old when the photograph was taken.

loss of perennial grass species and lowered basal cover, without much effect on the woody vegetation.

The effect of changes in the proportions of grass and woody plants on total available food is of particular interest here. There is ample evidence (West, 1947; Barnes, Chap. 1) to show that removal of the woody plants leads to a significant increase in the production of grass — by up to 400%. Unfortunately though, in these bush clearing trials no account was taken of the loss in browse which the increases in grass entailed. What evidence there is, however, suggests that in terms of utilizable dry matter the loss is much less than the increase in grass. Total seasonal production (including wood) is higher in a mixed tree/bush/grass community (Kelly, 1973) but only a proportion of this production is available to herbivores. Table 2.1 gives the biomass available to herbivores from the herbaceous and browse layers for two semi-arid ecosystems. It is apparent from these examples that production from the browse layer is one-third or less of that from the herbaceous layer.

On the basis of total food production, therefore, it would appear that in most instances bush-clearing and tree-removal will be beneficial. However, there are a number of considerations other than total production, all of which caution against indiscriminate clearing. Briefly they are as follows.

(1) Grasses under tree and shrub canopies (see Fig. 2.3) have a lower ratio of structural carbohydrates to green leaves and are more palatable than those which replace them when the trees are removed. The net effect of the changes in grass species composition following clearing is frequently undesirable, is sometimes of little significance (Barnes, 1972), but is seldom, if ever, an improvement. Furthermore, although production is lower under tree canopies the grass there remains green for longer into the dry period.

(2) The significance of browse production lies in the time of the year during which it occurs. As described earlier, the flush of new growth provides an invaluable source of high-protein food during the time of the year when grass quality constitutes a limiting factor to the sustainable level of herbivore biomass.

(3) Little is known about the resilience or the long-term stability of these ecosystems. Their highly variable and frequently adverse climatic conditions

TABLE 2.1

Peak biomass of herbaceous material and available browse in a savanna (Kelly, 1973) and an alluvial woodland (Goodman, 1975)

	Herbaceous material ($kg\ ha^{-1}$)	Available browse ($kg\ ha^{-1}$)
Savanna	1690	628 [a]
Alluvium	4113	780

[a] Total leaf and twig biomass <3 m, not all of which is available.

Fig. 2.3. The effect of tree-canopy cover on the species composition of the herbaceous layer. Under the *Acacia tortilis* tree is *Dactyloctenium giganteum*, with *Schmidtia pappophoroides* in the open (foreground).

have already been described, and the reduction in diversity of both the physiognomic structure and species composition of the vegetation, which results from clearing, may well reduce their long-term resilience. The persistence of woody vegetation during severe droughts dampens the extremes of environmental conditions and productivity.
In any event, a pure grass sward is unstable in this environment and requires a constant input in the form of mechanical or chemical bush-clearing, or controlled burning.

Animals

Details of the interactions amongst the herbivores, and between herbivores and their habitats, are dealt with later. The general characteristics of interest at this stage can be summarized rather briefly.

One can order the species of ungulates from those which occur as individuals or pairs, through smaller groups up to the large herds (Jarman, 1974). This trend corresponds fairly well with the gradient from woodland to plains, and a decrease in the proportion of browse in the diet. The seasonal limitation in food for grazing animals is a causal factor in the migration of these large herd species (Bell, 1970; Sinclair, 1975). The aggregation of animals in large herds has frequently given the impression of African plains teeming with game, yet more careful reading of some of the earlier accounts of these areas (e.g. Selous, 1881) suggests otherwise. Although such herds certainly occurred, they were relatively few and far between, and in the travels of these early writers long periods occurred when virtually no animals were seen.

The migrations tend to follow water and available food. In East Africa, they are regular and predictable (Bell, 1970) but in Southern Africa they are highly irregular and much less predictable. It is significant that this leads to a pattern of vegetation-use which is relatively light, on average, and has periods of fairly intensive use.

The species richness of large herbivores is relatively high in these ecosystems although, as will be seen later, the total biomass of animals is usually dominated by just a few of these species. Species diversity (α-diversity) and, more particularly species equitability (Whittaker, 1965) is consequently much lower than is commonly believed.

PAST AND PRESENT LAND-USE

Very little evidence is available concerning the pattern of land-use prior to the advent of Western Man, apart from that which can be deduced from accounts of the early explorers, as previously described. The semi-arid areas are, by definition, prone to droughts and both the distribution and number of animals were limited by available water. During the dry season they were concentrated in the vicinity of large water bodies and along the fringes of the larger rivers, but with the advent of the rains they dispersed from these areas, making use of the seasonal pans. Superimposed onto this annual pattern of wet season dispersal and dry season concentration were the more irregular migrations.

Prior to colonization from Europe the impact of Man was limited by disease and the availability of water. Vast areas within the semi-arid region could not be permanently settled, owing either to a lack of permanent water or to the fact that much of it was inhabited by the tsetse fly, which transmits trypanosomiasis — a disease carried by wild herbivores which is fatal to cattle.

The intrusion of modern civilization into this region has had three major effects. Firstly, the human population has suddenly and dramatically increased (Kay, 1972), owing to modern medicines. The resulting pressure on the land has forced people to live in the semi-arid areas on a permanent basis and cattle numbers, too, have increased as a result of modern husbandry methods and a campaign against the tsetse fly. The increase in cattle has been accompanied by a programme of game reduction or elimination, owing both to competition for food and to animal disease control. The result has been a virtual elimination of the herbivore browsing component and a big increase in both the proportion and actual number of grazers. Secondly, there has been a rapid development of artifical permant water points which have allowed, for the first time, year-round grazing and browsing. Finally, the introduction of fences, roads and human settlements have virtually eliminated the migrations by wild animals.

The net effect on the habitat has, in general, been a fairly predictable pattern of degradation (West, 1947; Brown, 1969; Kelly, 1973). Lack of controlled grazing has led to area-selective grazing and over-grazing of the palatable grass species, which has resulted in a loss of perennial grasses, increased inter-seasonal variation, a decrease in basal cover with a corresponding increase in soil erosion and, often, a scrub encroachment problem. This had led to further decreases in productivity and a vicious downward cycle has become the familiar pattern common to so many parts of Africa (see Figs. 2.4 and 2.5).

Attempts to alleviate the problem have included controlled burning, bush-clearing, fencing and associated controlled grazing techniques. There have been improvements, particularly where large areas are run as a single commercial enterprise, but even here the problems of attempting to increase productivity in an ecologically brittle area, persist. It is perhaps not surpris-

Fig. 2.4. Over-grazed *Colophospermum mopane* savanna. The photograph was taken in mid-rainy-season. No herbaceous growth had occurred and cattle were being forced to browse.

Fig. 2.5. Over-stocking with cattle, on the left, showing the effects of over-grazing on the grass layer, compared to unoccupied state land with a low density of game animals (Matabeleland Province, Rhodesia).

ing, therefore, that in the search for a solution, the utilization of adapted, natural populations of animals has been suggested. We proceed, therefore, to a consideration of game ranching.

THE THEORETICAL BASIS OF GAME RANCHING

Proponents of game ranching base their claims of higher sustainable secondary productivity on two main assumptions: (i) the theory of multiple species utilization of the vegetation, and (ii) superior adaptation of the animals to their environment (Dasmann and Mossman, 1960, 1962; Talbot and Talbot, 1962, 1963; Dasmann, 1964; Roth, 1964; Talbot et al., 1965; Brown, 1969). Each will be considered in turn.

Utilization of vegetation

According to the multiple species utilization theory a properly balanced wildlife ecosystem has a herbivore community such that the species involved exhibit non-overlapping, preferred diets, leading to an even pressure of utilization on all the available food resources. Reliable data to validate the theory are difficult to come by. However, we can examine it by considering the various ways whereby animal species differentially utilize vegetation.

There are three components to the variation in pattern of vegetation-use which can be attributed to differences between animal species in a mixed community of large herbivores. First, spatial separation of the herbivore species, owing to differences in basic habitat preferences. Second, temporal separation owing to differences in seasonal movement patterns, and third, where the species are sympatric, differences in preferred diet species and in the parts of the plants concerned.

Spatial or habitat separation

The marked spatial heterogeneity in vegetation structure in these semi-arid areas was described in the Introduction. Areas of a thousand hectares invariably have at least a few physiognomically different vegetation types which vary from open, more grassy communities at one extreme, to denser, more thicket-type vegetation at the other. The obvious association between plains animals such as wildebeest (*Connochaetes taurinus*) and open grassy areas, and between the more individualistic browsing animals such as kudu (*Tragelaphus strepsiceros*) and bushed areas, is evidence in favour of habitat separation. The degree of separation, though, is more difficult to establish, and in this regard we may refer to two examples. Lamprey (1963) made repeated observations of animal occurrence along transects through various vegetation types in the Tarangire Game Reserve in Tanzania. From these he derived a range of preferred habitat conditions for each species and the results (see Fig. 2.6) are an often-used example of habitat preferences. In the second study, Ferrar and Walker (1974) examined the habitat separation amongst 14 of the major herbivore species in Kyle National Park, Rhodesia. Separate observations were made on the occurrence of these species, and at each point of observation 26 habitat variables were measured. The analyses included a similar approach to that used by Lamprey (see Fig. 2.6) demonstrating a remarkably similar pattern of overall distribution across woodland, savanna and grassland. They also allowed for a multiple discriminant function analysis, which showed the major trends in habitat separation between the species (Fig. 2.7). It is of particular interest to note the variance around each species placement along each discriminant function.

Both these studies certainly demonstrate differences between herbivores in preferred habitats, but the striking feature about the results is not the separation which they show, but rather the high degree of overlap. Only a few, often minor, species show any clear-cut habitat separation. The point which begins to emerge at this stage, and which will become clearer later, is that the earlier theoretical accounts of African wildlife ecology wrongly emphasized habitat separation. What is important is the amount of overlap and flexibility in the animals' choice.

Temporal separation

Two examples again illustrate this component of species separation. The

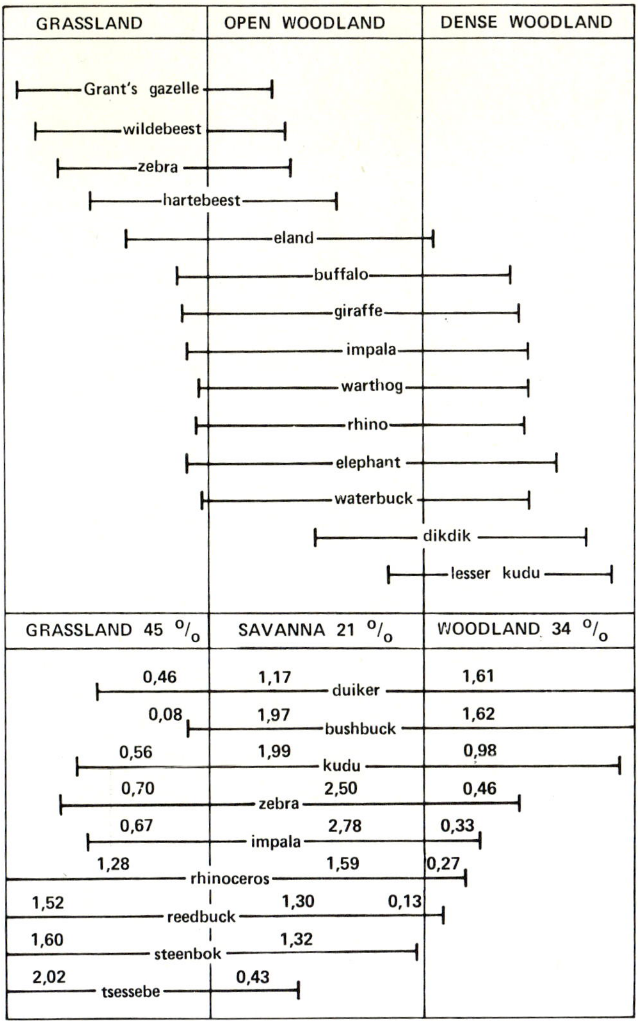

Fig. 2.6. Ecological separation — or overlap — in savanna habitat utilization by game animals. The top section is from Lamprey (1963, courtesy of East Afr. Wildl. J.), the lower section from Ferrar and Walker (1974, courtesy of South Afr. Wildl. Manage. Assoc.).

first (Anderson and Herlocker, 1973) was conducted in Ngorongoro Crater in Tanzania, and the results show (Table 2.2) for each of 12 different vegetation types, the months of the year during which they are occupied by the various herbivore species. Differences are certainly apparent, but at no time is any vegetation type used by only one species, although one is sometimes dominant, and the seasonal spread for each species is considerable.

In the second study, Hall-Martin (1974) considered a single species, the

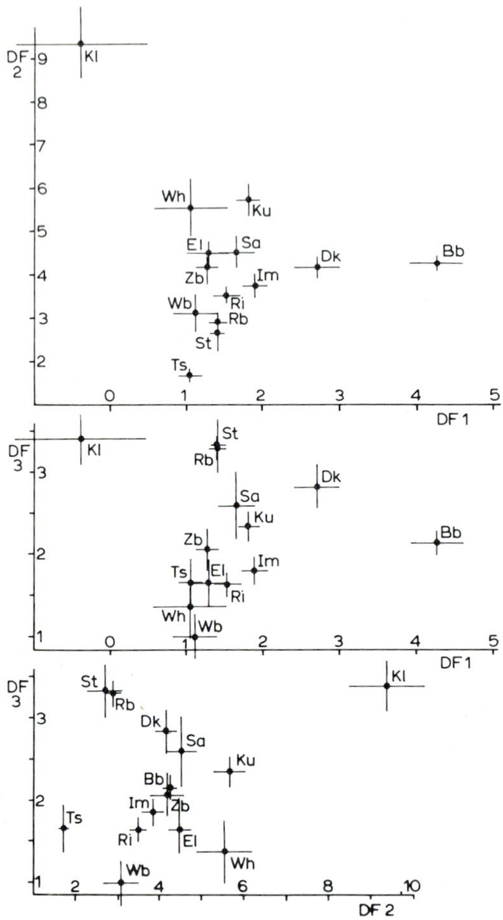

Fig. 2.7. Separation of herbivores according to a Discriminant Function Analysis of 26 habitat variables. The figure shows the location of species centroids or niche centres according to combinations of three discriminant functions, with two standard errors on each side of the centroids.
DF 1, A gradient from open, pure grassland to wooded shrubby patchy vegetation; DF 2, deep soil, herbaceous flat land to rocky, shrubby broken country; DF 3, short, sparse herbaceous layer to a tall dense herbaceous layer.
Bb, bushbuck; Dk, duiker; E, eland; Im, impala, Kl, klipspringer; Ku, kudu; Rb, reedbuck; Ri, rhinoceros; Sa, sable; St, steenbok; Ts, tsessebe; Wb, wildebeest; Wh, warthog; Zb, zebra.

giraffe (*Giraffe camelopardalis*), and examined its seasonal movement patterns within a ranch-sized game reserve in the lowveld of South Africa. Several different plant communities were recognized, and based on the densities of giraffe in each, there were significant differences between seasons in a number of them. Giraffe occurred, however, at all times in all communi-

... game utilization of vegetation types in Ngorongoro Crater (after Anderson and Herlocker, 1973)

Vegetation types	Animals (Jan. Feb. March April May June July Aug. Sept. Oct. Nov. Dec.)	Period
Cyperus/Aeschynomene swamp complex	Hippopotamus, zebra, wildebeest, rhinoceros, reedbuck, elephant	June – Oct.
Chloris/Cynodon grassland	Mainly larger ungulates, zebra, wildebeest	May – Oct.
Acacia xanthophloea/Achyranthes woodland	Mainly rhinoceros, elephant, hippopotamus, zebra, wildebeest, waterbuck	Jan. – Dec.
Sporobolus/Odyssea sparse very short grassland	Mainly Thomson's gazelle	June – Sept.
Sporobolus/Cynodon very short grassland	Mainly wildebeest, zebra, Thomson's gazelle	Jan. – June; Nov. – Dec.
Digitaria/Cynodon short grassland	Mainly wildebeest, zebra, Thomson's gazelle, kongoni	Jan. – July; Nov. – Dec.
Cynodon/Digitaria medium height grassland	Mainly wildebeest, zebra, eland, Thomson's gazelle	Jan.; May – Sept.; Dec.
Pennisetum/Cynodon/Andropogon medium height grassland	Zebra, wildebeest and gazelle	May – Oct.
Cynodon/Andropogon/Digitaria medium height grassland	Zebra, kongoni, eland, wildebeest, gazelle	May – Oct.
Andropogon/Digitaria/Themeda medium height grassland	Gazelle, zebra, wildebeest, kongoni, buffalo	June – Oct.
Themeda/Setaria tall grassland with *Lippia*	Mainly Grant's gazelle, zebra, eland, buffalo, elephant, kongoni	June – Oct.
Woodland/forest/bushland complex	Mainly zebra, buffalo, eland, Grant's gazelle, klipspringer, kongoni, elephant	Jan. – June; Nov. – Dec.

ties. Several other studies of individual species (e.g. Conybeare, 1975) demonstrate the same general pattern, namely that although seasonal preferences by game animals for different vegetation types certainly occur, the animals again show a high degree of overlap and flexibility in their choice.

Diet separation

The major split in composition of the available food is into grass and browse. The common use of the terms "grazer" and "browser" highlights this separation and promotes the concept of non-overlapping diets and the partitioning of available food resources.

Hoffman (1973) and Hoffman and Stewart (1972) have paid close attention to this aspect through an analysis of the stomach structures of African ruminants. They recognize three basic types of herbivore, each with subdivisions based on the feeding habits of the animals and on the internal structure of the stomach, and therefore its adaptation to breakdown and digestion of fibrous material: (1) bulk and roughage eaters, made up of (i) roughage grazers, (ii) fresh grass grazers dependent on water, and (iii) dry region grazers; (2) selectors of juicy, concentrated herbage, made up of (i) tree and shrub foliage eaters, and (ii) fruit and dicotyledonous herb selectors; (3) intermediate feeders, (i) preferring grasses (example impala, *Aepyceros melampus*), and (ii) preferring forbs, shrubs and tree foliage (e.g. eland, *Taurotragus oryx*).

Animals in the third category are clearly catholic in their diet but even the first two are noted as being only "selectors" of various foodstuffs and are not restricted to them. Further evidence from other detailed studies supports this view. Conybeare (1975) has shown that kudu in the Wankie National Park, although being predominantly browsers, consume significant amounts of grass during the early growing season — in one case the stomach contents contained 80% grass. Duckworth (1972) showed for the same area that buffalo (*Syncerus caffer*), generally regarded as being obligate grazers, consumed significant amounts of browse during the hot, dry season. Field (1975) has demonstrated a seasonal change in the diets of buffalo, eland, oryx (*Oryx gazella*) and cattle on a Kenyan ranch. All four feed in the herbaceous layer early in the season, but as the grass matures so eland concentrate more on browse.

There is no published quantitative evidence of separation between herbivores within the browse category. However, a fair amount has been done on separation in the grass category, mainly in East Africa (Gwynne, 1967; Gwynne and Bell, 1968; Bell, 1970; Field, 1972; Blankenship and Qvortrup, 1974; Owaga, 1975). In the first group of these studies, by Gwynne and Bell (1968), wildebeest, topi (*Damaliscus luntus*), zebra (*Equus burchelii*) and Thompson's gazelle (*Gazella thomsoni*) were compared with respect to the percentages of green leaf, sheath, grass stem and browse which were found in their stomach contents. Unfortunately, Bell's (1970) values for zebra do not sum to 100, but nevertheless the picture is clear (Table 2.3). Separation certainly occurs, but it is the degree of overlap which is most remarkable, being in excess of 75% for adjacent species. Owaga (1975) and Page and Walker (1978) showed the same degree of overlap between wildebeest and zebra in northern Kenya and Natal (South Africa), respectively.

Taylor and Walker (1978) compared the cattle and game sections of a lowveld ranch in Rhodesia, and examined the differences in grass species utilization. Cattle proved to be more selective than the combined game community. The two dominant species of grass made up 80% of the cattle diet, but only 59% of the game diet. The cattle section had 12 grass species with utilization values greater than 10%, as opposed to 20 such species in the game section. The multi-species game community therefore leads to a more even pressure of utilization on the different grass species, but even here there was a remarkable degree of overlap between cattle and game. The preferred grass species were the same for all herbivores, but some game animals are prepared to take other grass species in larger quantities.

Overall pattern of use

The three mechanisms which contribute to variation in vegetation utilization all demonstrate, individually, considerable flexibility and overlap between the animal species. The net effect of the three mechanisms interacting simultaneously must now be assessed. If the multiple-use theory is correct we should be able to detect an even pressure of use over all components of the vegetation, floristically and structurally. Few studies have been aimed at such an analysis, but the following are of particular interest.

Goodman (1975) examined the structure and composition of both the plant and animal communities in a segment of riverine habitat in the Sengwa wildlife area, Rhodesia. He measured both the biomass and percentage utilization of the plants, by species, in five different height layers, at each of the three seasons, and compared this to the structure and composition of the animal biomass. Considering any one season there was considerable variation in the pressure of use on the vegetation, both by species and by height

TABLE 2.3

Mean properties of different parts of plants occurring in the herbage intake of four ungulate species in the dry season, September 1967 (from Bell, 1970)

Ungulate species	Total dicotyledon	Total grass	Grass leaf	Grass sheath	Grass stem
Wildebeest	0	100	17.2	52.7	30.1
Topi	0	100	9.4	53.4	37.2
Zebra	0.1	99.9	0.2	48.7	57.0
Thomson's gazelle	38.75	61.25	2.88	37.5	20.88

layers, but an average for all three seasons produced a fairly even pressure, at least by height layers.

An interesting feature to emerge from this study is that a heterogeneous mixture of herbivores did not occur in any one particular plant community. The herbivore biomass was dominated by two species, impala and elephant (*Loxodonta africana*), and the several other species contributed insignificantly to the impact on the vegetation. It seems that particular habitat types are dominated by only a few herbivore species, often only one, and the multi-species "community" occurs only in terms of β-diversity, as a mosaic on a regional basis, with the species being only partially sympatric. Furthermore, the overall utilization of vegetation is by no means spatially even.

Spatial heterogeneity in utilization of vegetation was examined by Taylor and Walker (1978). The pattern and extent of vegetation use by game and cattle was compared by means of 80 vegetation transects. The variability and the range in overall utilization of the herbaceous layer between transects was greater in the game section (8—90%) than in the cattle section (4—60%). The cattle section was fenced into large paddocks and had more water points, but nevertheless it was clear that the dominant game species, impala, had an inherent behavioural tendency to remain on preferred sites and to graze them well below the level at which cattle moved onto other sites.

Conclusions concerning vegetation utilization

Assembling all that has been said about vegetation utilization, we can conclude that in a semi-natural, wildlife ecosystem, variation in utilization of vegetation is effected by spatial, temporal and dietary differences in the animal species concerned. In each of these, however, the degree of overlap is greater than the degree of separation and it is not the rigid aspect of "non-overlapping" diets, etc., that should be emphasized. Rather it is the enormous flexibility and overlap which is important, as this allows for utilization of all components of the vegetation (albeit somewhat uneven) and ensures that changes in the structure and composition of the vegetation can be accommodated. A decline in one vegetation component is accompanied by a shift to other components in the diet of those species feeding on it, or by a partial shift of the animals to other communities. Conversely, should one component (a species or height class) begin to expand, it will be given attention as a result of changes in the diet of some species and in the spatial distribution of others. It is this property of properly balanced wildlife communities which gives them superiority over cattle. In their comparative study of game and cattle, Taylor and Walker (1978) showed that, within the same plant community types, there was a significantly lower density of woody species in the game section than in the cattle section — demonstrating a control of woody species in the game section which was not being effected in the cattle section.

The differences between the two concepts of use has been illustrated diagrammatically by Walker (1976), using a single resource gradient, which could be a series of plant species, plant parts, habitats or time of occupation of a single habitat (Fig. 2.8). The area of each animal species block represents the amount of utilization, the length indicating range and the height indicating intensity of use. Section (a) depicts the non-overlapping concept, and (b) the flexible, overlapping viewpoint.

Other considerations

Apart from the pattern of vegetation utilization there are various other considerations related to the multi-species concept. Many of the difficult and expensive problems associated with cattle ranching stem from the fact that other animal species have been virtually eliminated. We have already noted the controlling effect of browsing animals on woody species density. The cost to farmers of controlling "thickening up" scrub on cattle ranches is a major problem. There are other, perhaps less obvious, advantages in trying to maintain as natural an ecosystem as possible, including other wildlife species.

The advantages of introducing or building up game bird populations are usually given in terms of sport and human benefit. However, they have other more fundamental functions. F. Meyer (Mlezu Agric. College, Rhodesia, personal communication 1973) records counting the remains of 5100 harvester termites (*Hodotermes mozambicus*) in the crop of a single guinea

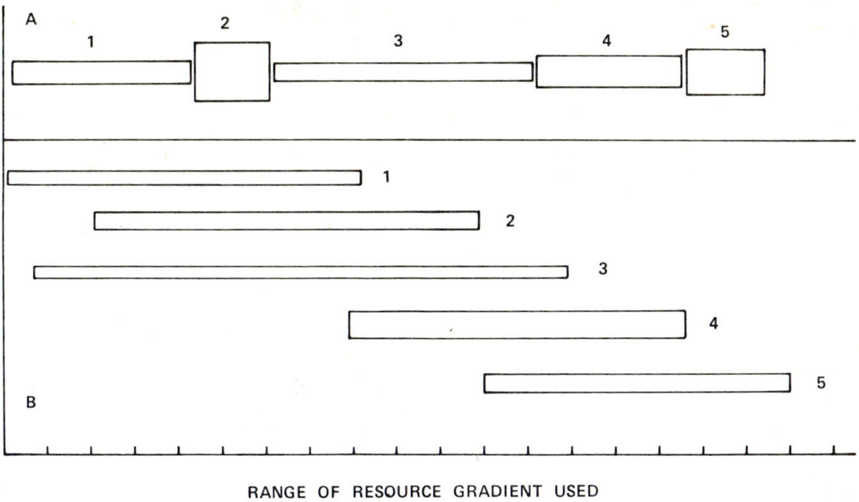

Fig. 2.8. The non-overlapping view of multiple species utilization of habitat (A) as opposed to the flexible, overlapping viewpoint (B). The area of the rectangles represents amount of use. All available evidence points to the latter as being more correct (after Walker, 1976, courtesy of Sthn. Afr. Grassl. Assoc.).

fowl (*Numicla mitrata*) shot on a Rhodesian ranch. Harvester termites are a major problem in semi-arid southern Africa due to their effect on the grass layer. In certain dry years they consume as much or more than all of the large herbivores. Game birds also have a significant effect on the soil surface, and their scratching in areas where flocks congregate leads to a noticeable improvement in the germination of grasses with awnless seeds, many of which are amongst the most desirable species (e.g. *Panicum* spp., *Urochloa* spp., *Dactyloctenium* spp.).

Scavengers and small predators are generally regarded as vermin by cattle ranchers. However, animals such as the jackal (*Canis mesomelas*) also have a useful function in that they exert a control on the populations of rodents, and help to prevent the spread of diseases by removing carcasses and sick animals. There is a need for reliable data comparing these advantages to the losses due to predation on calves, but already there are farmers who promote the ecosystem concept.

Adaptations and productivity of game animals

The ways in which game ranching and cattle ranching differ at the secondary production level again fall into three main areas, viz. adaptations of the animals to the physical environments, resistance to disease, and inherent potential productivity.

Adaptations to the physical environment

The species of African ungulates occurring naturally in the semi-arid regions have evolved a number of mechanisms adapting them to high temperatures and limited water supplies. Prevention of water loss is of prime importance and in a series of experiments in East Africa, Taylor and Lyman (1967) and Taylor (1969a, b, 1970a, b) have demonstrated the relative significance of these mechanisms in several species, including zebu and Hereford cattle.

Adaptive hyperthermia is one of the most important mechanisms found in a number of species. Taylor (1970a) showed that in Grant's gazelle (*Gazella granti*), Thomson's gazelle, oryx, wildebeest, zebu steers and buffalo, dehydration caused an increase in the temperatures at which the animals begin to pant or sweat. The magnitude of these effects was much greater in the species adapted to desert conditions. The maintenance of a body temperature at or even above the ambient temperature clearly reduces the requirements for evaporative cooling and is therefore an effective means of conserving body water. It is interesting, though, that when water is freely available, the animals maintain normal body temperatures, even under a heat load.

It is remarkable that species such as Grant's gazelle and oryx have the ability to maintain body temperatures of around 45°C for up to 6 h with no ill effects. It is particularly important that the brain temperature does not increase too much, and this is achieved by counter current cooling of arterial blood on its way from the heart to the brain. The external carotid artery passes through the cavernous sinus, where it ramifies into multiple small arteries. Cool venous blood from the nasal passages drains into the sinus, cooling the arterial blood (Taylor, 1970b).

The formation of dry faeces is another mechanism whereby antelope conserve water, and the moisture content of the faeces decreases with a decrease in available drinking water. The volume and concentration of urine is also affected by environmental conditions, and restriction of water to near minimal levels causes a two- to three-fold reduction in urine volume in eland, and a consequent approximate doubling in concentration. The same conditions have only a slight effect on the Hereford (Taylor and Lyman, 1967). However, where water is freely available, eland excrete a larger volume of urine per kg of body weight than do Herefords.

A further physiological mechanism in animals such as the eland is a narrow thermal neutral zone. At night the eland cools down rapidly, and in fact needs to increase respiration to keep warm. Its low morning temperature, though, is an advantage, as it warms up slowly during the day, and has a much shorter period during which it needs to keep cool. By allowing its body temperature to rise to almost the ambient temperature it stores a heat load which is dissipated later during the night (Taylor, 1970a).

Coat colour and specialized pelage are further means of reducing heat absorption. The light colours of most African ungulates reflect incident radiation, and their pelage keeps heat away from the body. Louw (1970) claims that the pelage temperature becomes higher than the ambient temperature, and therefore leads to a reverse heat flow. The pelage allows air to get to the skin, which causes cooling through cutaneous evaporation.

Finally, water conservation may be achieved by behavioural mechanisms. Avoidance of the mid-day sun and selection of hygroscopic food at night, coupled with the physiological adaptations, can render species like the eland and the oryx almost independent of surface water (Taylor, 1969a). It is nevertheless important to note that zebu cattle proved to be even more efficient than eland, wildebeest or buffalo in actual water conservation (Taylor, 1968), and although lacking the other physiological mechanisms, they are still very well adapted to semi-arid conditions.

Resistance to disease

Trypanosomiasis is the major disease limiting cattle ranching in the semi-arid areas. Wild herbivores are carriers of the disease but are not susceptible to it, and for this reason the tsetse-fly inhabited regions are suited only to game ranching. Game animals are susceptible to several diseases, notably anthrax, but are again only carriers of other diseases such as Foot-and-Mouth, Theilerias (buffalo) and Malignant Catarrhal Fever (wildebeest).

Their role as carriers of these diseases, which are lethal to cattle, has led to their elimination in many ranching areas.

Inherent productivity

Advantages claimed for wild herbivores as meat-producing animals are, in general, a rapid gain in weight, a higher reproductive potential than cattle, and better carcass qualities. The last of these is undisputed as game animals have a higher dressing-out percentage and about 2.5% fat, as opposed to up to 40% in cattle (De Vos, 1969). The first two, though, are not as clear. Skinner (1971) has examined the productivity of eland and has concluded that they are inferior to cattle in terms of efficiency of food utilization. Their narrow thermal neutral zone necessitates a greater consumption of food per kg of body weight. Taylor and Lyman (1967) showed that Hereford cattle excrete far less urea than do eland, and imply that this urea is recycled to protein in the Hereford, cutting down the dietary requirements.

Body size is an important consideration for, as Brody (1945) has shown, the smaller the animal the higher is its metabolic rate. A single animal weighing 456 kg is regarded by agriculturists as 1 animal unit, or livestock equivalent (LSE). It is incorrect, however, to assume that 10 impala (average weight of 45 kg) make up 1 LSE. The ratio should be computed as the 0.73 power of the values, which works out to 5.4 impala to 1 LSE. The average adult weight of most game species is less than that of cattle (c.f. Mentis, 1972) and a mixed community of wild herbivores therefore requires a larger food supply per unit of animal biomass than does a herd of cattle. Mentis (1977) discusses the basis of carrying capacity determination for African ungulates.

Reproductive rates of most game species are higher than those of cattle. Even larger animals such as the eland have a shorter gestation period and shorter post-partum anoestrus (Skinner, 1971). This may be offset in some species by restriction of breeding to a single season, but in general the average reproductive rate of a wild herbivore community is higher than that of a herd of cattle.

Turning aside from a purely theoretical comparison, there is little quantitative evidence which can be used to evaluate the relative productivities of cattle and game in the field. The references to high biomass by Talbot and Talbot (1963), Brown (1969) and others do not give details of the boundaries of the areas observed on a year-round basis. The waterhole phenomenon, where large numbers of animals congregate, can be extended to less-obvious congregation in preferred areas at various times, where huge biomasses may be recorded, but the animals concerned derive their food, in part, from other areas. Roth (1964) makes a case for higher productivity by game in the Rhodesian lowveld, based on the data of Dasmann and Mossman (1962) for the Henderson ranch. However, in their comparison of the game and cattle sections of Buffalo Range (also in the Rhodesian lowveld), Taylor and

Walker (1978) have produced evidence that overall percentage utilization of vegetation was slightly higher in the game section although the total animal biomass per unit area was about the same — if anything somewhat lower in the game section. Mentis (1977) and Mentis and Duke (1976) have similarly concluded that stocking capacities for wild herbivores in Natal in South Africa (which includes both semi-arid and higher rainfall areas) do not exceed those for cattle under present circumstances. There is a need for more quantitative data before any definite statements can be made on the comparative productivity of cattle and game.

Having considered the theoretical basis of game ranching it is now necessary to take a look at some of the more practical issues involved, before any conclusions or recommendations can be made.

PRACTICAL IMPLICATIONS IN GAME RANCHING

Game ranching, as envisaged here, follows the definition of Parker and Graham (1971); the husbandry of presently wild animals for the same ends as conventional ranching, i.e., the production of food and utilities as a means of occupancy of land by Man. It does not include the use of carcasses from game culling operations in National Parks and Game Reserves, which have very different primary objectives as forms of land-use.

There are a large number of registered game ranchers in Southern Africa, with fewer in East Africa. Some of the ranches concerned are not purely game ranches, and either have cattle and game sections or, less commonly, run cattle and game together. Economic analysis of these ranches are scarce, although in a few cases data on cropping levels and meat sales have been published. Johnstone (1971, 1973), in particular, has detailed the activities on Rosslyn Ranch in Rhodesia, and has demonstrated the economic and ecological advantages of using the land for game rather than cattle. However, personal observations have shown that in practically all cases, including Rosslyn Ranch, the income from meat production has had to be supplemented by income from safari hunting. Considering the animal biomass which is carried on the land, this situation should not exist, and the reasons are worth examining.

A major drawback to commerical game ranching at present is the marketing system and the prices paid for game meat. Veterinary restrictions prevent game carcasses from being processed in the same abattoirs or distribution centres as domestic animal carcasses. This results in uncertain supplies, and combined with a lower demand for game meat by the public, has kept prices well below those paid for beef. An added problem is that game ranches are generally far from major centres and transport of fresh carcasses is therefore difficult and costs are high. As a result, it is common practice in Southern Africa to cut the meat into small strips and to market it as salted, air-dried sticks, locally known as "biltong".

The methods of game cropping have, until very recently, been wasteful. Night shooting using powerful lights has proved to be the most efficient way of obtaining animals without unduly increasing their wariness. Even with good hunters, though, the percentage of spoiled meat is fairly high, and reduces the amount of utilizable meat production per unit area of land. This particular problem is likely to disappear as the use of opaque plastic sheeting increases (cf. Densham, 1974). By this method animals can be relatively easily captured and slaughtered at a convenient time, with minimal hardship to the animals, and virtually no wastage.

A fundamental problem in game ranching is the reduced possibilities for management. Because they are wild, game animals are not easily handled and therefore many standard practices on cattle ranches are not possible. For example, Skinner (1971) has shown how castration of young eland bulls enhances their carcass quality. If, in addition to castration of males, game animals could be dosed to clear them of internal parasites, immersed in acaricide solutions (via plunge-dips) to prevent debilitation and disease from external parasites, de-horned, treated for wounds, and so forth, their productivity would be increased. Furthermore, wild herbivores cannot be herded or moved on a regular basis, and therefore improved range management through the use of rotational or short-duration grazing is not possible. Game animals, to their own individual advantage, operate on a free-range system.

CONCLUSIONS

Considering both the theoretical basis and the practice of game ranching we may conclude that the claims made by its early proponents are at least partly correct, but in general were somewhat over-optimistic. This no doubt stemmed from a desire (perhaps subconscious) on the part of these early workers to use game ranching as a means of effecting wildlife conservation. As more quantitative data become available, some of the theories need to be modified. The example given in this chapter is the change in emphasis from the non-overlapping, separation view of multi-species utilization to the view that it is a flexible, overlapping system which in fact occurs, and which is important in the overall stability of these semi-arid ecosystems. As census techniques improve, the numbers of animals counted tend to moderate and the density and biomass claimed for game animals becomes less dramatic (cf. Robinette et al., 1974). Nevertheless, game ranching has a number of definite advantages, and it is useful to attempt a summary of its pros and cons.

On the positive side, a properly balanced wildlife community makes more efficient and more even use of the available food resources than does any mono-specific herbivore community. The vegetation in most of the semi-arid areas of East and Southern Africa consists of a mixture of trees and/or shrubs and grasses, and the presence of a significant proportion of predominantly browsing animals is particularly important in effecting efficient utilization, and in maintaining stability of the vegetation. Where the introduction of a comprehensive network of water supplies is either impossible or prohibitively expensive, the use of adapted game species is the only viable form of ranching. This also applies to those areas infested with tsetse fly, although their absolute superiority over cattle in this respect appears to be in some doubt (Cunningham, 1968).

The two major disadvantages of game ranching are, firstly, the lower food conversion rates of wild herbivores and, secondly, the reduced options in management. From all that has been said previously it seems unlikely that game ranching will ever replace cattle ranching in areas where the latter is not restricted by water or disease.

The evidence reviewed here indicates that perhaps the optimum form of land-use in these areas is a mixture of cattle and game. The findings of Taylor and Walker (1978), that game animals took a wider range of plant species than cattle, but that cattle spread their grazing more evenly over the range, are of particular interest in this regard. Research is still required to determine the optimum structure of the vegetation for maximum utilizable plant production. When this is known for each ecosystem, maximum secondary production will then be obtained by determining the optimum structure of the mixed game and cattle community which will make most efficient use of the plant production.

RECOMMENDATIONS

If game ranching is to play a significant role as a form of land-use in the region, its development needs both short- and long-term objectives. The short-term is concerned with making the best use of what is available. The long-term concerns the development of new domestic species.

Short-term improvements

The most common problem on game and game/cattle ranches is an imbalance in the composition of the herbivores. There is a general need to increase the ratio of browsing to grazing and the aim should therefore be to increase, where the habitat is suitable for them, the proportion (if not the actual numbers) of the larger, predominantly browsing, animals such as eland, kudu and giraffe. It is a basic principle of sound range management that the stocking rate be adjusted to suit the carrying capacity of the range. As shown earlier, the amount of browse on offer to the animals is considerably less than the amount of grass, and the correct approach is therefore to begin by determining the stocking rate of browsing animals. The optimum number of predominantly browsing animals is determined by dividing the carrying capacity of the browse, in livestock equivalents, by the average proportion of the animals' diet which is made up of browse. The remaining portion of the total amount of required food for this number of animals must then be deducted from the grazing carrying capacity. The number of animals, expressed in livestock equivalents, is then made up to the latter figure with predominantly grazing animals, either cattle, game or a mixture of both.

Determination of the composition and stocking rate of the herbivores in this way is obviously only approximate. But provided that the derived composition is not too far removed from that which would in fact give an overall equal utilization of available vegetation, the ecological amplitude of the animal species concerned will allow for the necessary adjustments in their selected diets. For this reason it is advantageous to include some of the more euryphagous species, such as impala, in the community.

In addition to even utilization of the plant species, it is necessary to utilize each area of the ranch according to its optimum level for sustained maximum production. This can be achieved to some extent by an even and adequate distribution of water points and by placing mineral salt licks in non-preferred areas. The latter technique has been highly successful on some ranches but has had little effect on others. The distribution of cattle is easily controlled by paddocking, but wild game animals are not amenable to herding or being moved. For this reason, the best procedure is to encourage free movement of game by maintaining all water supplies and by using a system of fencing which allows easier movement across them by game animals.

Integrated cattle and game ranching is presently restricted by veterinary problems, but it is unlikely that these will prevent the development of game/cattle ranches as envisaged here, for the following reasons. The most important game species as regards disease are buffalo and wildbeest, the former being carriers of Theilerias, and Foot-and-Mouth, and the latter of Malignant Catarrhal Fever. Both these species are predominantly grazers and their inclusion into a cattle/game enterprise is therefore undesirable. The most desirable species are the larger animals at the browsing end of the spectrum, together with species such as the wart-hog (*Phacochoerus aethiopicus*), which is highly palatable and concentrates on a different food source from other herbivores. These species do not pose a great disease threat. Furthermore, if game animals are being introduced for the first time, or re-introduced after previous complete removal, they can be immunized by vaccination while being handled during the introduction stage. Eland, for example, are susceptible to Brucellosis (contagious abortion), of the same sub-type affecting cattle, and can be vaccinated with the same vaccine (J. Condy, Veterinary Dept., Rhodesian Ministry of Agriculture, personal communication, 1974).

Game cropping should be carefully planned for each species on an age and sex basis, bearing in mind that frequency and intensity of cropping significantly affects maximum sustained yield (cf. Walters and Brady, 1972). The practical difficulties associated with hunting (which lead to wasted carcasses and unwarranted efforts in achieving desired offtake levels) are being overcome with the introduction of the capture techniques using plastic sheeting.

Long-term development

The evidence presented earlier concerning the adaptations and productivity of game animals was somewhat conflicting. Definite advantages in optimum use of the vegetation would accrue from the use of game, but the two major disadvantages seem to be, firstly, generally inferior food conversion ratios and, secondly, the difficulties associated with management. Both of these problems could be at least partially solved by a long-term policy aimed at the development of new domestic species.

Game domestication has not yet been tried on a large scale as a serious, major programme. The evidence in favour of using game animals, particularly in integrated cattle and game ranches, warrants such a programme. Species which are considered desirable on the basis of their food preferences, carcass size, water relationships, disease resistance and perhaps other factors, should be captured in large numbers and included in carefully designed selection and breeding programmes. Only with large numbers will it be possible to apply the rigorous culling, based on food conversion and growth rates, which is necessary to select the desirable gene combinations and still maintain a viable herd.

Concurrent with the selection programme the animals would be tamed and subjected to the normal management activities of a ranch, such as dehorning and internal and external parasite control. They should be constantly handled in order to allow for easy control of their movement. The latter development is necessary for the achievement of controlled grazing and browsing programmes. After a few generations of such selection and taming the herds would no longer be the same as the wild species, but the castrated "steers" would be approaching their genetic potential for meat production.

REFERENCES

Anderson, G.D. and Herlocker, D.J., 1973. Soil factors affecting the distribution of the vegetation types and their utilization by wild animals in the Ngorongoro Crater, Tanzania. J. Ecol., 61: 627—651.

Anonymous, 1963. Rainfall map No. 12A. In: Atlas of the Federation of Rhodesia and Nyasaland. Federal Government Printer, Salisbury. Rhodesia.

Barnes, D.L., 1972. Bush control and veld productivity I. Mod. Farming Cent. Afr., 9(6): 10—19.

Bell, R.H.V., 1970. The use of the herb layer by grazing ungulates in the Serengeti. Br. Ecol. Soc. Symp. No. 10: 111—124.

Blankenship, L.H. and Qvortrup, S.A., 1974. Resource management on a Kenya ranch. J. South. Afr. Wildl. Manage. Assoc., 4: 185—190.

Brody, S., 1945. Bioenergetics and Growth. Rhinehold, New York.

Brown, L., 1969. Wildlife versus sheep and cattle in Africa. Oryx J. Fauna Preserv. Soc., 10: 92—101.

Conybeare, A., 1975. Notes on the feeding habits of kudu in the Kalahari sand area of Wankie National Park, Rhodesia. Arnoldia (Rhodesia), 7: 1—7.

Cunningham, M.P., 1968. Trypanosomiasis in African wild animals. East Afr. Agric. For. J., 33: 264—365.

Dasmann, R.F., 1964. African Game Ranching. Macmillan Co.

Dasmann, R.F. and Mossman, A.S., 1960. The economic value of Rhodesian game. Rhod. Farmer, 30: 17—20.

Dasmann, R.F. and Mossman, A.S., 1962. Road strip counts for estimating numbers of African ungulates. J. Wildl. Manage., 26: 101—104.

Densham, W.D., 1974. A method of capture and translocation of wild herbivores using opaque plastic material and a helicopter. Lammergeyer, 21: 1—25.

De Vos, A., 1969. Ecological conditions affecting the production of wild herbivorous mammals on grasslands. Adv. Ecol. Res., 6: 137—183.

Duckworth, B.J., 1972. The distribution and movement of buffalo (*Syncerus caffer caffer*) herds in the Kalahari sand area of Wankie National Park. Proj. Rep., Cert. Field Ecol., University of Rhodesia.

Ferrar, A.A. and Walker, B.H., 1974. An analysis of herbivore/habitat relationships in the Kyle National Park, Rhodesia. J. South. Afr. Wildl. Manage. Assoc., 4: 137—147.

Field, C.R., 1972. The food habits of wild ungulates in Uganda by analysis of stomach contents. East Afr. Wildl. J., 10: 17—42.

Field, C.R., 1975. Climate and the food habits of ungulates on Galana Ranch. East Afr. Wildl. J., 13: 203—220.

Fogg, G.E., Stewart, W.D.P., Fay, P. and Walsby, A.E., 1973. The Blue-Green Algae. Academic Press, London and New York.

Glover, J., Robinson, P. and Henderson, J., 1954. Provisional maps of the reliability of annual rainfall in E. Africa. Q.J. R. Meteorol. Soc., 80: 602—609.

Glover, P.E., Glover, J. and Gwynne, M.D., 1962. Light rainfall and plant survival in East Africa. II. Dry grassland vegetation. J. Ecol., 50: 199—206.

Goodman, P.S., 1975. The relation between vegetation structure and its use by wild herbivores in a riverine habitat. M.Sc. Thesis, University of Rhodesia.

Gwynne, M.D., 1967. Selective intake of vegetation by cattle in semi-arid areas. Annu. Rep. East Afr. Agric. For. Res. Organ. 1967, pp. 78—80.

Gwynne, M.D. and Bell, R.H.V., 1968. Selection of vegetation components by grazing animals in the Serengeti National Park. Nature (London), 220: 390—393.

Hall-Martin, A.J., 1974. A note on the seasonal utilization of different vegetation types by giraffe. S. Afr. J. Sci., 70: 122—123.

Hoffman, R.R., 1973. The Ruminant Stomach. E.A. Literature Bureau, Nairobi.

Hoffman, R.R. and Stewart, D.R.M., 1972. Grazer or browser: a classification based on the stomach-structure and feeding habits of East African ruminants. Mammalia, 36: 226—240.

Jarman, P.J., 1974. The social organization of antelope in relation to their ecology. Behaviour, 48(3, 4): 215—267.

Johnstone, P.A., 1971. Wildlife husbandry on a Rhodesian game ranch. Int. Conf. Behav. Ungulates Relat. Manage., University of Calgary.

Johnstone, P.A., 1973. Evaluation of a Rhodesian game ranch. Part 1: Operations, productivity and economic appraisal over six years 1967—72. Symp. Wildl. Conserv. Utiliz. Afr., Pretoria.

Kay, G., 1972. Distribution and Density of African Population in Rhodesia. University of Hull, Dep. of Geography. No. 12 (Misc. Series).

Kelly, R.D., 1973. A comparative study of primary productivity under different kinds of land-use in south-eastern Rhodesia. Ph.D. thesis. University of London.

Kelly, R.D. and Walker, B.H., 1976. The effects of different forms of land-use on the ecology of a semi-arid region in south-eastern Rhodesia. J. Ecol., 64: 553—576.

Lamprey, H.F., 1963. Ecological separation of the large mammal species in the Tarangire Game Reserve, Tanganyika. East Afr. Wildl. J., 1: 63—92.

Loope, W.L. and Gifford, G.F., 1972. Influence of a soil microfloral crust on selected properties of soils under pinyon-juniper in south-eastern Utah. J. Soil Water Conserv., 27: 164—167.

Louw, G.N., 1970. Physiological adaptation as a criterion in planning production from wild ungulates. Proc. S. Afr. Anim. Prod., 9: 53—56.

Mentis, M.T., 1972. A review of some life history features of the large herbivores of Africa. Lammergeyer, 16: 1—89.

Mentis, M.T., 1977. Stocking rates and carrying capacities for ungulates on African rangelands. S. Afr. Wildl. Res., 7: 89—98.

Mentis, M.T. and Duke, R.R., 1976. Stocking capacities of natural veld in Natal for large wild herbivores. Sthn. Afr. Wildl. Res. Assoc., 6: 65—74.

Owaga, M.L., 1975. The feeding ecology of wildebeest and zebra in Athi-Kaputei plains. East Afr. Wildl. J., 13: 375—383.
Page, B.R. and Walker, B.H., 1978. Feeding niches of four large herbivores in the Hluhluwe Game Reserve, Natal. Proc. S. Afr. Grassl. Assoc., 13: 117—122.
Parker, I.S.C. and Graham, A.D., 1971. The ecological and economic basis for game ranching in Africa. Br. Ecol. Soc. Symp. No. 11: 393—404.
Phillips, J., 1965. Fire — as master and servant: Its influence in the bioclimatic regions of Trans-Saharan Africa. Proc. 4th Ann. Fire Ecology Conf. Tall Timbers, Florida, pp. 7—109.
Robinette, W.L., Loveless, C.M. and Jones, D.A., 1974. Field tests of strip census methods. J. Wildl. Manage. 38: 81—96.
Roth, H.H., 1964. Game utilization in Rhodesia. Mammalia, 30: 397—427.
Rushworth, J.E., 1975. The floristic, physiognomic and biomass structure of Kalahari sand shrub vegetation in relation to fire and frost in Wankie National Park, Rhodesia. M.Sc. thesis, University of Rhodesia.
Scott, J.D., 1972. Veld burning in Natal. Tall Timbers Fire Ecol. Conf., 11: 33—51.
Selous, F.C., 1881. A Hunter's Wanderings in Africa. Richard Bentley and Son, London.
Shields, L.M., Mitchell, C. and Drouet, F., 1957. Alga and lichen stabilized surface crust as a soil nitrogen source. Am. J. Bot., 44: 489—498.
Sinclair, A.R.E., 1975. The resource limitation of trophic levels in tropical grassland ecosystems. J. Anim. Ecol., 44: 497—520.
Skinner, J.D., 1971. Productivity of the eland: an appraisal of the last five years' research. S. Afr. J. Sci., 67: 534—539.
Talbot, L.M. and Talbot, M.H., 1962. Food preferences of some East African wild ungulates. East Afr. Agric. For. J., 27: 131—138.
Talbot, L.M. and Talbot, M.H., 1963. The high biomass of wild ungulates on East African savannas. Trans. North Am. Wildl. Conf., 28: 465—476.
Talbot, L.M., Payne, W.J.A., Ledger, H.P. and Talbot, M.H., 1965. The meat production potential of wild animals in Africa. A review of biological knowledge. Commonw. Agric. Bur. Tech. Commun. No. 16.
Taylor, C.R., 1968. Hygroscopic food: a source of water for desert antelopes. Nature (London), 219 (No. 5150): 181—182.
Taylor, C.R., 1969a. The eland and the oryx. Sci. Am., 220: 89—95.
Taylor, C.R., 1969b. Metabolism, respiratory changes and water balance of an antelope, the eland. Am. J. Physiol., 217: 317—320.
Taylor, C.R., 1970a. Dehydration and heat: effects on temperature regulation of East African ungulates. Am. J. Physiol., 219(4): 1136—1139.
Taylor, C.R., 1970b. Strategies of temperature regulation: effect on evaporation in East African ungulates. Am. J. Physiol., 219(4): 1131—1135.
Taylor, C.R. and Lyman, C.P., 1967. A comparative study of the environmental physiology of an East African antelope, the eland, and the Hereford steer. Physiol. Zool., 40(3): 280—295.
Taylor, R.D. and Walker, B.H., 1978. A comparison of vegetation use and condition in relation to herbivore biomass on a Rhodesian game and cattle ranch. J. Appl. Ecol., 15: 565—581.
Walker, B.H., 1976. An assessment of the ecological basis of game ranching in southern African savannas. Proc. Grassl. Soc. S. Afr., 11: 125—130.
Walters, C.J. and Brady, P.J., 1972. Periodic harvest as a method of increasing big game yields. J. Wildli. Manage., 36: 128—134.
West, O., 1947. Bush encroachment in relation to veld management. Rhod. Agric. J., 44: 488—495.
West, O., 1965. Fire in vegetation and its use in pasture management, with special reference to tropical and subtropical Africa. Mimeographed publication No. 1/1965. Commonw. Bur. Pastures Field Crops. Hurley, Berkshire, England.
Whittaker, R.H., 1965. Dominance and diversity in land plant communities. Science, pp. 250—260.

Chapter 3

MANAGEMENT OF THE SEMI-ARID AND ARID RANGELANDS OF AUSTRALIA

A.D. WILSON and R.D. GRAETZ

CSIRO, Division of Land Resources Management, Deniliquin, N.S.W. (Australia)

INTRODUCTION

The arid and semi-arid lands of Australia cover the vast central area of the continent, extending to the western and southern coasts and comprising

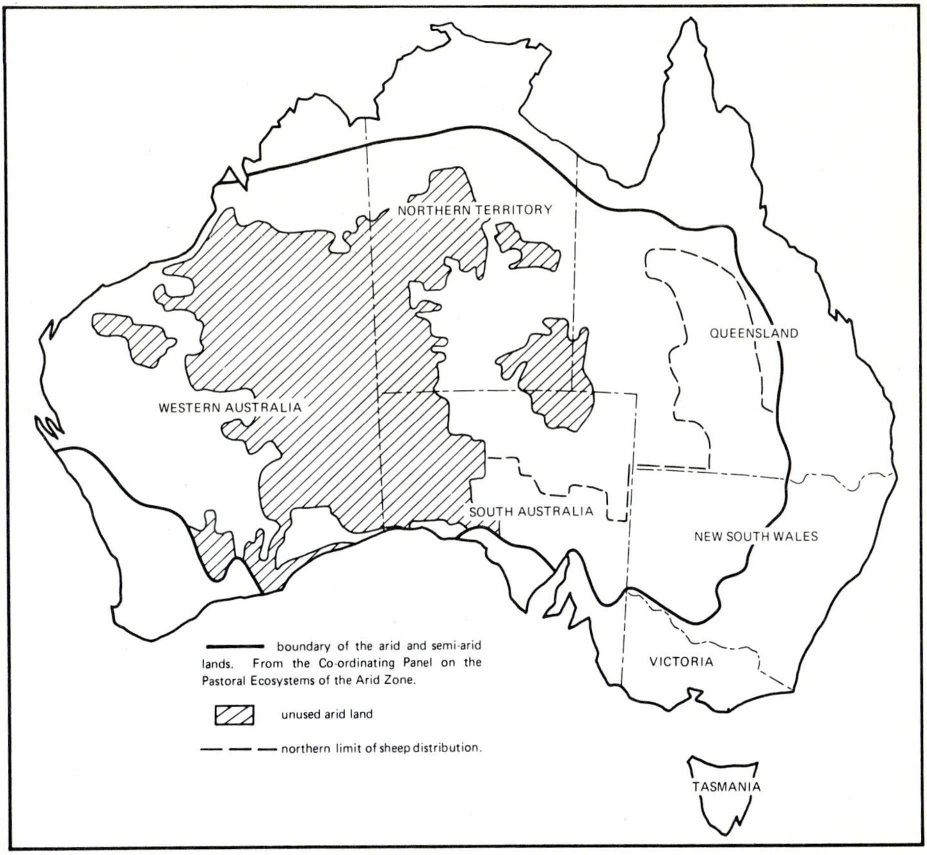

Fig. 3.1. A broad-scale map of land use in the arid lands of Australia.

5.5×10^6 km^2, or nearly 70% of the total land surface of Australia. The higher rainfall margin to this zone is defined as the boundary between the land used solely for extensive grazing and that used for dry-land cropping (see Fig. 3.1). This is similar to the boundary presented by Perry (1968) and Moore (1975), but has been modified according to changes made by the Panel on Arid Zone Pastoral Ecosystems (Division of National Mapping, Canberra). The lower rainfall limit for cropping is principally determined by rainfall effectiveness and soil type, and approximates to the medium rainfall isohyets of 250 mm in the southern winter rainfall zone and 650 mm in the northern summer rainfall zones. The latter, however, is somewhat fluid as agricultural experience in this area is young and intensive improvement is proceeding. A large portion of the arid lands (35%), comprising 2.0×10^6 km^2 is too dry or barren for livestock grazing and remains unmanaged and hence outside the consideration of this chapter.

Within such a large area there is naturally great variation in climate, vegetation and land-use, so that for the discussion of management the area has been divided into five major types — the woodlands, chenopod shrublands, *Acacia* shrublands, tussock grasslands and hummock grasslands. These are based on the ecological characteristics of the rangelands as used for extensive grazing and differ slightly from the strict vegetation classification of Moore and Perry (1970), because of an emphasis on the herbage within grazing height. These types are sufficiently uniform to allow discussion of management principles, but each contains land systems and units of much variation. Consequently the detail given here will be insufficient for the consideration of management at the land-holding or locality level.

A DESCRIPTION OF THE ARID LANDS

Climate

No continent is drier than Australia, for no continent is more dominated by the sub-tropical high pressure belt (Gibbs, 1969). It is therefore probable that no continent has less reliable rainfall, for there is an almost universal association of high variability with low annual totals. Rainfall characteristics are paramount in the prescription of management for arid lands.

The precipitation classification presented here (Fig. 3.2) is that of Gaffney (1971) and it is based on a combination of median annual rainfall and seasonal incidence. The seasonal incidence is determined from the ratio (greater/lesser) of the median summer rainfall (November—April) and the median winter rainfall (May—October). The classification is synoptic but it correlates well with the other atmospheric parameters of temperature and humidity deficit as well as with the arid land boundaries.

The main features of this classification for the arid lands are the marked

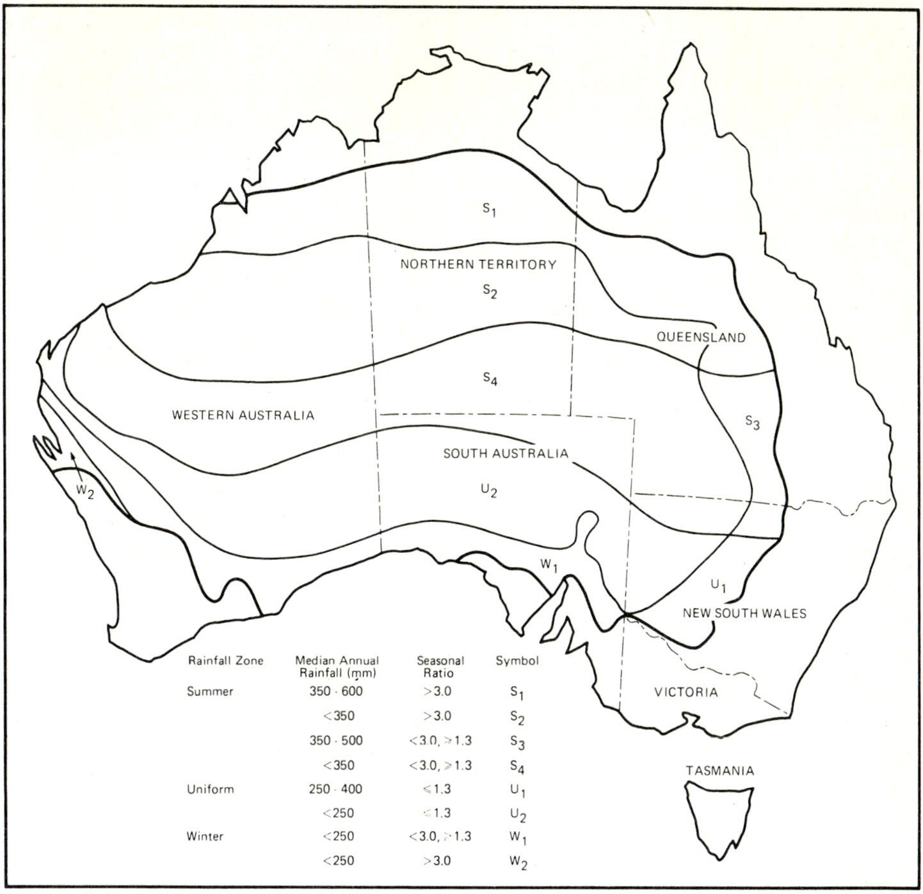

Fig. 3.2. Seasonal rainfall zones in the arid lands of Australia.

wet summer and dry winter of northern arid lands and the comparatively small area in the south that has either a uniform incidence, or a winter maximum. The significance of this for animal production will be discussed later.

The next most important parameter in the appraisal of the rainfall regime for the region is the variation from year to year, and we find that variability in annual rainfall follows closely the main rainfall incidence zones of Fig. 3.2, which in turn reflects the nature of the atmospheric circulation patterns responsible. It suffices to say that most of the arid lands are characterized by high rainfall variability with a large central area roughly coincident with the Lake Eyre basin and which is occupied by both cattle and sheep grazing enterprises, that experiences very high to extreme variability.

A necessary broad-scale analysis, complementary to that of annual rainfall variability, is that of rainfall deficiency or probability of drought. Such analyses, using various indices (e.g. Gaffney, 1975), credit the arid lands with a high incidence of droughts. The area of greatest drought-tendency extends from the north-west coast of Western Australia across the continent to south-west Queensland. Most of the winter rainfall rangelands of the south-east of the continent experience a moderate to high risk of drought.

All of the above-mentioned rainfall characteristics influence the management strategies for the vegetation and the animals in the arid lands; for although the rainfall is both low and variable winter temperatures are not low enough to inhibit plant growth. Each rainfall tends to produce a finite period of plant growth; winter rainfall is not accumulated in the soil store. Thus there is no regular plant growth rhythm and, as a consequence, no seasonal livestock grazing strategies so common in South Africa and North America. The probability of drought — i.e. periods greater than a year without any significant periods of plant growth — is so high for most of the rangelands that it has been suggested by Perry (1974) that all management be based on drought levels, but with the flexibility to take advantage of seasonal responses of the vegetation.

Relief and drainage

The arid lands of Australia are characteristically flat or of low, gently undulating topography with relief being generally less than 50 m. It has been estimated that some 60% of the arid lands can be classified as "smooth plains" (Anonymous, 1972). The lowest point on the continent is Lake Eyre, the focus of the largest area of co-ordinated interior drainage with an elevation of —15 m (Mabbutt, 1969) and the highest elevations of 1500 m are to be found amongst the peaks of Macdonnell Ranges in central Australia. The increased rainfall yield from these and other ranges is small but the runoff from them is often of considerable importance in increasing forage production in run-on areas for distances of hundreds of km. More detailed accounts of the geological history and geomorphology are available in Mabbutt (1969), Mabbutt and Sullivan (1970) and Dury (1968).

Soils

The soils of the arid lands, and of Australia in general, are distinctive when compared with those of the Northern Hemisphere (Hubble, 1970). Their most characteristic properties are a general low nutrient status, with extremely low levels of phosphorus and nitrogen (Charley, 1972), which is coupled with very poor physical attributes such as low macro-porosity and surface sealing. Only about 5% of the Australian continent has dominant soils which are entirely restricted to the non-arid lands; most soils in mesic Australia also occur in the arid interior or show some features of aridity. The use of the terms "desert" or "arid" soils is inappropriate for these and other reasons (Jackson, 1962). A broad-scale map and description of the soils of the arid lands is given in Hubble (1970) and Perry (1970).

Vegetation

The flat or gently undulating landscapes of infertile soils are clothed in a flora of characteristic composition and structure. This flora consists mainly of sclerophyllous species which have diversified greatly from the ancient and more widespread stock. Such genera are *Eucalyptus*, *Eremophila*, *Acacia*, or to a lesser extent, the grass genera *Dicanthium*, *Themeda*, *Heteropogon* and the cosmopolitan group of halophytic shrubs *Atriplex*, *Kochia* and *Rhagodia*. It is thought that sclerophylly is as much a product of the widespread low phosphate status of the soils as aridity (Beadle, 1966; Beard, 1975).

Of the endemic genera of the arid lands about 40 are specific to it; these genera are mainly in the families Chenopodiaceae, Asteraceae and Cruciferae. Endemic and widespread tree species are, for example, *Acacia aneura* (mulga), *A. harpophylla* (brigalow) and *Eucalyptus oleosa* (mallee) (Moore and Perry, 1970).

The flora has been sorted into vegetation formations through climatic—edaphic interactions, with fire as a component since the arrival of the aboriginal man (20 000 years) and non-indigenous herbivores in the last 150 years. The major vegetation formations are mapped in Fig. 3.3. Each of these vegetation types and their response to pastoral use will be dealt with in later sections; a more extensive description may be found in Moore et al. (1970) and Perry (1970).

Fauna

The Australian—New Guinea area has been isolated by deep seas from all other land masses since the Cretaceous period. As a consequence the Australian fauna, particularly the vertebrate fauna, reflects this isolation in being represented by only a few basic groups (Orders), with some of these showing spectacular radiation to fill available niches. Within the continent itself it is difficult, if not incorrect, to describe a peculiarly arid fauna. Many of the large marsupials, e.g. the kangaroos, that inhabit the arid lands have very wide distribution and are largely concentrated in the arid lands because of heavy hunting pressures in the cultivated areas. Kangaroos in particular have increased in number since the advent of the European pastoralist who established permanent water supplies and eliminated the natural predator, the dingo. An excellent summary of the interaction of kangaroos and domestic stock has been given by Newsome (1975).

There is a large exotic component to the fauna of the arid lands, for large populations of feral camels, goats, foxes and pigs exist in various areas. These escapees from domestication have had nowhere near the devastating effect as the deliberate introduction of the European rabbit (Leigh, 1974). The numbers of the latter have now been biologically controlled by the introduction of myxomatosis (Rolls, 1969).

In summary, the native faunal population of the arid lands has been

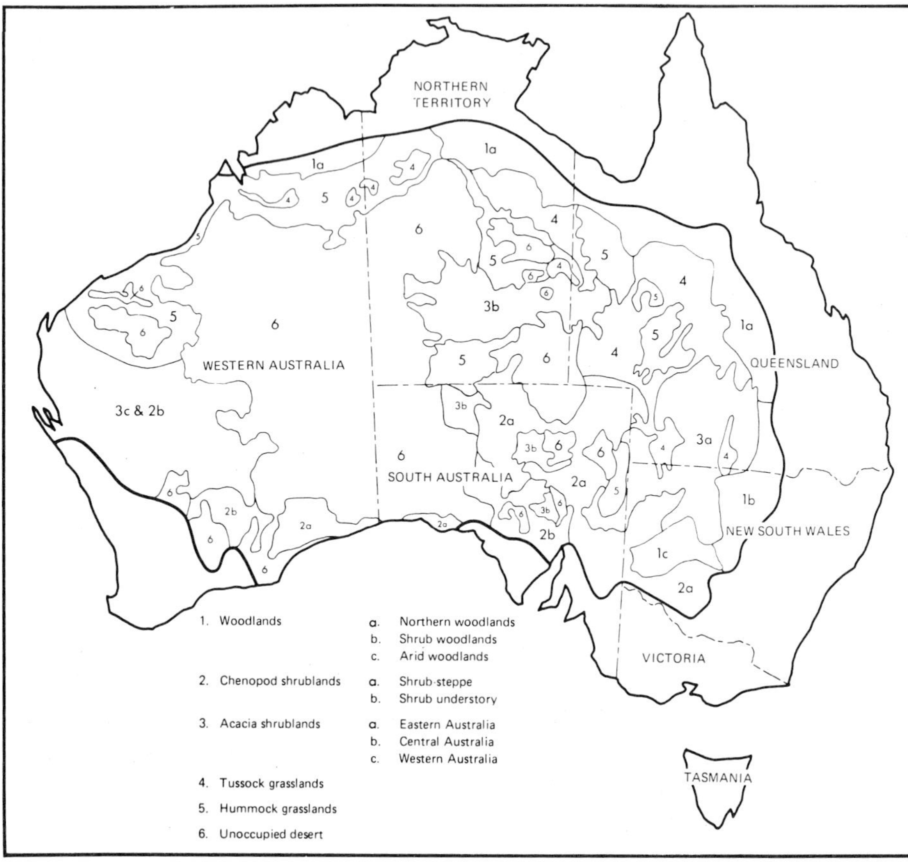

Fig. 3.3. The major rangeland types in the arid lands of Australia.

altered by the arrival of domestic stock; the large marsupials have increased whilst many small marsupial rats and mice have declined or become extinct through habitat destruction. Wildlife is generally regarded by pastoralists as a competitor with, or in the case of foxes and pigs, predators of, livestock. It is not, as yet, seriously regarded as a harvestable resource either for protein or hunting for recreation. However, this should develop within the next decade or so.

HISTORY AND STRUCTURE OF LAND-USE

Early settlement

The arid and semi-arid lands were almost wholly occupied within the 50-year period 1850—1900, a relatively short period of time compared to the occupation of similar vast areas in the United States of America. This occupation was associated with the grazing of cattle and sheep and was at first in large holdings along the permanent streams. It was fuelled by large profits made by the first settlers from the sale of wool, the sale of livestock and subsequent development, sub-division and resale of the land.

In the eighteen-fifties and eighteen-sixties the main expansion was into western New South Wales, central Queensland and northward from the agricultural areas in South Australia. In the eighteen-seventies and eighteen-eighties occupation extended to the drier far west of New South Wales and Queensland and the occupation of the Northern Territory and the arid areas of Western Australia was begun. In most areas occupation was at first with cattle, and these were subsequently displaced by sheep as development of watering points and fencing were made. The peak use of these lands was reached by the eighteen-nineties in the east, but not until the nineteen-twenties in the Northern Territory and the west and so it was in the east that the problems of over-stocking first arose (Barnard, 1969).

These problems were most severe and best illustrated by events in western New South Wales. The eighteen-eighties saw the development of high expectations, both of the land itself because of relatively good seasons and of the profits that were possible from wool-growing, sheep sales and capital gains. This was followed, in the eighteen-nineties, by a run of poor drought seasons and falling prices, which led to widespread land degradation and financial collapse (King, 1957). The problem was also compounded by the expansion of the European rabbit (*Oryctolagus cuniculis*) to plague proportions which occurred at the same time. Sheep numbers, which reached 15.5 million in 1891, fell to less than 6 million in 1900 and this was associated with extensive soil erosion and the abandonment of some 2 million ha. Subsequent legislation to reconstruct the pastoral industry in this area recognized the variability of the climate and the need to control livestock numbers (King, 1957). At present the rangelands support about 5 million cattle, largely in the northern areas, and 30 million sheep, mainly in the south, and this stock represents about 20—25% of the national herds and flocks. The absolute number of animals in the rangelands has remained fairly static over the last 50 years, whereas the number in the non-arid lands has increased considerably through pasture improvement by plant introduction and the extensive use of fertilizers and trace elements.

The Australian rangelands are organized and managed on a very different basis from most other arid land areas. Firstly, and of most importance, all the livestock enterprises are geared to a monetary export market-oriented system which contrasts markedly with the subsistence systems of most of the world's arid lands. Secondly, the livestock industry is organized on an extensive, sedentary basis; there is no nomadism, no herders and virtually no seasonal grazing.

Modification of vegetation and land resources

The intensity of use characteristic of the early stages of settlement described above was based on an ignorance of the climatic variability and the fragility of most of the plant communities. The combined and cumulative effects of over-grazing and of the introduction of the European rabbit have produced significant changes in the botanical composition of most of the southern rangelands. In the light of this it is fortunate that most of the occupied arid land is of low relief and thus the erosion hazard is also lowered. There are no accurate estimates of the degree of degeneration (in terms of ecological change), but an overall estimate of 65% moderate and severe degeneration for the shrublands, 30% for the low woodlands and 15% for the grasslands has been made (Newman and Condon, 1969).

It is not surprising that in the degraded areas there has been a decline in the number of livestock carried (Ealey, 1967; Perry, 1968), but with the exception of a few specific localities numbers have now been stable for decades. This supports the conclusion that most of Australia's arid and semi-arid grazed areas are now in a stable equilibrium with pastoral use (Newman and Condon, 1969).

Land tenure and size of holdings

The individual unit of the pastoral industry is the sheep or cattle "property" or "station", which is an area leased from the state on a 35—40 year tenure and managed by one man. The size of these units may range from 25 000 to 100 000 ha.

The principal form of land tenure in the semi-arid regions is leasehold, varying in tenure from 30 years to perpetuity (Heathcote, 1969). However, the conditions attached to those leases, particularly on size limitations, vary considerably from state to state. Thus, in New South Wales and the more intensively grazed areas of Queensland (sheep areas), there is a limitation in the area that may be leased by one person to approximately twice a "living area" or "home maintenance area", a living area being variously defined as 4—6000 sheep in New South Wales and 8—12 000 sheep in Queensland (Mawson et al., 1974). These restrictions were designed to limit the aggregation of land into large company holdings even though such limitations do not appear in the best interests of good land-management in areas of low and variable rainfall.

The present average size of holdings and the number of livestock carried is shown in Table 3.1. In the areas where property size is not limited by statute, size may vary widely from these averages. The leases in all states also impose some limitations on the stocking intensity, and require lessees to control noxious weeds. However, in practice stock numbers are only loosely controlled and depend on the development of goodwill between the lessee and the Lands Boards.

TABLE 3.1

The average size of pastoral holdings and the number of livestock carried (cattle carried on sheep properties have been added to sheep numbers on an equivalent basis of one beast = eight sheep)

State and district	Property size	Stock number
New South Wales, Western	11 000	5700 sheep
Queensland, Central	16 000	7500 sheep
Queensland, Western [a]	77 000	1930 cattle
South Australia, Central	38 000	5200 sheep
South Australia, Northern [a]	301 000	1630 cattle
Northern Territory, Central	384 000	4000 cattle
Northern Territory, Barkly Tableland	638 000	17 000 cattle
Western Australia	176 000	12 600 sheep

Source: Bureau of Agricultural Economics.

[a] Statistical divisions for cattle overlap those of sheep areas. Properties carrying cattle only are larger and carry more cattle than values shown.

Structural limitations to intensification

Productivity, both per animal carried and numbers carried per unit area, declines within the rangelands as both aridity and incidence of summer rainfall increases. Generally the distance to markets and from sources of supply increases also. As a consequence livestock operations become more and more extensive in nature. The sheep-breeding and wool-growing enterprises of the south-eastern areas operate within fenced, well-watered paddocks, with a consequent good control of mating and disease. At the other extreme, the cattle industry of the Northern Territory operates with little control of mating and a minimum of fencing and husbandry; compensating for the low productivity by an increased size of operation and financial returns and a minimizing of inputs. As a result the type of management input which is both feasible and economic to apply changes from one region to another. In the more intensive, better-watered margins of the zone some modification may be possible (Moore et al., 1970) while livestock husbandry including parasite control, seasonal breeding and weaning is standard practice. In the more extensive regions, management is limited to genetic improvement (e.g. introduction of Zebu type cattle) and control of stock numbers.

Other aspects which affect the type of management applied are the transport facilities and the degree of interchange between sheep and cattle.

Transport

Because of the extensive nature of arid and semi-arid grazing enterprises and remoteness from markets, transport is important in determining the type of operation and its economic viability, as well as determining the level of resource conservation possible through the transport of livestock out of drought-stricken areas. The development of railways, and more recently "Beef Roads" in Australia has allowed the arid and semi-arid lands to develop as breeding areas for the more humid zones, which is a logical integration of their respective industries. It is estimated that 190 000 cattle and 2.3 million sheep are presently bred in the semi-arid zone and transported to other zones for breeding or fattening (Wilson, 1978); the type and direction of these movements has been outlined elsewhere (Ricketts, 1969). It is suggested that some of the land degradation during the early phases of settlement arose from an inability to transport stock to markets in more favoured areas; a disability that has now been largely removed. Nevertheless, the cost of transport is significant, particularly for the cattle industry, and these areas will not be immune from future problems arising from a shortage of fossil fuels.

Interchange of cattle and sheep

Throughout the semi-arid lands of Australia there is only one major factor limiting the interchange of sheep and cattle, although there are a number of minor factors arising more from the structural development of properties into one form of production or the other. The major factor is the climatic limitation which restricts Merino sheep, the predominant breed in Australia, to areas with at least some winter rainfall. The productivity of these sheep, both in terms of wool growth and fertility, declines steadily from 4.1—4.5 kg wool/sheep and lamb and a breeding rate of 60—70% lambs per ewe mated in western New South Wales (latitude 32°S) to 3.2—3.6 kg and 20—40% in northern Queensland (latitude 21°S) for the same breed and strain of sheep (Brown and Williams, 1970). This decline is attributed to both nutrition and heat stress, but whatever the reason it effectively limits the northward extent of sheep grazing, which has been, overall, the more profitable enterprise in the arid regions.

This northward limitation of sheep grazing has subsequently been formalized by the erection of an extensive "dog" fence across South Australia, around New South Wales and northward into Queensland (Fig. 3.1). The property holders within this fence contribute to its upkeep and to the eradication of the native wild dog, the dingo (*Canis familians dingo*), which is an important sheep predator. Within the fence sheep are predominant, but there are, nevertheless, large numbers of cattle within this zone as well. These areas have become known as the "sheep areas", but as far as can be determined the emphasis on sheep is not based on any biological advantage (Wilson, 1976), but has arisen from their greater income and reduced income fluctuation from seasonal variability (McKay, 1973). The dependence of this difference on economic factors was illustrated by the sharp increase in cattle numbers in the early nineteen-seventies as a response to a price differential in favour of cattle. Other factors which tend to formalize the separation of sheep and cattle industries is the standard of fencing, the supply of water, the type of facilities and the expertise which develops to favour the industry initially adopted in that region.

PRESENT MANAGEMENT OF THE ARID AND SEMI-ARID LANDS

In the following sections we outline the ecological characteristics and current land-use of the five major rangeland types and recommend optimal conservative management strategies for each. Of necessity this must require appreciable generalization, for while the variance of slope, soil and vegetation over scales of 1 km or so have importance at the level of the individual manager—property, such detail must be ignored in what follows.

Woodlands

The woodlands are composed of areas of semi-arid and arid low woodlands and shrub woodlands that extend around the higher rainfall fringes of the rangeland zone (see Fig. 3.3). These may be separated into three types — the sub-tropical woodlands of the Northern Territory and northern Queensland, the shrub woodlands of southern Queensland and northern New South Wales and the arid woodlands of central and southern New South Wales.

(i) Northern woodlands. The northern woodlands are dominated by various species of Eucalypt (*E. dichromophloia*, *E. melanophloia*) and other trees (e.g. *Melaleuca* spp., *Bauhinia cunninghamiana*) and the herbage layer by *Aristida* spp. with the addition of other grasses such as *Chrysopogon fallax* and *Triodia mitchellii*. A more detailed description is presented by Moore et al. (1970). They are of relatively low productivity and nutritive value and are grazed solely by cattle at low intensities (one beast to 20—50 ha) despite their relatively high rainfall. This may be attributed to the poor quality of the grass, which may be deficient in protein, sulphur and phosphorus for the greater part of the year, and the low forage production arising from poor soils and competition from the tree layer. In the more northern areas the cattle are grazed extensively with little management or control, except for the provision of artesian water from deep bores. The cattle are mustered only for branding, sale and some tick control. The principal improvements that can be effective under this regime are the introduction of heat and tick-resistant cattle, and the provision of supplements, usually urea and molasses, which reduce the protein and sulphur deficiences.

No form of pasture management is practised, although natural rotation occurs through the grazing of the woodlands in the wet season and other pasture lands in the dry season — a feature which has remained unchanged from pre-settlement times (Perry, 1960). Little is known of the degradation that may have occurred in these areas, but because it is a harsh environment for cattle, it is generally assumed that vegetation changes have been slight.

(ii) Shrub woodlands. The shrub woodlands of southern Queensland and northern New South Wales are dominated by *Eucalyptus populnea* with an understory of shrubby species of *Eremophila*, *Dodonaea*, *Cassia* and other genera. The herbage layer contains a number of palatable grasses such as *Eragrostis* spp., *Bothriochloa decipiens*, *Chloris acicularis*, *Stipa variabilis*, *Neurachne mitchelliana*, unpalatable species of *Aristida* and some chenopod forbs (*Bassia* spp.) in the south (see Moore et al., 1970). The rainfall of 400—600 mm is spread over all seasons, but has a pronounced summer dominance in the north. These are grazed by sheep for breeding and wool-growing, but cattle are also common, the stocking intensity varying from one sheep to 2 ha to one sheep to 8 ha, depending on rainfall and tree density. The latter are grazed year-long but are more intensively managed and husbanded than the cattle of the northern areas.

The principal problems of vegetation management are the reduction of tree and shrub density and the control of erosion. Research in southern Queensland, in an area of higher rainfall but the same vegetation type, has shown that herbage growth may be increased three-fold by the removal of trees and shrubs that are inedible to sheep and cattle (Walker et al., 1972). They concluded that a density of six trees or 360 shrubs per ha is the upper limit before herbage growth is reduced (see Figs. 3.4 and 3.5). Unfortunately clearing is expensive in relation to the expected financial returns from the cleared land and the woody plants quickly regenerate after clearing by regrowth from the bases unless specific control measures are undertaken. Experimental work indicates that regeneration may be controlled in average years by heavy sheep-grazing, or in years of above average rainfall by burning (Moore and Walker, 1972). However, controlled burning for this purpose is rarely practised and further research and extension is required to promote its adoption. Similar control of regrowth may be achieved in the more southerly regions by grazing with goats (G.N. Harrington, personal communication, 1976) although there is no organized goat industry in Australia and its adoption is unlikely in the near future. The grazing of goats may also have application for the initial clearing of shrubs, provided that the unpalatable species *Eremophila sturtii* and *E. mitchellii* are absent (Wilson et al., 1976). Apart from the management of the shrub and tree layer, management for improved herbage composition is by the control of grazing pressure. Heavy grazing and the absence of fire is thought to promote increases in *Aristida* spp., which are unpalatable and whose sharp seeds become embedded in the wool and skin of sheep, but such changes are not substantiated experimentally. More

Fig. 3.4. Semi-arid shrub woodland. Herbaceous species much reduced by a dense growth of unpalatable shrubs and trees.

severe grazing may lead to dominance by *Bassia birchii*, which is inedible to sheep. Its presence is an indication of severe degradation and erosion. In small areas the introduction of cultivars of the exotic *Cenchrus ciliaris* has been successful, particularly on soils of higher phosphate status, and offers possibilities for improved productivity (Christie, 1975).

Erosion is also a major concern, particularly on undulating brown soils of northern New South Wales (Condon, 1961). The slope and structure of these soils make them particularly susceptible to erosion and it is recommended that particular care be taken to avoid over-grazing these soils in drought times and to site watering points and shearing sheds on more stable

Fig. 3.5. Semi-arid shrub woodland. Open woodland with a good cover of herbaceous species.

areas to avoid problems arising from concentrations of livestock that occur at these points. The combination of both erosion and high shrub density has given rise to a marked decline in the carrying capacity of the whole Cobar district, which is estimated to be a reduction from one sheep to 2.5 ha to one sheep to 4 ha (Anonymous, 1969). In this area reclamation is a major requirement of management. Reduction in livestock alone is insufficient to achieve regeneration and there is a requirement for some form of mechanical treatment, such as contour furrowing (Cunningham, 1967). Another alternative is the intensive development of the flat run-on areas for crops or forage crops, using the run-off water derived from the eroded ridges.

Continuous year-long grazing is adopted in most areas although some movement of sheep is associated with the weaning of lambs and the saving of feed for the breeding flock. Investigation of the autumn and spring deferment of *Chloris* pastures (Biddiscombe et al., 1956) showed no advantage in either sheep productivity or pasture composition over continuous grazing. Grazing intensity (stocking rate), however, was important and this remains the most important factor in good pasture management.

(iii) Arid woodlands. The arid low woodlands of western New South Wales are dominated by the tree *Casuarina cristata* and the herbaceous layer by the grasses *Stipa variabilis*, *Enneapogon avenaceus* and edible chenopod forbs. These woodlands receive 250—300 mm of rainfall, without a marked seasonal incidence and are grazed almost entirely by sheep at an intensity of one to 4—6 ha. Nutritive value is high, with protein content adequate at all times of the year (Wilson, 1974a) and on average the wool yield per sheep is amongst the highest of any region in Australia (Brown and Williams, 1970). In addition, the plant composition and soils are stable under grazing (Condon, 1961) so that the principal management requirement is the control of sheep numbers, according to seasonal conditions. The number of sheep is usually restricted to that which can be carried safely in 75% of years and in drought years the numbers are reduced by sale or agistment.

Stock watering points are provided by bores or surface catch-tanks within each paddock and their position is usually determined by the features of the land. Sheep need 6—7 l of water daily on these pastures in summer (Wilson, 1974b) and will need to drink once daily. On this basis they are able to range evenly over the vegetation which is within a 4 km range of water and most properties are now developed to meet such standards. However, problems still arise in drought years when some water sources become exhausted and stock are concentrated onto more permanent waters. Naturally this further accentuates the problem of feed shortage and over-grazing.

Another form of arid woodland in the southern arid winter-rainfall region is the Mallee shrublands, dominated by *Eucalyptus oleosa* and *E. dumosa*. These have a herbaceous layer of the hummock grass *Triodia irritans*, or *Stipa variabilis* and *Atriplex stipitata*. In the higher rainfall fringes the land may be cleared for wheat growing, but in their native state they are of low productivity (one sheep per 12 ha), or in areas with *Triodia*, may be ungrazed altogether. The gradual build-up of litter encourages the spread of uncontrolled fires in these communities on a 15—20 year cycle. After burning, the herbaceous growth is enhanced for 2 or 3 years, sufficient to increase the carrying capacities to one sheep to 5 or 6 ha for this time. More frequent burning on a controlled basis, if sufficient fuel is present, could be desirable for this reason. However, except for the areas with *Triodia*, burning is not practised and it requires some field experimentation to determine its feasibility and value.

The chenopod shrublands

The chenopod shrublands, the "saltbush" and "bluebush" country, swing in a large, discontinuous arc across the southern portion of the continent (Fig. 3.3), where though summer rainfall may account for 50—70% of the annual total, the winter rainfalls are the most reliable and effective.

The main distribution lies between the latitude of 24°—35°. The rainfall regime of the areas shows trends in both an E—W and N—S bearing. Median annual rainfalls range from 400 mm in the east to 125 mm on the western edges of the distribution whereas the seasonal incidence changes from winter to summer and the variability increases in the north-west of the continent where chenopods occur either as isolated homogeneous stands or as the understory of *Acacia* shrublands.

The shrublands are largely restricted to lands of low relief; the featureless plains or gently undulating lowlands where relief is generally less than 30 m (Mabbutt and Sullivan, 1970). The soils range from the heavy grey, brown and red clays of the alluvial fans and occluded drainage lines, to the desert loams and calcareous soils of the undulating lowlands (Oxley, 1979). It has been estimated that this low shrubland in all its forms occupies some 451 000 km^2 or about 6% of the total land surface area of Australia.

The dominants of these low shrubland communities are all members of the family Chenopodiaceae, characteristically being xeromorphic halophytes with semi-succulent, reflectant and usually hairy leaves, occurring in extensive uniform stands that are often mono- or bi-specific in dominants (Graetz, 1975). The shrubs are generally of 1.5 m or less in height and, coupled with flat or very gently rolling landscapes, produce panoramic views that despair travellers; "the everlasting sameness of the never ending plains".

The most widespread and economically important genera, with the number of endemic species in parentheses, are: *Atriplex* (40), *Bassia* (80), *Chenopodium* (20), *Enchylaena* (1), *Maireana* (Syn. *Kochia*) (40), *Rhagodia* (15), and *Salsola* (1). The most important species of pastoral and ecosystem stability value are *Atriplex vesicaria* and *A. nummularia* (commonly called "saltbush"); *Maireana* (Syn. *Kochia*) *sedifolia*, *M. pyramidata* and *M. astrotricha* (variously called "bluebushes").

It can be seen from Fig. 3.3 that the boundary of the distribution of sheep largely coincides with the northern edge of the shrub steppe (2a) or shrub understory (2b). These areas have traditionally been sheep areas since their earliest occupation. In the last 5 years an increase in cattle numbers, at the expense of sheep, has occurred for reasons of differential economic return rather than ecological ones. The density of sheep ranges from one sheep to 1.5 ha in the high rainfall (400 mm) areas of the Riverine Plain of New South Wales which carries about half of the sheep on shrublands, to one sheep to 25—30 ha on the western fringes in South Australia and Western Australia.

Animals introduced by European Man have been grazing most of these shrublands for over 100 years and unfortunately some areas have been degraded and eroded through injudicious stocking (Ratcliffe, 1936; Newman and Condon, 1969). This largely occurred in the early days of settlement or, less widely, in later periods of economic depression.

At the time of initial European settlement the shrublands, particularly the shrub steppe (see Fig. 3.3), had high densities of shrubs which competitively excluded or suppressed the grass and forb species. The latter component increased as the shrubs were themselves suppressed by grazing. This was the case in the wettest end of the distribution, the Riverine Plain, where areas of permanent disclimax grasslands were created ca. 1850 (Williams, 1974) as in the more xeric parts, the Nullabor Plain, as described by Adamson and Osborne (1922).

In general, sheep concentrate their grazing on the grasses and forbs which grow between or beneath the shrubs after significant winter or summer rainfall. The shrubs are of lower acceptability, but become the principal or even the sole dietary component when rainfall is low and grasses are absent. This change in preference forms a natural rotation of grazing pressure and removes the need for imposed grazing systems. At moderate stocking rates the shrubs regrow and regenerate in the wetter years when the animals graze mainly on the grasses (Graetz and Wilson, 1979). In parts of the Western Australian shrubland this does not apply and protection from grazing for periods of up to 2 years after good germinating rains is required to ensure the establishment of *Atriplex* seedlings (Wilcox, 1963).

Nevertheless, over-grazing and loss of these perennial shrubs does occur. The shrub *Atriplex vesicaria*, which is of widespread occurrence (see Figs. 3.6 and 3.7) does not regrow when defoliated to less than 20—30 leaves (Leigh and Mulham, 1971) and regeneration is then dependent on seedling establishment, which occurs only in favourable years. Over-grazed shrublands are replaced by various communities including perennial grasslands, annual grass/forb communities and *Bassia* communities, depending on locality; some of which provide good forage and soil protection, but others are inedible. However, without exception the disclimax communities have a lower drought grazing capacity and hence carry a more variable number of sheep than the climax shrublands.

These chenopodiaceous shrubs are more stable in growth and nutritive value than the associated grasses and forbs. Variability with season and maturity is low (Wilson, 1974a) and whilst the more productive forage is obtained from the grasses and forbs when available, the presence of shrubs sets minimum levels of protein and energy for the whole community. For *A. vesicaria* this is known to be a minimum of 11% crude protein, a digestibility of 52% and and a digestible intake of 550 g of dry forage daily for Merino sheep, which provides an adequate maintenance ration (Wilson, 1966; Graetz and Wilson, 1979).

Fig. 3.6. Chenopod low shrubland. Typical saltbush (*Atriplex vesicaria*) country (200 mm annual rainfall) with the scattered shrubs (50 cm high) and intervening forbs and grasses.

Fig. 3.7. Degraded chenopod shrubland. Here the perennial shrubs have been removed through over-grazing. There is an increase in the biomass of grasses and forbs but evidence of wind erosion is also apparent.

They do, however, have one disability, in that they are all high in soluble ash (chlorides of sodium and potassium), which may amount to as much as 25—30% of their dry matter. These salts are excreted in the urine of grazing animals and increase their water requirements several fold. In dry summers, when *Atriplex* is the major component of the diet, water intakes in sheep are as high as 12 l per day, compared to 3.5 l per day on grassland (Wilson, 1974b). This has the disadvantage of increasing the amount of water that must be supplied to livestock, decreasing the distance that stock may graze away from water points and decreasing the amount of salt that may be tolerated in water obtained from underground sources. The high salt intakes do not appear to otherwise affect the health of the stock, although there is a suggestion that it may reduce the number of lambs born (Potter and McIntosh, 1974).

These shrublands are valuable grazing areas that give a stable and high level of productivity for sheep, provided that they are stocked conservatively. This is usually achieved by conservative set-stocking, but in some cases it is achieved by periodic adjustment of stock numbers according to rainfall.

Acacia *shrublands*

The *Acacia* shrublands occur in a wide belt through the centre of Australia, in low rainfall country that receives both winter and summer rainfall (Nix and Austin, 1973) (see Fig. 3.3). They are characterized by a layer of tall shrubs, principally *Acacia aneura* (Mulga), but including many other species such as *A. sowdenii*, *A. kempeana*, *A. georginae* and *A. cambagei*. These shrubs vary in density from 1 ha^{-1} in the more arid regions to 1000 ha^{-1} in the higher rainfall areas of Queensland. In South Australia and parts of Western Australia they may have a chenopod low shrub understory but in general possess a ground layer of short-lived perennial grasses (Perry, 1970).

In Queensland and northern New South Wales the Mulga lands are mostly grazed by sheep, but this is intermingled with cattle and becomes entirely cattle in the more arid regions (250 mm and below). These may be grazed at rates of 3—4 ha per sheep, or equivalent, on a continuous grazing basis, although actual rates vary from this according to rainfall, land system and land condition. Degradation has occurred on the more unstable land systems, usually associated with greater slope and shallow soil, and takes the form of soil loss, loss of herbage and a change in the density of trees. The latter varies from the undesirable elimination of trees in the more arid region to the development of dense thickets in the higher rainfall areas (Boyland, 1974).

The area is of variable rainfall and drought management is a major concern. General drought management consists of rotating the use of the major pasture types to make best use of those that provide feed only after rain, the reservation of others that maintain feed quality in dry times and the avoidance of heavy grazing on the more unstable land systems. The use of tree leaves (topfeed) also plays a major role. The more accessible foliage is eaten first, and then the trees are cut or pushed to provide a maintenance ration for sheep. Its nutritive value may be enhanced by supplementation with molasses and phosphorus. The mulga is usually cut to allow regrowth from lateral branches which are high enough to prevent sheep browsing (Everist, 1972). Drought management also includes the reduction of livestock numbers by sale or agistment, as the purchase of fodder for feeding is uneconomical and promotes resource degradation. Such a policy may be aided in future by the development of soil water-balance models which can predict changes in sheep numbers, wool growth and lambing percentages 6 months in advance (Easter, 1975).

The management of tree and shrub density is important so that a balance between the tree and herbage strata is maintained. In the drier regions seedling regeneration of most species is an episodic event in response to above-average rainfall (Cunningham and Walker, 1973) and some protection from sheep grazing may be necessary to ensure adequate recruitment. However, in the higher rainfall areas the high tree density reduces the growth of ground forage. Densities of 640 and 140 ha^{-1} reduce herbage growth by half compared to 40 trees ha^{-1}, and even then there is some reduction. Mechanical clearing of these dense stands is practised in some areas, but care must be taken to avoid problems of increased erosion and increased activity by termites which harvest dry forage and build compact nests which reduce water penetration. On balance it is clear that an open shrubland is the most desirable community, retaining stability and drought reserve with reasonable forage yields, but such an optimum is difficult to attain in practice (Dawson et al., 1975).

The type of grazing on cleared land may also influence the type of forage plants obtained. It is desirable to retain a high proportion of valuable perennial grasses such as *Monachather paradoxa* and *Thryridolepis mitchelliana* and a low proportion of less desirable grasses such as *Aristida* spp. It is generally considered that *Aristida* is encouraged by heavy continuous grazing but grazing systems to encourage the better grasses have not yet been devised. On the other hand, continuous sheep grazing reduces the regrowth of trees and undesirable shrubs and the present trend towards an increase in cattle at the expense of sheep could have widespread and undesirable consequences (Dawson and Boyland, 1974).

The Mulga zone in Western Australia may be divided into a number of topographic vegetation types. Along the streams and around the depressions of internal drainage there are belts of chenopod shrub country, which have been mentioned in the previous section. Flanking this is the "Wandarrie Country", a perennial and short grass—forb pasture based on the grasses *M paradoxa*, *Eragrostis* spp. and *Eriachne helmsii*. Both of these areas are highly productive, but have suffered extensive degradation through heavy grazing, resulting in a reduction of perennial grasses and edible shrubs and extensive erosion. Flanking these there are short grass (*Aristida arenaria*)—forb pastures and other species on the alluvial plains which provide good grazing only after rain. Interspersed between these productive areas are the hill country, which is of low value but provides some reserve forage, and the spinifex grasslands (*Triodia basedowii* and *Plectrachne melvillei*) on sand plain country which is virtually useless (Wilcox and Speck, 1963).

These lands receive 200—250 mm of rainfall which may fall in either winter or summer although the winter rains produce more forage. They are generally grazed by sheep, although cattle are important in the lower rainfall areas. The principal problems facing these industries are the decreased carrying capacity associated with land degradation and low fertility in their breeding flocks which is associated with poor nutrition. Overall carrying capacities are now of the order of 15—20 ha per sheep compared to 6—8 ha per sheep in the early nineteen-thirties. However, the Wandarrie and chenopod communities are capable of stocking rates in excess of 6 ha per sheep (Wilcox, 1963, 1972).

The need for improved management in this particular zone has led to the development of government-sponsored schemes (in conjunction with the landholders), which are illustrated by the Gascoyne catchment scheme (Wilcox and McKinnon, 1974). This involves the mapping of districts and properties on a land system basis of pasture types at a scale of 1 : 50 000 and a classification of each land unit into condition classes based on pasture deterioration and the degree of erosion. This is followed by the determination of carrying capacity and the completion of property management plans which are developed in conjunction with the landholders.

The management plans are based on the need to restrict overall animal numbers and the encouragement of regeneration through the protection from grazing during periods of chenopod shrub regeneration or seeding by the desirable grasses. These procedures will restore the capacity of these pastures to carry sheep in dry seasons and encourage the growth of grasses that respond to rainfall in both seasons (Wilcox, 1963). Stock may be excluded altogether from the worst areas.

This type of overall management planning which includes land system mapping, rangeland management, livestock husbandry and economic planning, involving all government agencies that are concerned with a particular area should be a blueprint for future management of Australia's arid and semi-arid lands.

The Mulga lands of central Australia are similar in structure to the more arid Mulga lands of Queensland. These receive 200—300 mm of rainfall and are grazed by cattle at 35—40 ha per beast. The pastures include some perennial grasses, such as *Eragrostis eriopoda*, but are generally composed of short-lived grasses such as *Enneapogon* spp., *Aristida* spp. and forbs (*Bassia* spp., etc.). The landscape also includes extensive areas of hills and spinifex sand plains and dunefields that remain virtually unused (Perry, 1962).

As in Western Australia a management system has been devised based on land system mapping (Condon et al., 1969). In this system one of the important and well-studied pasture lands is chosen as a reference and the carrying capacity of other pasture lands in normal seasons is determined from it by the adoption of rating factors for soil, topography, tree density, rainfall and condition. Similar assessment of drought grazing capacity are determined according to the presence of topfeed and the retention of feed quality by the herbage species. Allowance is also made for the proportion of each pasture land that is within walking distance of water. Together these factors specify the number of cattle that can be carried without damaging the resource, particularly in drought times when most damage occurs. This may be coupled with the management of particular land systems, such as the siting of water sources on stable land systems and the regeneration of the more eroded areas, and may include improvements in cattle husbandry.

The tussock grasslands

The tussock grasslands are composed of perennial, summer growing grasses of the genus *Astrebla*, which all have the common name of "Mitchell Grass". The grasslands are supported on treeless, or almost treeless plains that border some inland rivers and ranges. The genus *Astrebla* is represented by four species, all of which are endemic to Australia (Orr, 1975).

The grasslands lie in an arc around the northern and north-eastern summer rainfall part of the zone (Fig. 3.3) though *Astrebla* as a species has a much wider occurrence than this as small isolated pockets on flood plains, etc. These grasslands commonly intergrade with the chenopod sub-shrubs particularly on the extensive gibber plains. In all it has been estimated that they occupy an area of some 450 000 km^2, all of which is utilized for pastoral purposes.

The Mitchell grasses grow in discrete tussocks 25—50 cm high and 10—30 cm in diameter and show extraordinary persistence under heavy grazing and frequent droughts. The former is partly due to a low acceptibility to animals, for the leaves are fibrous and non-succulent when mature, and the latter to a well-developed root system. In pristine state the grassland consists of an even, though sparsely distributed stand, of one or more *Astrebla* species occupying a basal cover of $\leqslant 4\%$ (Orr, 1975). As a minor component other perennial grass genera occur, e.g. *Eragrostis*, *Dichanthium*, *Aristida* and *Panicum*, along with annual grasses and forbs in post-rain periods. Summer rainfall promotes the growth of *Astrebla*, the other grasses and the few perennial herbs present; winter rainfall promotes the growth of annual forbs which, when present, despite their lack of bulk, increase the annual productivity considerably.

The northern *Astrebla* grasslands (Western Australia, Northern Territory and north Queensland) are used for cattle grazing exclusively while sheep predominate throughout the central and southern Queensland distribution. Despite the inhospitable nature of these hot and shadeless plains, covered with a forage of low nutritive value for several months of the year, the Mitchell grass plains support several times the intensity of stock of any of the adjacent land types. Carrying capacities range from one sheep to 2 ha down to one sheep to 5 ha.

What little change in, or deterioration of, the *Astrebla* grasslands that has been noted has variously been attributed to long-term rainfall changes and/or excessive grazing and trampling. Newman and Condon (1969) have emphasized the resilience of this community under all except gross disturbance. Nevertheless, many isolated areas of this grassland have been completely eliminated by over-stocking, particularly along the alluvial plains of the southern edges of the distribution. In all cases replacement with annual grasses or forbs results, which markedly affects animal production characteristics. However, even when devoid of vegetative cover, the flat topography and the fine-textured soils characteristic of these grasslands are very resistant to erosion.

The Mitchell grasslands are somewhat analogous to the chenopod low shrublands with respect to the reciprocal interaction of diet on nutrition with domestic stock, and hence have similar ecologically based management strategies. *Astrebla* species have low acceptibility to livestock even when green and actively growing, their preference being the annuals and ephemerals (Weston and Moir, 1969). Pure stands of these grasses have little value for the "fattening" of livestock, their low protein content makes them barely a maintenance ration. In the dry winter months when all the other forage species are gone the dormant but perennial *Astrebla* provides the only forage present.

In terms of animal production the *Astrebla* is also analogous to the chenopod shrubs in acting as a seasonal forage buffer in a largely monsoonal environment — a yearly drought reserve. But unlike the chenopods these grasses show a marked change in quantity, through haying-off and dormancy, and in quality. Dormant Mitchell grass, though alive, constitutes a poor maintenance ration for stock and they are often in a negative nitrogen balance at the end of the winter before the arrival of the summer rains (Weston and Moir, 1969). The use of urea and molasses as nitrogen and energy supplements results in improved animal production overall.

The tussock grasslands are perhaps the most robust and stable of all the rangeland types. Originally this natural and extensive grassland was held to be a fire disclimax but this is now disputed. Rather it may be surmised that low rainfall effectiveness exacerbated by heavy-textured soils performs the same role as fire in preventing the invasion and growth of woody species. The recruitment of new individuals to this very long-lived grassland is a rare and spasmodic event apparently related to exceptional rainfall events (Williams and Roe, 1975). Their robustness is aided by their low nutritive value; the grazing animals themselves die before the grassland is eliminated.

The hummock grasslands

The hummock grasslands are a very widespread and distinctive arid community. It is generally known as "spinifex" country, whereas in fact the communities have as dominants one or more species of the genera *Triodia* and *Plectrachne*, with the genus *Spinifex* not being represented at all.

The hummock grasslands is the most extensive arid plant community occupying about one-third of the arid lands (Leigh and Noble, 1969; Perry, 1970). Most of the land designated as unoccupied desert in Fig. 3.3 is spinifex country. In most cases the communities are treeless, or almost so, but in the southern, less arid edges of the distribution, the spinifex may occur as an understory in *Eucalyptus* and *Acacia* shrublands. Thus, though this formation is distinctive, evaluation of its pastoral use requires detailed consideration of site and floral characteristics and its relationship to other community types. *Triodia* and *Plectrachne* hummock grasslands have been described generally by Specht (1972) and Perry (1970).

In a hummock grassland the individual plants are up to 1—2 m high (including the seed head) and up to 1 m in diameter. In some *Triodia* spp. the centre of the hummock dies and the plants become annular rings 0.5 m wide and 0.5 m high, often exceeding 5 m in diameter. This distinctive form is unique to the Australian arid lands (Beard, 1975).

The hummocks are well spaced and the ground between them is bare, except for short periods after rainfall when there may be a cover of annual grasses or ephemerals. The supporting soils are usually the deep sands with minimal profile development on the red sands and the red clayey sands of the great sand plains. In other areas it may be restricted to the rocky hill-slopes with skeletal soils.

The most detailed ecological studies of *Triodia* and *Plectrachne* hummock grasslands are those by Winkworth (1967), who investigated their botanical composition, structure and biomass in relation to climate, soil water and soil nutrients; Winkworth (1967) concluded that the vegetation was a stable edaphic climax subject to, and adapted to, repeated burning. Burbridge (1943) has described the successional changes which follow burning of *Triodia pungens* communities in Western Australia.

The aridity, remoteness and poverty of these grasslands for livestock production are the reasons for the small amount of research information on them. Although the grazing value for domestic livestock is poor (Perry 1960; Leigh and Noble, 1969), some *Triodia* spp. are acceptable to cattle (Chippendale, 1968) and small areas of hummock grassland are in the mediocre to fair condition classes (Newman and Condon, 1969; Boyland, 1974).

presumably because they are adjacent to more productive forage types which are less drought resistant. However, the substantial areas of hummock grassland distant from better classes of country will forever be unoccupied land.

All species of "spinifex" burn readily even when green and burning to alter botanical composition or to produce green active shoots of improved nutritive value is practised on the occupied margins of this country. In fact, these grasslands are the only Australian arid community wherein fire has been investigated and recommended as a management tool.

These techniques particularly have been developed in the Pilbara and Kimberley districts of Western Australia (Wilcox, 1972). This region is characterized by summer cyclonic rains that give a short growing period of 6 to 8 weeks (December—March) only. Injudicious grazing in the past and burning in winter had all but eliminated more desirable perennial grasses and encouraged complete dominance of "soft" spinifex (*T. pungens*) with a consequent lowering of the carrying capacity of that country, from 5 ha to 10—15 ha per sheep.

The contrasting changes in sheep and kangaroo populations, the sheep declining and the kangaroos increasing, that followed this rather drastic vegetation change have been described by Newsome (1975).

Suijdendorp (1969) has demonstrated that burning should be in early summer to reduce competition from soft spinifex which is killed at this time. Deferment of grazing after the burn and opening rains for at least 6 weeks, or until the plants have matured, should follow. Under this regime the unpalatable grasses are maintained for a much longer period in the *Triodia* grass communities, and can even emerge as the dominants in the pure grass stands. Pure soft spinifex pastures also require burning to maintain the spinifex tussocks in a palatable and manageable condition. Suijdendorp (1969) has demonstrated that these changes have profound beneficial effects upon the carrying capacity of this community and upon its ability to support lactating ewes and their progeny. Pure soft spinifex pastures are quite inadequate for the latter purpose (Williams and Suijdendorp, 1968).

CONCLUSIONS

Throughout the extensively grazed arid and semi-arid lands of Australia the principal management decision is the selection of the correct number of grazing animals for the area of land available. The desire of managers to carry more animals must be tempered by the need to keep them in good condition for meat or wool production and the need to stock conservatively as an insurance against low forage production in drought years. Drought strategy must protect the basic breeding herd or flock and conserve the basic vegetation and soil resources which vary considerably in resilience from one rangeland type to another. This basic management decision on stocking rates over-rides all other decisions.

Other lesser management options arise in better seasons when regeneration or improvement in vegetation conditions may be possible. These include the revegetation of eroded areas and the management of seedling grasses or shrubs to protect those that are desirable and discourage those that are undesirable. The control of these undesirable plants is usually undertaken at the seedling stage when grazing or fire have their greatest effect. Control of botanical composition by chemical or mechanical means is technically feasible, but not adopted because of expense. Organized rotations of grazing for vegetation management are rarely adopted because of their noticeable lack of effect on either vegetation or animal production.

These lands have been altered by the grazing pressure, more noticeably in some areas than others, but in general they have not suffered to the same extent as the arid lands of the old world. Nevertheless, improvement is possible and the main need here is the extension and adoption of existing knowledge, particularly on the monitoring of rangeland condition, so that future improvements or declines in condition can be determined. One particular scheme in Western Australia (Wilcox and McKinnon, 1974) has been mentioned and this provides a general formula for other states and localities. Within such improvement schemes the close co-operation and involvement of research, extension and administrative personnel and agencies is essential, for without such co-operation the best of schemes will fail.

The economic and administrative structures also have their part to play in promoting or restricting good management practices. The major effects of these factors have been mentioned in a historical sense, but the effects of present factors such as overseas markets, government economic policy for primary industries and land tenure provisions are often overlooked. While some specific research has been undertaken on these topics, its application to the promotion of good land management is still in its infancy.

The unique conditions of the arid and semi-arid lands in Australia, such as the topography, climate, tenure and economic structures, are unlikely to find exact counterparts overseas. Consequently overseas readers will recognize the problems of extrapolating this information to other countries. Ecological differences within Australia itself are sufficient to restrict the transfer of information from one region to another. Nevertheless, some concepts have more general application and their parallels will be found within other chapters of this book.

REFERENCES

Adamson, R.E. and Osborne, T.G.B., 1922. On the ecology of the Ooldea district. Trans. R. Soc. S. Aust., 46: 539—563.

Anonymous, 1969. Report. Inter-Departmental Committee on Scrub and Timber Regrowth in the Cobar-Byrock District and Other Areas of the Western Division of N.S.W. Government Printer, Sydney.

Anonymous, 1972. Hydrology of smooth plainlands of arid Australia. Australian Water Resources Council, Hydrological Series No. 6, Canberra.

Barnard, A., 1969. Aspects of the economic history of the arid land pastoral industry. In: R.O. Slatyer and R.A. Perry (Editors), Arid Lands of Australia. A.N.U. Press, Canberra, pp. 209—228.

Beadle, N.C.W., 1966. Soil phosphate and its role in molding segments of the Australian flora and vegetation, with special reference to xeromorphy and sclerophylly. Ecology, 47: 992—1007.

Beard, J.S., 1975. The evolution of Australian desert plants. In: D.W. Goodall (Editor), Evolution of Desert Biota. University of Texas, Austin, pp. 51—64.

Biddiscombe, E.F., Hutchings, R.J., Edgar, G. and Cuthbertson, E.G., 1956. Grazing management of natural pasture at Trangie, New South Wales. Aust. J. Agric. Res., 7: 233—247.

Boyland, D.E., 1974. Vegetation. In: Western Arid Region Land Use Study Part I. Division of Land Utilization Tech. Bull. No. 12. Queensland Department of Primary Industries, pp. 47—70.

Brown, G.D. and Williams, O.B., 1970. Geographical distribution of the productivity of sheep in Australia. J. Aust. Inst. Agric. Sci., 36: 182—198.

Burbridge, N., 1943. Ecological succession observed during regeneration of *Triodia pungens* R. Br. after burning. J. R. Soc. West. Aust., 28: 149—156.

Charley, J.L., 1972. The role of shrubs in nutrient cycling. In: Wildland Shrubs — Their Biology and Utilization. USDA Forest Service, Gen. Tech. Rep. INT-1, pp. 182—203.

Chippendale, G.E., 1968. The plants grazed by Red Kangaroos, *Megaleia rufa* (Desmarest), in Central Australia. Proc. Linn. Soc. N.S.W., 93: 98—110.

Christie, E.K., 1975. A study of phosphorus nutrition and water supply on the early growth and survival of buffel grass grown on a sandy red earth from south-west Queensland. Aust. J. Exp. Agric. Anim. Husb., 15: 239—249.

Condon, R.W., 1961. Erosion and management problems in western New South Wales. J. N.S.W. Soil Conserv. Serv., 17: 86—102.

Condon, R.W., Newman, J.C. and Cunningham, G.M., 1969. Soil erosion and pasture degeneration in central Australia. III. The assessment of carrying capacity. J. N.S.W. Soil Conserv. Serv., 25: 225—250.

Cunningham, G.M., 1967. Furrowing aids revegetation at Cobar despite the worst drought on record. J. N.S.W. Soil Conserv. Serv., 23: 192—202.

Cunningham, G.M. and Walker, P.J., 1973. Growth and survival of Mulga (*Acacia aneura* F. Muell. ex Benth.) in western New South Wales. Trop. Grassl., 7: 69—77.

Dawson, N.M. and Boyland, D.E., 1974. Resource use. In: Western Arid Region Land Use Study Part I. Division of Land Utilization Tech. Bull. No. 12. Queensland Dep. of Primary Industries, pp. 104—129.

Dawson, N.M., Boyland, D.E. and Ahern, C.R., 1975. Land management in south-west Queensland. Proc. Ecol. Soc. Aust., 9: 124—141.

Dury, G.H., 1968. An introduction to the geomorphology of Australia. In: G.H. Drury and M.I. Logan (Editors), Studies in Australian Geography. Heinemann, Melbourne.

Ealey, E.H.M., 1967. Ecology of the euro, *Macropus robustus* (Gould), in north-western Australia. I. The environment and changes in euro and sheep populations. CSIRO Wildl. Res., 12: 9—25.

Easter, C.D., 1975. Some agronomic aspects of factors underlying production in the New South Wales grazing industry. Q. Rev. Agric. Econ., 28: 177—199.

Everist, S.L., 1972. Australia. In: Wildland Shrubs — Their Biology and Utilization. USDA Forest Service, Gen. Tech. Rep. INT-1, pp. 16—25.

Gaffney, D.O., 1971. Seasonal rainfall zones in Australia. Working Paper No. 141. Bureau of Meteorology, Melbourne.

Gaffney, D.O., 1975. Rainfall deficiency and evaporation in relation to drought in Australia. Unpublished manuscript. Bureau of Meteorology, Melbourne.

Gibbs, W.J., 1969. Meteorology and climatology. In: R.O. Slatyer and R.A. Perry (Editors), Arid Lands of Australia. A.N.U. Press, Canberra, pp. 33—54.

Graetz, R.D., 1975. Biological characteristics of Australian *Acacia* and chenopodiaceous shrublands relevant to their pastoral use. In: D.N. Hyder (Editor), Arid Shrublands. Society of Range Management, Denver, Colorado, pp. 33—38.

Graetz, R.D. and Wilson, A.D., 1979. An assessment of herbivore diets in the chenopod shrublands. In: R.D. Graetz (Editor), Studies of the Australian Arid Zone IV. Management of Chenopod Shrublands. CSIRO, Melbourne.

Heathcote, R.L., 1969. Land tenure systems: past and present. In: R.O. Slatyer and R.A. Perry (Editors), Arid Lands of Australia. A.N.U. Press, Canberra, pp. 185—208.

Hubble, G.D., 1970. Soils. In: R.M. Moore (Editor), Australian Grasslands. A.N.U. Press, Canberra, pp. 44—58.

Jackson, E.A., 1962. Soil studies in Central Australia; Alice Springs—Hermannsburg—Rodinga Areas. CSIRO Soil Publication No. 19, Melbourne.

King, C.J., 1957. An outline of Closer Settlement in New South Wales. Government Printer, Sydney.

Leigh, J.H., 1974. Diet selection and the effects of grazing on the composition and structure of arid and semi-arid vegetation. In: A.D. Wilson (Editor), Studies of the Australian Arid Zone. II. Animal Production. CSIRO, Melbourne, pp. 102—126.

Leigh, J.H. and Mulham, W.E., 1971. The effect of defoliation on the persistence of *Atriplex vesicaria*. Aust. J. Agric. Res., 22: 239—244.

Leigh, J.H. and Noble, J.C., 1969. Vegetation resources. In: R.O. Slatyer and R.A. Perry (Editors), Arid Lands of Australia. A.N.U. Press, Canberra, pp. 73—92.

McKay, L.E., 1973. Some aspects of diversification of production in the pastoral zone of the Australian sheep industry. Q. Rev. Agric. Econ., 26: 140—148.

Mabbutt, J.A., 1969. Landforms of arid Australia. In: R.O. Slatyer and R.A. Perry (Editors), Arid Lands of Australia. A.N.U. Press, Canberra, pp. 11—32.

Mabbutt, J.A. and Sullivan, M.E., 1970. Landforms and structure. In: R.M. Moore (Editor), Australian Grasslands. A.N.U. Press, Canberra, pp. 27—43.

Mawson, W.F.Y., Hunter, H.C.D., Robinson, I.B. and Dawson, N.M., 1974. Current land use. In: Western Arid Region Land Use Study Part I. Division of Land Utilization Tech. Bull. No. 12. Queensland Department of Primary Industries, pp. 84—103.

Moore, R.M., 1975. Australian arid shrublands. In: D.N. Hyder (Editor), Arid Shrublands. Society of Range Management, Denver, Colorado, pp. 6—11.

Moore, R.M., Condon, R.W. and Leigh, J.H., 1970. Semi-arid woodlands. In: R.M. Moore (Editor), Australian Grasslands. A.N.U. Press, Canberra, pp. 228—245.

Moore, R.M. and Perry, R.A., 1970. Vegetation. In: R.M. Moore (Editor), Australian Grasslands. A.N.U. Press, Canberra, pp. 59—73.

Moore, R.M. and Walker, J., 1972. *Eucalyptus populnea* shrub woodlands. Control of regenerating trees and shrubs. Aust. J. Exp. Agric. Anim. Husb., 12: 437—440.

Newman, J.C. and Condon, R.W., 1969. Land use and present condition. In: R.O. Slatyer and R.A. Perry (Editors), Arid Lands of Australia. A.N.U. Press, Canberra, pp. 105—132.

Newsome, A.E., 1975. An ecological comparison of the two arid zone kangaroos of Australia, and their anomalous prosperity since the introduction of ruminant stock to their environment. Q. Rev. Biol., 50: 389—424.

Nix, H.A. and Austin, M.P., 1973. Mulga: A bioclimatic analysis, Trop. Grassl., 7: 9—21.

Orr, D.M., 1975. A review of *Astrebla* (Mitchell grass) pastures in Australia. Trop. Grassl., 9: 21—36.

Oxley, R.E., 1979. The perennial chenopod pasture lands of Australia. In: R.D. Graetz (Editor), Studies of the Australian Arid Zone IV Management of Chenopod Shrublands. CSIRO, Melbourne.

Perry, R.A., 1960. Pasture lands of the Northern Territory, Australia. In: Land Research Series No. 5. CSIRO, Melbourne.

Perry, R.A., 1962. Present and potential land use of the Alice Springs area. In: Land Research Series No. 6. CSIRO, Melbourne, pp. 258—278.

Perry, R.A., 1968. Australia's arid rangelands. Ann. Arid Zone, 7: 243—249.

Perry, R.A., 1970. Arid shrublands and grasslands. In: R.M. Moore (Editor), Australian Grasslands. A.N.U. Press, Canberra, pp. 246—259.

Perry, R.A., 1974. Strategies available for managing multispecific communities in Australia. In: Plant Morphogenesis as the Basis for Scientific Management of Range Resources. USDA Miscellaneous Publication 1271, pp. 124—139.

Potter, B.V. and McIntosh, G.H., 1974. Effect of salt water ingestion on pregnancy in the ewe and on lamb survival. Aust. J. Agric. Res., 25: 909—917.

Ratcliffe, F.N., 1936. Soil drift in the arid pastoral areas of South Australia. Pamphlet No. 64, CSIRO, Melbourne.

Ricketts, W.J., 1969. The role of transport in the utilization of arid land resources. In: R.O. Slatyer and R.A. Perry (Editors), Arid Lands of Australia. A.N.U. Press, Canberra, pp. 142—157.

Rolls, E.C., 1969. They All Ran Wild: The Story of Pests on the Land In Australia. Angus and Robertson, Sydney.

Specht, R.L., 1972. The Vegetation of South Australia. Government Printer, Adelaide.

Suijdendorp, H., 1969. Deferred grazing improves soft spinifex association. J. Agric. West. Aust., 10: 487—488.

Walker, J., Moore, R.M. and Robertson, J.A., 1972. Herbage response to tree and shrub thinning in *Eucalyptus populnea* shrub woodlands. Aust. J. Agric. Res., 23: 405—410.

Weston, E.J. and Moir, K.W., 1969. Grazing preferences of sheep and nutritive value of plant components in a Mitchell grass association in north-western Queensland. Queensl. J. Agric. Anim. Sci., 26: 639—650.

Wilcox, D.G., 1963. The pastoral industry of the Wiluna-Meekatharra area. In: Land Research Series No. 7. CSIRO, Melbourne, pp. 195—212.

Wilcox, D.G., 1972. The use of native pastures by sheep in Western Australia. In: J.H. Leigh and J.C. Noble (Editors), Plants for Sheep in Australia. Angus and Robertson, Sydney, pp. 45—54.

Wilcox, D.G. and McKinnon, E.A., 1974. A report on the condition of the Gascoyne catchment. Departments of Agriculture and Land and Surveys, Western Australia.

Wilcox, D.G. and Speck, N.H., 1963. Pastures and pasture lands of the Wiluna—Meekatharra area. In: Lands of the Wiluna—Meekatharra Area, Western Australia. Land Research Series No. 7. CSIRO, Melbourne, pp. 162—177.
Williams, O.B., 1974. Vegetation improvement and grazing management. In: A.D. Wilson (Editor), Studies of the Australian Arid Zone. II. Animal Production. CSIRO, Melbourne, pp. 127—143.
Williams, O.B. and Roe, R., 1975. Management of arid grasslands for sheep: plant demography of six grasses in relation to climate and grazing. Proc. Ecol. Soc. Aust., 9: 142—156.
Williams, O.B. and Suijdendorp, H., 1968. Wool growth and wethers grazing *Acacia aneura—Triodia pungens* savanna and ewes grazing *Triodia pungens* hummock grass steppe in the Pilbara district, W.A. Aust. J. Exp. Agric. Anim. Husb., 8: 653—660.
Wilson, A.D., 1966. The value of *Atriplex* (saltbush) and *Kochia* (bluebush) species as food for sheep. Aust. J. Agric. Res., 17: 147—153.
Wilson, A.D., 1974a. Nutrition of sheep and cattle in Australian arid areas. In: A.D. Wilson (Editor), Studies of the Australian Arid Zone. II. Animal Production. CSIRO Melbourne, pp. 74—84.
Wilson, A.D., 1974b. Water consumption and water turnover of sheep grazing semi-arid pasture communities in New South Wales. Aust. J. Agric. Res., 25: 339—347.
Wilson, A.D., 1976. Comparison of sheep and cattle grazing on a semi-arid grassland. Aust. J. Agric. Res., 27: 155—162.
Wilson, A.D., 1978. Future resource management in the arid rangelands of Australia. J. Aust. Inst. Agric. Sci., 44 (in press).
Wilson, A.D., Mulham, W.E. and Leigh, J.H., 1976. A note on the effects of browsing by feral goats on a Belah (*Casuarina cristata*)—Rosewood (*Heterodendrum oleifolium*) woodland. Aust. Rangel. J., 1: 7—12.
Winkworth, R.E., 1967. The composition of several arid spinifex grasslands of Central Australia in relation to rainfall, soil water relations and nutrients. Aust. J. Bot., 15: 107—130.

Chapter 4

MANAGEMENT OF SEMI-ARID ECOSYSTEMS IN ISRAEL

I. NOY-MEIR and N.G. SELIGMAN

Department of Botany, Hebrew University, Jerusalem,
and *Department of Range and Fodder Crops, Volcani Center, Agricultural Research Organization, Bet-Dagan (Israel)*

INTRODUCTION

A wide belt of semi-arid climate arches along the northern margins of the Arabian Peninsula, from the Persian Gulf to northern Sinai. Here we refer to the western-most corner of this belt, between the Jordan Valley and the Mediterranean, and between the isohyets of 100 and 300 mm mean annual rainfall (Fig. 4.1). The total area is about 5000 km^2.

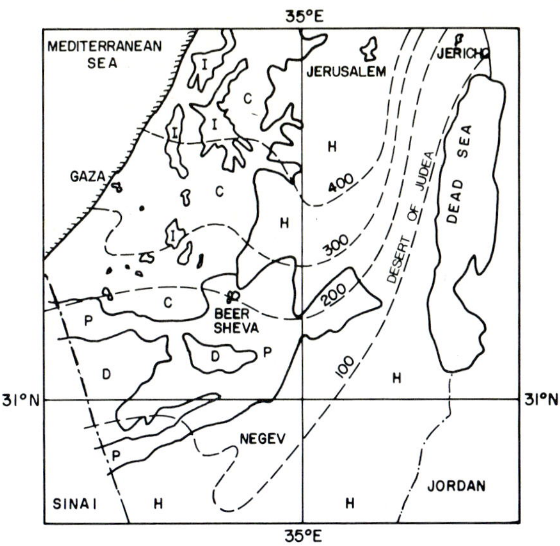

Fig. 4.1. The semi-arid zone in southern Israel. The area discussed in this chapter is bounded by the 100 and 300 mm isohyets in the south and north, and by the Sinai and the Judean Desert in the west and east. Land-use types are marked by: H, hilly, shallow soil, mainly pasture; P, loessial plains with sporadic cultivation; D, sand dunes, partly stabilized; C, cultivated under dryland conditions or occasional supplementary irrigation; I, irrigated land. (After Atlas of Israel, 1970. Survey of Israel, Jerusalem and Elsevier, Amsterdam; pre-1948 boundaries only are marked, without reference to present or future boundaries.)

In the main features of climate, soil, vegetation and land-use, this sector is fairly typical of the belt as a whole. But there are some special features which distinguish it from the main body of the Syrian Desert. Firstly, the climatic gradients are steeper, so that the semi-arid belt is narrow (10—70 km rather than hundreds of km) and much closer to the sub-humid Mediterranean cultivation lands on one side and to the arid deserts on the other. This proximity has important biological and historical effects. Secondly, the semi-arid climate is modified by a certain maritime influence, which is absent in the Syrian Desert. In both these features there is some similarity with the semi-arid lands of coastal Egypt and Cyrenaica.

A geographical term which covers most of the subject region as defined here is Négev, which in biblical Hebrew meant "dry land" or "South land". The semi-arid parts towards the Jordan Valley and the Dead Sea, which were in the east, rather than in the south, in relation to the settled land, were known as Midbár Yehudá (Desert of Judaea) or simply "the midbár". The original meaning of the latter word was probably "grazing land" or "unsettled land", rather than "arid desert" as it is used today.

This chapter includes first, a brief description of the environment and biota of the region and of its history; then, several systems of land-use and management are described and discussed: the "traditional" system of livestock grazing and dry-land farming which has existed here for 5000 years; a modern intensive dry-land farming system which has developed recently; and new farming—grazing systems which may be developing.

ENVIRONMENT AND BIOTA

Climate

The climate can be described as semi-arid Mediterranean, or BShs (dry, steppe climate, warm, dry season in summer) in the Köppen system (Evenari et al., 1971). Rainfall is restricted to the cooler half of the year (October—April) and concentrated mainly in winter (December—February). There is considerable variability between years in the total rainfall (30—40% coefficient of variation) and in the distribution within the rainy season (Table 4.1).

Conditions at Beer-Sheva (Fig. 4.2; Katznelson, 1959) are fairly typical for the region. Three major trends can be noted within the region:

(a) a decrease in rainfall from north-west to south-east;

(b) a decrease in winter temperatures by about 5°C as altitude increases from sea-level (in the west) and 300 m below sea-level (in parts of the east) to 700 m in parts of the central range;

(c) a maritime effect in the west: higher air humidity and dew incidence, smaller diurnal and annual temperature amplitudes.

TABLE 4.1

Variability of mean monthly rainfall and number of rainy days at Beer Sheva (1921—1951) (after Katznelson, 1956)

Month	Rainfall		Rainy days	
	mm	cv	No.	cv
September	0.2	539	+	—
October	4.2	181	1.7	100
November	24.9	97	3.6	70
December	40.5	82	6.7	60
January	46.4	82	6.2	40
February	43.5	71	6.7	43
March	27.4	85	7.4	39
April	8.5	129	1.1	82
May	2.8	268	+	164
Total	198.4	30	33.4	26

cv, coefficient of variation (%).

Two extreme conditions are biologically important. Night surface temperatures near 0°C occur in most parts of the region on 5 to 25 nights per year on average, but may be more frequent and severe in some years. Very dry and hot weather with easterly to southerly winds (H[1]: sharav) occurs on 30—50 days between March and October; in spring this condition accelerates desiccation of annual vegetation and the end of the growing season.

The most common winds, bringing most of the rain, as well as raising dust storms, are from the south-west to north-west.

There seems to be no evidence of any drastic changes of climate in the last 8000 years, though some fluctuations may have occurred.

Geomorphology and soils

From west to east, the region can be divided broadly into the Coastal Plain, the Central Range and the Rift Valley. However, right in the middle of the semi-arid region, the drainage system of Nahal Beer-Sheva (Wadi Saba') has cut through the central range, reduced it to some low hills and created the large Beer-Sheva-Arad Plains, which virtually separate the Judaean Mountains from those of the Negev.

The Coastal Plain is covered with aeolian sediments of silt (loess), fine sand, coarse sand and various mixtures. The soils are generally deep and of low salinity (Table 4.2, Dan et al., 1962).

---

[1] Abbreviations used in the text for non-English terms: H, Hebrew; A, Arabic.

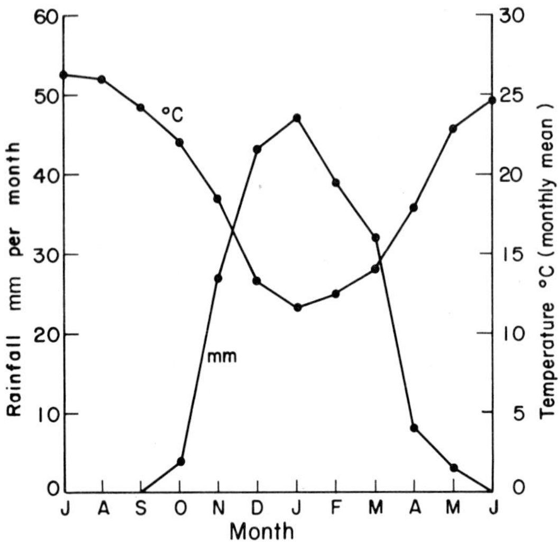

Fig. 4.2. Climagram of Beer Sheva. Monthly means, average for years 1931—1960. (After Ganor et al., 1973).

The Central Range consists of Upper Cretaceous and Tertiary calcareous rocks, varying from soft marls and chalks to hard limestones and dolomites. These are covered by generally shallow and stony silty soils, with many rock outcrops. The ranges are dissected by a dense network of wadis (A: dry watercourses) with silty or gravelly beds. The silty soils of the old wadi-terraces and of wider plains have accumulation layers of $CaCO_3$, NaCl (and sometimes $CaSO_4$) at a depth varying from 80 to 20 cm, depending on the dryness of the site.

The lower levels of the Rift Valley are dissected plains with marly and gravelly sediments. The transition to the Central Range is marked by steep escarpments, incised by deep canyons.

Hydrology

The drainage basin of Nahal Besor (Wadi Shalala) and its many tributaries occupies most of the region. In the far west there are sand plains and dunes without well-defined surface flow patterns. The eastern part is drained to the Jordan and Dead Sea, through many parallel wadis.

Permanent springs, and short permanent streams fed by them occur mainly along the eastern escarpments (notably Wadi Qelt and Jericho); elsewhere, there are only short sections of the Besor with permanent surface water.

The loessial, marly and rocky surfaces covering most of the area start producing runoff even from small storms of 3—5 mm (Evenari et al., 1971).

TABLE 4.2

Major soil types in the semi-arid region of Israel (after Dan et al., 1962; Danin et al., 1975)

Landscape	Soil type	Parent material	Rainfall (mm)	Texture	Depth (in cm) of		
					Rock	$CaCO_3$	NaCl
						Accumulation layer	
Hills	Brown rendzina	Hard limestone + chalk	250<	Loam	10—40	(±40)	—
	Brown lithosol	Hard limestone + chalk	100—250	Loam	10—40	20—40	—
	Light rendzina	Soft chalk + mark	250<	Loam	10—40	20—40	—
	Rendzinic lithosol	Soft chalk + mark	100—250	Loam	10—40	0—30	20—40
Plains	Loessial burozem	Loess (aeolian)	200—300	Silty sandy loam	(>100)	60—100	(>100)
	Loessial serozem	Loess (aeolian)	100—200	Silty sandy loam	(>100)	20—60	40—80
	Sandy burozem	Sand (+ loess)	100—250	Silty sand	(>100)	(>100)	—
	Sand dunes	Sand	100—150	Sand	(>100)	—	—
Wadis	Loessial alluvium	Loess		Silt + fine sand	(>100)	—	—
	Coarse alluvium	Gravel + sand		Gravel + sand	(>100)	—	—

Falls of 10 mm or more in a day usually cause flow in most wadi channels while in the largest wadis, flash floods may occur one to 10 times a year. These floods carry considerable volumes of water to the sea, but much larger amounts infiltrate in the gravelly beds.

Fairly shallow aquifers, which have long been tapped by wells, usually occur locally, associated with gravelly deposits in wadi beds and sand dunes near the coast. Over the rest of the area, evapotranspiration balances infiltration in most years, so that the soil is permanently dry beyond the root zone, and there is no deep drainage and no groundwater recharge.

Vegetation

Phytogeographically, the semi-arid part of Israel is regarded as part of the Irano-Turanian Region (Zohary, 1962), which due to the steep climatic gradient is reduced here to a narrow belt between the Mediterranean and Saharo-Arabian Regions. Typical Irano-Turanian species (e.g. *Artemisia herba-alba*, *Noaea mucronata*)[2] are indeed prominent in the flora and vegetation of this belt, but always in mixtures with plants of one of the two adjacent regions (Table 4.3, Danin et al., 1975). Saharo-Arabian elements are predominant in the vegetation of warm and dry habitats in the east and south and on sand in the west.

The typical vegetation form is a sparse shrubland dominated by low (10—50 cm) chamaephytes on slopes and plains and medium tall (1—2 m) shrubs in wadis. Trees are very scarce, but there are interesting populations of *Acacia raddiana* in the western plains (and in wadis) (Halevy and Orshan, 1972) and of *Pistacia atlantica* in the Negev Highlands (Danin and Orshan, 1970) which might possibly be remnants of larger populations in the past. In silty wadis and on plains (and on some slopes) there is a dense and highly diverse ephemeral vegetation of annuals and herbaceous perennials, which appears for 2 to 5 months in winter and spring, if rainfall permits.

The vegetation has been considerably affected and modified by human activities (cultivation, livestock grazing, woodcutting) over several millenia (see below).

Fauna

The fauna of the semi-arid region in Israel is mainly Mediterranean and Irano-Turanian in its affinities. It is replaced by a desert fauna (Saharo-Arabian) only below the 100 mm rainfall line (Shkolnik, in Evenari et al., 1971).

Ants are the most conspicuous group among the invertebrates, in species

[2] All plant names are according to Zohary (1976).

TABLE 4.3

Dominant perennial plant species in major habitats in the semi-arid region of Israel (after Danin et al., 1975; Zohary, 1962)

	Slopes						Plains						Wadis		
Rainfall (mm)	>250		150—250		<150		>200		<200						
Substrate	H	C	H	C	H	C	L [a]	SL [a]	L	SL	S	D	L	G	S
Sarcopoterium spinosum	+														
Echinops spinosus		+													
Artemisia herba-alba		(+)	+		+					(+)					
Noaea mucronata			+		(+)					+					
Reaumuria hirtella				+		+									
Salsola vermiculata ssp. *villosa*				+		+									
Anabasis articulata					+					(+)					
Achillea santolina							+								
Anabasis syriaca							+		(+)						
Hammada scoparia									+						
Artemisia monosperma								+		(+)	+	(+)			
Stipagrostis scoparia												+			
Thymelaea hirsuta			(+)										+		
Atriplex halimus													+	+	
Retama raetam											(+)			+	+

[a] Mostly after cultivation.

Substrates: H, hard limestone and chalk; C, soft chalk and marl; L, loess; SL, sand covering loess or chalk; S, sand (stable); D, sand dunes; G, gravel.

diversity, population density and possibly in ecological importance; though isopods, land snails and grasshoppers are also common. Among vertebrates, lizards and rodents are the dominant groups in terms of diversity, density and biomass. Populations of large mammals have been severely reduced by human pressure, but viable populations of gazelles (*Gazella gazella*) have survived long enough to benefit from recent conservation efforts. Some predators (wolf, jackal, fox, hyaena, caracal) also still survive in small numbers. Several species of large herbivores and carnivores seem to have become extinct in historical times.

HISTORY

Persistence amidst change

Our region happens to be at one of the ancient cross-roads of human civilizations. It has therefore been subject to fairly frequent and dramatic political changes (Table 4.4). Often these have caused, directly or indirectly, changes in the economy and in patterns of settlement and land-use.

However, despite this turbulent history, some main characteristic features of the land-use system can be traced back continuously for more than 5000 years. The inhabitants of the subterranean villages of the Chalcolithic Beer-Sheva culture (Josien, 1955; Perrot, 1955) had an economy that combined small livestock grazing with dry-land winter grain cropping. These have remained the two dominant economic activities in the Negev from then until recently. Throughout history there have been fluctuations in their relative importance, and other activities have been important for some periods.

During all this period, two major groups in the population of the region can be distinguished mainly by habitation and mobility: tent-dwellers (more or less nomadic) and house-dwellers (sedentary). In some periods, but not always, this corresponded with an ethnical and political dichotomy. There was also economic specialization, but it was probably never as clear-cut as i sometimes presented: tent-dwellers tended to put more emphasis on livestock (but grew crops as well; see Genesis, 26, 12), house-dwellers put a more consistent effort in cropping (but also raised large herds; see Samuel I 25, 2).

What were the ecological implications of the changing political relations between nomads, settled people and governments which ruled the area (or claimed to do so)? Some basic situations may be recognized.

The limes of empires

Over long periods of its history, the Negev was a frontier province (Latin limes) of a large empire: Egyptian, Assyrian, Persian, Hellenistic, Roman-

Byzantine, etc. It was important in guarding the Fertile Crescent and the Asia-to-Africa land route against the desert tribes. Two different policies were used by the imperial governments to facilitate their military tasks: (a) alliances with local nomadic tribes, employing them as auxiliary troops, guides and scouts in return for seasonal grazing rights in settled areas, (b) encouragement of a southward extension of settlement and agriculture, including land grants to soldiers and ex-soldiers. It seems that often both policies were pursued simultaneously, but under Roman and Byzantine rule the latter became dominant. In these periods, settled agriculture in the Negev reached its peak. It extended well beyond the semi-arid zone into the arid Negev Highlands (less than 100 mm), by developing to perfection techniques of runoff farming which had been used earlier by Nabataeans and Israelites (Evenari et al., 1971). Any nomads that did not settle must have been under strict control.

However, this dense settlement could prosper only as long as the central government was both interested and strong enough to support it (also economically) by keeping its army in the limes province.

Weak empires and nomad invasions

In other periods, the Negev was nominally part of a large empire, the government of which was either too weak or too indifferent to control it effectively. Usually nomadic tribes gained the real power and their system of land-use became dominant in the region. The sedentary population abandoned, or retreated to a few fortified towns. This had occurred sometimes even during Roman and Byzantine rule, as well as many times earlier, probably as far back as the Early Bronze age.

The longest period of nomad dominance began soon after the Arab conquest, when the Negev suddenly lost its importance as a frontier. With short intermissions, the nomads continued to gain ground throughout the periods of Crusader, Mameluk and Ottoman rule (until about 1900). During the latter period, several waves of Bedouin tribes invading from the Arabian Desert pushed sedentary farming back to the 400—500 mm isohyet in the coastal plain and the Jordan Valley. Only in the Judaean Mountains (Jabal al-Khalil) a chain of strong villages managed to keep a frontier at about 300 mm, by alternately fighting the Bedouin, allying with them or paying them the "brotherhood tribute".

Settling nomads and local kingdoms

Nomadic tribes from the arid desert, which invaded the semi-arid region (and stayed there long enough before being displaced by a new wave) tended to adapt gradually to the new environment. Migrations became shorter and less frequent, camels were replaced by sheep and goats and grain cultivation

TABLE 4.4

A summary of the history of the semi-arid region of Israel

Years	Period	Political	Economic activities		Technological innovations	Form of settlement
			Major	Secondary		
Before 10 000 B.C.	Palaeolithic		Hunting-gathering			Nomadic
10 000—5000	Neolithic		Hunting-gathering	Livestock, grains	Sheep and goat domestication, early agriculture	Mainly nomadic (oasis town at Jericho)
5000—3000	Chalcolithic	(Beer-Sheva culture)	Grains and livestock		Wells	Semi-nomadic and subterr. villages
3000—1200	Bronze	Canaanite towns (Egyptian hegemony)	Grains and livestock		Camels	Walled towns, tents
1200—600	Early Iron	Israelites, Kdm. of Juda (since 1000)	Grains and livestock		Cisterns, early runoff agric.	Homesteads, towns and tents
600—330	Late Iron	Persian Empire	Grains and livestock			Decreasing rural settlement
330—160	Hellenistic (early)	Ptolemaeid + Seleucid Empires	Grains and livestock			Decreasing rural settlement

B.C. 160—0	Hellenistic (late)	North: Judaea South: Nabataea	Grains and livestock	Runoff agric. and caravan trade	Development of runoff	Increasing rural settlement
A.D. 0—330	Roman	Roman Empire (south: Nabataean until 106)	Grains and livestock	Runoff agric. and military	agriculture techniques	Growth of towns
330—634	Byzantine	Byzantine Empire	Grains and livestock	Runoff agric. and military		Growth of towns
634—1100	Early Arab	Ommayads, Abbasids, Fatimids	Livestock and grains	—		Decrease in settlement
1100—1260	Crusader	Crusader Kdm., Ayyubids	Livestock and grains	—		Decrease in settlement
1260—1517	Mameluk	Mameluk Kdm. (Egypt)	Livestock and grains	Raids, caravans		Mainly tents
1517—1917	Ottoman	Ottoman (Turkish) Empire	Livestock and grains	Raids, caravans		Mainly tents (1903 Beer-Sheva refounded)
1917—	Modern	1917—1948 British mandate, since 1948 Israel (in part 1948—1967 Jordan)	Grains and livestock and irrigat. agric.	Industry	Pipelines, N-fert., mechanization, veterinary care	Bedouin settling, cooperative villages, towns growing

became more regular and extensive (e.g. most recently in the last century). Sometimes in history the loose tribal system then evolved into a more centralized kingdom. Whenever the semi-arid region was ruled by such a locally-based government (rather than by a remote imperial one), the conflict between nomads and settlers, farming and stock-raising did not seem to be a major issue, both elements being ethnically related and involved in the same political structure.

The first documented occurrence of such a process in the Negev was about 3000 years ago. The Israelites arrived as tent-dwelling, semi-nomadic pastoralists, and gradually shifted to permanent settlement and cultivation; but parts of the population (e.g. the tribe of Simeon) seem to have remained tent-dwellers for a long time. For 400 years, the Negev and its population (both nomadic and settled) were an important part of the economic and political base of the Kingdom of Judaea, which in turn had a vital interest in the stability and prosperity in the Negev. This situation, as well as the development of techniques of cistern-building and runoff farming, favoured a considerable extension of agriculture in the semi-arid and arid zones under the Judaean kings. In contrast to the Roman—Byzantine period, agriculture in the Negev in this period was not concentrated around a few large towns but apparently dispersed over smaller villages and hamlets or homesteads (H: haṣér); and pastoralism probably retained a much larger role.

A similar process apparently occurred again between 300 and 100 B.C. The Nabataeans arrived in the Negev as nomadic pastoralists from the desert. They developed there a prosperous economy based on runoff agriculture and long-range trade (in addition to livestock), under an independent kingdom with its base in the arid zone itself.

THE OLD CROP—LIVESTOCK SYSTEM

Definition

One system of land-use can be clearly recognized as prevailing in the Negev from the Chalcolithic age (3500 B.C.) until the early 20th century (and in part until today). It may be defined by the following characteristics:

(1) grazing by mobile herds of small livestock (sheep and goats) is combined with cultivation for winter cereals (wheat and barley) where and when suitable;

(2) there is no significant use of irrigation water, fertilizer and supplementary feeds;

(3) grain and livestock products are used primarily for subsistence of the local population, but in favourable years surplus is exported.

The following sections describe some of the features of this traditional crop—livestock system, which can generally be traced through all or most of the periods of its existence. The description is based on a variety of sources: for the earliest periods, the Bible and archaeological data (e.g. Negev, 1972; Aharoni, 1974); for the Roman period, the Mishna, [3] recent works on agriculture in the Mishna and Talmud [4] (Felix, 1963; Ben-David, 1974), archaeological data; for the Bedouin in the 19th century, writings of explorers (particularly Musil, 1908); and in recent years mainly studies by Aref (1934), Epstein (1939), Marx (1967) and personal observations (Noy-Meir, 1975 and unpublished, 1974—1975).

Land classification and tenure

Two main categories of land are recognized: sown (seedland) and unsown land. Practically all the unsown land is grazing land. The seedland may be cultivated and sown every year, in rotation or only in favourable years. Arable land that has not been actually ploughed and sown for some years is considered as grazing land.

The distribution of the two types depends mainly on physical and partly on socio-economic factors. The best land ($\sim$20%) has practically always been seedland: all deep soils (>50 cm) in areas with more than 250 mm rainfall, and most deep soils in wadi-beds in the 150—250 mm belt. Much land has never been cultivated, being too rocky, stony, sandy, shallow (<30 cm), saline or dry ($\sim$50%). Marginal seedland has been cultivated in periods when socio-economic conditions in the region favoured the expansion of agriculture, and was abandoned in other periods. This includes stony and/or slightly saline soils of medium depth (30—50 cm) and above 200 mm rainfall; some loessial and sandy plains in the 100—250 mm belt; wadis in arid areas below 150 mm (down to 50 mm).

In some areas receiving runoff water, in terraced wadis or depressions, fruit trees (almond, fig, olive, etc.) and vines were planted. Such orchards were probably most extensive in the Roman—Byzantine period (Evenari et al., 1971); many small ones can still be seen in the Negev today.

In general, sown and planted land has been held and owned by individuals or by families while grazing land has been used in common by whole villages, tribes or nations. The acquisition of both types of land by such groups has usually been by force, or by default. The acquisition of fields in new or marginal land by individuals within a group may be by organized distribution, by actual cultivation, or by declaring ownership in front of witnesses and marking the boundaries with stones, or with bulb plants (e.g. *Urginea maritima*) where stones are scarce. Thereafter the field was regarded as private property, which could be inherited, sold, or leased. The normal

[3] Mishna: the body of Jewish traditional law, collected and edited in the second century.
[4] Talmud: Commentary and extension to the Mishna, recorded in the fourth to fifth centuries.

size of the area owned or cultivated by a family was probably 5 to 10 ha; some larger holdings were cultivated and harvested with hired labour, or leased out. Communal ownership of arable land (A: mashá), with an annual redistribution of strip-fields by lot, was practised in villages throughout the country in the last centuries, but not by the Bedouin in the Negev. State ownership of semi-arid lands was widely effective only in the Roman, Byzantine (and perhaps the Persian) period, when large estates were managed by the crown, or granted to individuals or churches. Under Ottoman rule, most of the Negev was theoretically crownland, but only after 1900 did the Turkish government (and later the British and Israeli governments) try to realize this claim in any substantial part of the area.

Private tenure of seedland had its limitation in the Negev as in the rest of the country. The Law of Moses and the Ten Rules of Joshua (Talmud Babli, Baba Kama 80) do not allow the owner to prevent certain uses of the field by the general public; picking of green ears and fruits for immediate consumption, gathering of wood and wild plants, free passage (except when the crop is standing). These rules probably date back to the Israelite conquest 3200 years ago, and have mostly survived to the present in the common law of Bedouin and Arab peasant societies.

The common grazing lands of the semi-arid zone were in some periods divided into large territories, in each of which a tribe or a village had priority rights. This may have been the situation in Biblical times, as indicated by geographical names like the "Negev of . . ." (name of tribe: Yehuda, Qeini, Kaleb, etc.) and the "Midbar of . . ." (name of village: Zif, Maón, Teqóa). In the recent past, each of the Bedouin tribes (A: 'ashíra) had a territory within which its members had grazing priority. Tribes belonging to the same alliance (A: s̱af or qabîla) also granted mutually free grazing in their territories, particularly when drought hit some areas worse than others. Tribes of different alliances could be refused, or special arrangements could be made. Boundary disputes, or a massive intrusion by a strange tribe, often caused fighting. At some times in the past, tribal territorial systems seem to have broken down, resulting in completely irregular grazing-use in the whole region.

The crops and their management

(1) The plants. The winter cereals, wheat and barley, have always been the main crops in the semi-arid zone. Their presumed ancestor species grow wild in adjacent, slightly more humid areas, and their growth cycle is fairly well adapted to the climate.

The main wheat species in Chalcolithic times was emmer (*Triticum dicoccum*), but later it was replaced by varieties of soft and hard wheat (*T. aestivum, T. durum*). Wheat grain was reserved for human consumption and was the main source of food energy as well as an important source of protein for most of the population. Chaff and straw was fed to livestock.

Barley (*Hordeum sativum*; ssp. *distichum*, later mainly ssp. *vulgare*) was regarded as inferior to wheat as human food, but in many periods barley was used for making bread and brewing beer. Otherwise, barley grain was fed to animals used for work and transport.

In all periods, both wheat and barley were grown in the Negev, their proportions varying in space and time according to ecological and economic factors. Barley is ready for harvest 6—7 weeks earlier than wheat, thus is less susceptible to drought failure and preferred in dry (or shallow) sites. Wheat can make better use of moisture in deeper layers and can give higher yields in favourable sites and years. However, in much of the area either of the two can be grown with about equal expected success. Thus a high relative demand for wheat (such as caused by a dense human population with a high standard of living) would cause an increase in its proportion in the sown area; the opposite would be caused by a high demand for barley (sparse or poor human population, but large populations of camels or horses engaged in transport or military activities). The proportion sown to wheat may have fluctuated in historical times between 20 and 70%; in most periods it was probably well below 50%.

Pulses (*Lens* and *Vicia* spp.) have had a small but persistent role in the agriculture of the semi-arid zone from the Chalcolithic until the present. They were grown mainly above 200 mm rainfall or in wadis, accounting for 2—5% of the cultivated area at any time.

Sorghum (*Sorghum bicolor*) has probably been sown in the Negev (1—3% of area, mainly in the north-west) only since the 19th century, as a summer cereal which uses moisture stored in land that was fallowed in winter.

(2) Crop rotation. A two-year rotation was common in the traditional agriculture: one year winter cereal, one year ploughed field fallow (A: krab, H: bur), which may be sown to a summer crop.

The ploughed fallow conserves part of the rainfall and most of the natural nitrogen inputs and usually allows a higher yield in the following crop. Due to the large year-to-year and site-to-site variation in the Negev, the rotation there seems to have been less regular, and more opportunistic than in other areas. Fields with a good moisture supply (in wadis) may have been sown almost every year, while marginal fields were ploughed only once every 2, 3 or more years, depending on early rains. In a few fields (about 5%) wheat or barley were grown in rotation with pulses (in winter) or with summer crops (sorghum, watermelons) after winter fallow.

(3) Cultivation. Fields were ploughed with a simple wooden or iron plough, drawn by a camel, ox or donkey, which breaks up the soil to a depth of 10—20 cm. This shallow cultivation is not very effective in weed control, but quite adequate for other purposes; in particular it breaks the hard crust of

the loess soil and creates a micro-topography, which minimizes runoff and maximizes infiltration. In some soils, where the salt accumulation layer is as shallow as 30—40 cm, it is actually preferable to deep ploughing.

Most fields are ploughed between September and November, after stubble grazing and before the rains. In years with good early rains, marginal fields may be ploughed for immediate sowing as late as January.

(4) Sowing. The amount of wheat and barley sown per unit area in the Negev was between 50 and 80 kg seed ha^{-1}, somewhat less than in higher rainfal areas.

The time of sowing was, and still is, one of the most difficult decisions in semi-arid agriculture. Early sowing (in October—November, before or jus after the first rain) allows early canopy development, long growing season and high yields — if good rains continue at short intervals; but a long pause in rains which may occur in early winter will cause damage or complete loss of the crop, requiring a second sowing. Late sowing (only after stored soi moisture is sufficient to assure a continuous growth, i.e. usually Decembe or January) minimizes this risk but reduces the potential growing season and yield.

The dilemma could not be solved by any fixed sowing date or rule, and various combined or flexible strategies were used (Ecclesiastes 11, 4—6)[5] In recent Bedouin agriculture, some fields in favourable sites (in terms o rainfall or runoff supply) were sown in November and early December in any case; most other fields (and certainly marginal ones) were sown only i and when 50—100 mm moisture had been stored, which could be as late a the end of January.

(5) Harvest and after-harvest treatment. Barley is harvested in March—April and wheat in May. Either of two methods was used: reaping with a sickle, o pulling by hand. The former leaves a stubble 20—30 cm high (and some ears), the latter leaves practically none, all the straw being gathered in. The choice varied between individuals, depending on their convenience (and available labour), and on their preferences of stubble for summer grazing o straw for autumn and winter feeding.

The stubble may be ploughed in directly (to maintain soil fertility) o after burning (to facilitate ploughing) or left standing for grazing until the autumn. The relative merits of these alternatives are discussed in the Mishna with a tendency to encourage the latter "so as not to deprive man (of left

---

[5] "He who watches the wind will not sow, and he who keeps his eye on the clouds wi not reap . . . . . . Sow your seed in the morning, and let your hand not rest until evening For you do not know which will succeed, this or that, or whether both alike will b good". "Morning" and "evening" probably refer to early and late in the season.

over grains) and beast''. This has also been the common practice in the recent Bedouin system.

(6) Fertilization. There is no evidence for the use of manure in Biblical times, but in the Roman (Mishna) period its role in improving yields was already highly appreciated. It seems that its application was mainly in selected fields, and it is not clear whether it extended to the dry land in the Negev.

In the recent period, spreading of the manure from pens and stables was occasionally practised in villages on the margins of the semi-arid zone, but rarely (if at all) in the Bedouin system. Throughout the region, however, much manure and urine is returned to the fields by livestock which graze and camp on the stubble for 3 to 5 months in summer.

Livestock and grazing management

(1) The animals. Sheep and goats (together: A: ghanam, H: ṣon) were the dominant livestock. They were probably domesticated in adjacent areas and are biologically adapted to the semi-arid winter rainfall environment.

The sheep (*Ovis aries*) is of the fat-tailed Awassi breed. It is reared for meat, milk (and milk products) and wool, but its specific advantage is as meat producer for marketing, due to the ability of the lamb to grow to almost full size (30—50 kg) in 3 to 5 months. For the rest of their annual cycle sheep can survive on poor pasture, but for successful lambing and lactation and lamb growth it requires a high level of nutrition, as supplied by abundant green herbaceous pasture (spring pasture, A: rabí'a). This requirement severely curtails sheep production in drought years and in arid or stony areas with poor annual vegetation.

The goat (*Capra hircus*) excels the sheep in being able to survive and produce also on shrub vegetation of arid and stony hills, being less dependent on annual spring pasture. It is reared for milk, meat and hair (the material of the nomad's tent). In terms of kid growth-rate the black goat is inferior to the Awassi sheep, and this is only partly compensated for by a higher twinning potential. But its milk production is higher; goat milk, cheese, liquid butter (A: samne) and yoghurt (A: leben) are important in the subsistence diet of the Bedouin, and surplus products may be marketed. The black goats of the Bedouin in Sinai are a distinct breed which differs from the black mountain (Mamber) goat of the Mediterranean zone by smaller size, lower metabolic requirement and lower required drinking frequency (Shkolnik et al., 1972). In the semi-arid zone there is a transition between the dwarf desert and the mountain populations.

The proportions of sheep and goat numbers in the semi-arid zone have varied around a rough average of 1 : 1, depending on both environment and economy. The proportion of sheep is higher (up to 75%) in the cooler and less dry parts of the semi-arid zones and in areas with extensive plains (or large wadis) and dense annual vegetation. Goats predominate (up to 95%) in the more arid or hot parts, in stony ranges with mostly shrub vegetation, and in areas where drinking points are sparse. When the human population was near subsistence, goats were more abundant; the proportion of sheep tended to rise with the standard of living and the demand for meat.

The camel (*Camelus dromedarius*) appeared in the region only about 3500 years ago. Since then it has been the main animal of the true desert nomads, who have invaded the Negev from time to time. It also had a more permanent place in the economy of semi-nomads and villagers of the semi-arid zone, for transport, ploughing, milk, hair and occasionally meat. But its numbers were only 5 to 15% of the sheep and goat populations. Camels graze preferentially on shrubs (including thorny and salty ones).

In most periods there is evidence that cattle were also raised in the semi-arid zone and used its pastures. Their numbers were probably smaller than those of camels and they were kept mainly for ploughing.

The donkey has been the most important home-transport animal in the region for at least 4000 years, as well as being the poor man's ox for ploughing. It also uses some plants which are not grazed by sheep and goats.

Horses were sometimes kept in the Negev for military purposes or as racing horses, always with liberal supplements of grain.

(2) Grazing. (This section is based in part on observations of Bedouin herds by H. Levi (unpublished data, 1975).)

Sheep and goats usually graze in mixed flocks, which may vary in size from 10 to 500 head (typically about 100), belonging to one or several owners. They are always accompanied by a shepherd, who may be a member of the family (often a young girl or boy, a woman or an old man) or hired professional (male) shepherd. The animals are taken out daily from the temporary or permanent night-camp, unless the weather is extremely cold and wet. The grazing usually begins within an hour of sunrise (later in winter, earlier in summer) and ends at dusk. Within these 10—14 h there are about 6—8 h of grazing periods. In hot weather there is a 4—5 h rest period about mid-day; otherwise, there is often a rest period in mid-morning.

In summer, and on mostly dry pasture, the animals are watered every day or two (goats up to 4 days), if possible in the morning (08.00—10.00 h). This limits the maximum range of grazing away from water points to about 5—8 km (goats up to 20 km). Drinking frequency may be decreased (and range increased) as the moisture-content of the pasture increases. When grazing on abundant green spring pasture (from January to April in a good year) sheep and goats need not drink at all, though they may be taken to water occasionally.

The daily grazing route depends on the spatial relations of night-camp, water point and pastures of different types, and on the shepherd's personal considerations. The distance walked in a day may be between 4 and 20 km and the direct distance between the farthest points 1 to 8 km. The extent to which schedule and route are controlled by the shepherd or by the herd may vary considerably according to the experience and personality of the shepherd.

The speed of movement is inversely related to the density and quality of the pasture, but the herd always moves on after consuming only a small proportion of the available biomass in any one place. Strong species preferences are generally evident, but in any grazing period several species are grazed. The herd usually moves and grazes in a loose front 50—200 m broad and 20—50 m deep.

There are some behavioural differences between sheep and goats in the mixed flock. The goats are better walkers and climbers and normally lead the flock. In hilly terrain, the sheep tend to move in the valleys while the goats spread out on the hills. In hot weather, sheep often stop grazing and huddle while goats continue to graze.

Camels graze separately from small livestock, and over larger distances, being less tied to night-camps and watering points. Large groups are accompanied by a herder but odd camels and donkeys are left to graze by themselves (and frequently get lost).

(3) Supplementary feeding. In the traditional system, locally-produced forage was used to supplement livestock when pasture was insufficient: barley grain, straw and chaff of wheat, barley and pulses. Working cattle, donkeys and camels had priority over sheep and goats. The latter received only the surplus of straw and chaff, and only in periods of severe pasture shortage in early winter and in drought years. Barley grain was hardly, if ever, fed to sheep and goats, as the price of animals and animal products was rather low compared with the price of grain. The marginal liveweight gain per kg barley fed can at most be about 400 g in the most extreme circumstances (growing lamb with optimal biological efficiency, no alternative feed); 250 g kg^{-1} is probably a more realistic maximum estimate. Thus supplementation is economically sensible only if the price of 1 kg of live lamb or sheep is more than that of 4 kg of grain. From fragmentary data in the Bible and the Mishna it may be deduced that the value of a sheep was equivalent to 50—120 kg barley, i.e. 1.5 to 3 kg barley per kg of liveweight. A similar price ratio was still common about 1900.

(4) Yearly grazing schedules and migration patterns. The most important part of the grazing year is the period of grazing green (mainly annual) pasture in late winter and spring. Most of the potential for milk production by sheep and goats and for weight gain by lambs and kids, occurs between December and May. The actual production achieved depends on the existence of adequate green pasture (A: rabi'a) in this period. Rabi'a conditions end abruptly with the drying-out of the herbaceous vegetation. This occurs

usually about April, with differences of up to only 4 weeks between regions, habitats and years. The beginning of the rabi'a is more variable between areas and years. The norm is January—February but it can be any time between November and March, depending on rainfall timing, runoff concentration, winter temperature, soil fertility and seedling biomass. Stock migrations in winter and early spring are aimed at maximizing the period of production by the herd by taking advantage of early green pastures: on the warmer eastern slopes and far-western plains; in large wadis where even a small early rain can cause germination; and in any areas which happened to get early local rainstorms (sometimes far south in the desert). As rabi'a conditions become general in February, herds spread back to all other areas.

After most of the pasture is dry, the herds concentrate on the barley stubble, and from May also on wheat stubble, in the semi-arid zone. Some grazing on late-green shrubs and perennials and on dry annuals in adjacent uncultivated land may occur simultaneously. As stubble quantity and quality are depleted in July—August, more time is spent on the rangeland. Some herds move north and west to graze on stubble and dry pasture in higher-rainfall areas.

Between October and December pasture is scarce and poor in quality throughout the region. The remnants of dry herbage and of shrub foliage are usually sufficient to keep sheep and goats alive, but early-born kids and lambs (from November) are endangered. Herd owners try to bridge this critical period until the new rabi'a either by moving to any area with slightly better pastures or by supplementing at home with straw and chaff.

In a drought year this pattern is modified in several ways. Failed wheat and barley crops are grazed-off, in whole, usually between March and May Migrations become more frequent and extended, both within and outside the Negev. In a bad drought almost all herds may leave the Negev north wards and move to the coastal plain, the mountains and the Jordan Valley and even to Transjordan or Egypt, if political conditions allow this; this occurred in Biblical times too (e.g. Genesis 12: 10).

(5) Animal husbandry and herd demography. There are few or no attempts to control mating. Sheep mate mainly between June and September and lamb between November and February. With goats the season is somewhat more extended in both directions and more dependent on seasonal conditions; occasionally, more than one pregnancy a year can occur. In a good year, up to 90% of adult female ewes and goats (and also some of the yearling females) may conceive, but in a drought the proportion can go down to 40%. Twinning is rare in Awassi sheep, but can reach 30—40% in the local goat breeds (which even produce occasional triplets). Poor nutrition, hard weather and diseases can cause heavy losses of lambs and kids about the end of pregnancy or just after lambing, particularly in November and December Considerable attention and care is given to newborn and suckling lambs and

kids. For instance, some may be nursed in "incubators" made of shrubs or dug in the ground.

Most male lambs are sold or slaughtered between the age of 2 and 5 months (male goats between 4 and 8 months), preferably before they lose too much weight due to deteriorating nutrition in summer. Most young females are kept in the herd. Ewes and goats are sold between the age of 5 and 8 years, depending on their reproductive performance. Under normal conditions, mortality of adult animals (due to diseases, predation, etc.) is fairly low, but in severe and prolonged drought it may reach 30—50%.

The demography of the livestock population is thus strongly dependent on rainfall and pasture conditions. In a very good year, which follows a year with fair pasture conditions in the mating period, the net increase in the female herd (reproduction less mortality and replacement) can be 20% for sheep and 30% for goats. In a drought which follows upon a dry year, there can be a net decrease of 30—40%. It is likely that in the traditional system, occasional catastrophic losses in drought years balanced, in the long term, population increases in favourable years, and prevented excessive over-population and over-grazing.

Actual numbers and densities of livestock in the semi-arid region in the traditional system are difficult to estimate. There is no quantitive evidence at all before the 20th century and all estimates before the nineteen-sixties are very inaccurate. Using some rough estimates from the period 1920—1940 (Aref, 1934; Shmueli, 1970), with some extrapolation backwards to account for recent trends (see below), it may be estimated that around 1900 there were between about 80 000 and 100 000 sheep and goats (+10—15 000 camels and a similar number of donkeys) in the semi-arid zones of the Negev and the Judaean Desert (Aref, 1934; Shmueli, 1970). These numbers which correspond to an average density of 16—20 sheep and goats km^{-2} may be taken as an indication of the situation in the past 3000 years. However, there would have been considerable variation in time and space around this average. A recent survey of the region indicated an average carrying capacity of 17 sheep km^{-2}, rising to 60 sheep km^{-2} on the northern loess plains (Seligman et al., 1962).

Extrapolation from recent observations (Noy-Meir, 1975) indicates that the local instantaneous density may have varied between 2 km^{-2} (in a drought summer in poor pastures at the arid margins) and 80 km^{-2} (in good annual pastures in spring), and up to 200 km^{-2} on stubble in early summer.

Use of wild plants and animals

Apart from grazing by livestock, the semi-arid vegetation was used by Man for other purposes. Quantitatively most important was (and in places still is) the gathering of firewood. The wood and roots of the large shrubs *Thymelea hirsuta* and *Retama raetam* were preferred, but in their absence all small shrubs and half-shrubs, in fact any combustible plant materials, were used. Dry dead shrubs and branches were obviously gathered first, but when none were available, live branches or whole shrubs were taken.

Parts of many plants were used as human food; leaves (e.g. *Atriplex halimus*, many annual forbs), bulbs (*Erodium hirtum*, *Scorzonera* spp., etc.) and fruits (Bailey and Danin, 1975). The time spent gathering and the quantities were generally small (except in emergency, Job, 30, 4) [6].

The contribution of wild plants to the human diet was small in terms of calories and protein; but it might have had some importance in diversifying the diet and providing certain minerals or vitamins. Many plants were used medicinally (e.g. *Artemisia* spp.).

Similarly, hunting was only an opportunistic occasional activity rather than a concerted effort. Large game (gazelle and ibex) has probably been scarce already for thousands of years. Hares, porcupines, partridges (also their eggs) and other small mammals and birds could more frequently and easily be obtained.

ECOLOGICAL EVALUATION OF THE TRADITIONAL SYSTEM

Any evaluation of the ecological effects of the traditional cropping—grazing system over the last 5000 years is necessarily based on sparse evidence. A number of statements can nevertheless be made with reasonable confidence.

Production and human welfare

Under this system, the resources of the semi-arid zone have supported a human population of 25 000—100 000 (5—20 km^{-2}), through a livestock population of 80 000—100 000 sheep and goats (16—20 km^{-2}) and an average annual grain production of 15 000—50 000 ton (300—400 kg ha^{-1} of actually cultivated land). In most periods the standard of nutrition and general welfare of this population does not seem to have been lower than in adjacent, more humid areas.

Stability

Between-years variation in rainfall caused considerable fluctuations in production and, in extreme droughts, migration of livestock and people out of

---

[6] "who pick saltbush (*Atriplex*) from the bush, and the root of broom (*Retama*) is their bread". Since the roots of *Retama* could hardly have been eaten even by desperate outcasts, two alternative interpretations have been suggested: (a) read "for their warmth" instead of "is their bread" (the roots of *Retama* are excellent firewood); (b) "the root of broom" alludes to the root-stock of a *Cistanche* (Orobanchaceae), which parasitizes *Retama* and other desert shrubs, and is indeed edible.

the area. Despite this short-term variability the system seems to have had reasonable long-term stability (or rather resilience in the sense of Holling, 1973). There is no evidence, at any time in the written history of the region (3000 years), of changes in land-use, or in population, or social—political changes, that were caused by cumulative ecological deterioration within the region itself. Rather, such changes can almost always be attributed to known external (often international) economic or political factors. The resilience and, indeed, the longevity of the system may be related to the flexibility inherent in the crop—livestock combination, with migratory options for stock and to other regulatory feedbacks, such as stock mortality in droughts. These allowed the fluctuations in the state of the semi-arid ecosystem with its human population to be contained within a well-bounded domain, without irreversible catastrophes, for at least 3000 years.

Effects of the transition

This ecosystem was almost certainly different from that of pre-Chalcolithic times. The transition from a sparse population of hunters—gatherers to a denser population of farmers—pastoralists would have involved changes of a trend type in vegetation, fauna and soil (even without postulating a climatic change). The major change must have occurred between 3500 and 1000 B.C., but some processes may have taken much longer to complete.

(1) Effects on vegetation. The greatest effect of the human population on the vegetation has been through cultivation. The shallow ploughing method does not cause immediate total eradication of shrubs and geophytes. In marginal land, which is only intermittently cultivated, many shrubs remain in the fields and recolonization after abandonment is fairly rapid. The regrowth may differ in composition from the original one. *Anabasis syriaca* and *Achillea santolina* apparently tend to increase and dominate in areas subject to intermittent shallow ploughing (Danin et al., 1975). But many other species of the original vegetation (including the endemic and large-flowered species of *Iris* section *Oncocyclus*) survive in such habitats.

The situation is different in areas which have been almost continuously cultivated for millenia, in the western loess plains and in parts of the Beer-Sheva Plains. There the shrub populations (originally mainly *Noaea mucronata*) have become extinct, or survived only in small enclaves of stony ground and steep banks. Even after 30 years of abandonment there is little recolonization by shrubs. The regrowth vegetation is dominated by herbaceous plants, mostly annuals but also some perennials (e.g. *Leontice*). Some of these survive the cultivation stage as weeds, while others re-invade from uncultivated areas. This rather diverse flora includes species which are common in fields and abandoned fields in the humid Mediterranean zone to the north. It seems that these originate from the semi-arid (Irano-Turanian) region and invaded the Mediterranean region in the agricultural period, rather than vice versa (Danin et al., 1975).

Firewood gathering must have further depressed populations of shrubs and trees, in cultivated land as well as in adjacent non-arable land, where human pressure was most continuous. There can be little doubt that trees were more abundant in the Negev in the pre-agricultural period, but it is not certain that they were dense and widespread enough to describe the vegetation then as *Acacia* savannas and *Pistacia atlantica* woodlands. It seems perhaps surprising that even relict populations have survived over 5000 years. A Bedouin tradition bans cutting of mature *Acacia* and *Pistacia* trees (A. Danin, personal communication, 1975); this may be an old rule aimed at protecting a valuable fodder resource. It does not prevent destruction of tree seedlings by grazing stock, which is probably the main cause of a slow (rather than catastrophic) decline of tree populations in the Negev since the Chalcolithic. Conditions for successful establishment of tree seedlings are rare in the Negev anyway (due to the consistently long summer drought) and not much of a grazing pressure may be needed to block recruitment and replacement of the old population.

The effects of 5000 years of grazing by domestic livestock on the vegetation are complex and not easy to assess. There seems to be a gradient of long-term grazing pressure from the grazing land nearest to cultivation, settlement and permanent wells (along the north-western margins and around the Beer-Sheva Plains), towards the more remote areas in the east and south where grazing pressure has been rather light. Despite the confounding effects of habitat factors (in particular a corresponding rainfall gradient) and of recent decades (see below), vegetation changes along this gradient can give an indication of the long-term effects of the traditional grazing system.

(a) Total shrub cover and density have been reduced by a factor of 2 to 5 (particularly in the Judaean Desert and around Beer-Sheva).

(b) Some of the most palatable and susceptible shrub species (notably *Artemisia herba-alba;* also *Salsola vermiculata*?) have been almost eliminated in the most heavily grazed areas. The few almost totally unpalatable shrubs (*Anabasis syriaca*, *Thymelea hirsuta*) may have increased in density, but the distribution of most shrub species has probably not changed much.

(c) Annual grasses (e.g. *Stipa capensis*) and forbs (e.g. *Reboudia pinnata*) may have increased in areas where grazing reduced shrub cover. Most of the annuals of the semi-arid region are rather tolerant of grazing pressure (by rapid growth in a short active life-cycle). Only in local conditions of heavy grazing has the composition of the annual vegetation been significantly modified by a selection in favour of low, prostrate species with small seeds and against tall, erect and large-seeded species. The productivity of the vegetation was probably only slightly reduced by direct short-term effects of grazing (leaf-area reduction effects; Noy-Meir, 1975); long-term effects through soil fertility (see below) and perhaps seed reserves may have been more significant.

(d) The perennial herbaceous plants which are at all common in the region have properties which reduce relative grazing pressure on them: thistles (species of *Carlina*, *Atractylis*, *Echinops*), low growing grasses (*Poa bulbosa*, *Cynodon dactylon*), geophytes (or others) containing substances which make them unpalatable at least when green (*Asphodelus microcarpus*, *Urginea maritima*, *Peganum harmala*). Tall perennial grasses (*Dactylis glomerata*, *Hyparrhenia* spp., *Stipa* spp., *Oryzopsis* spp.) are rare, but do occur in remote areas and in micro-refuges (between rocks, inside shrubs) elsewhere. While these grasses (and palatable perennials in general) must have been more common in the pre-pastoral period, the hypothesis that the vegetation then was a perennial grassland (Boyko, 1949) is still speculative.

(e) Small areas of accumulation of livestock excretions (at camp sites and near caves) have become dominated by a distinct nitrophilous vegetation, mainly annuals (e.g. *Chenopodium murale*, *Malva parviflora*); sometimes a concentric zonation can be observed. This persists for decades (or centuries?) after the site is abandoned.

In addition, the various species of domestic livestock have many subtler effects on the vegetation, which are only partly understood. (For instance, seeds of many leguminous plants — trees such as *Acacia*, shrubs such as *Retama* and annuals such as *Trigonella* — are dispersed by livestock which eat them; apparently they are also made more germinable by passing through the intestines of livestock.) In many ways, the present vegetation of the semi-arid region is well adapted to co-exist with a domestic livestock population, of the density that was typical of the traditional system (until recently). It is difficult to assess whether this adaptation evolved over the last 5000 years, or to what extent wild herbivore populations previously had the effects now performed by domestic ones.

(2) Effects on fauna. The occasional hunting by farmers and pastoralists probably exerted more pressure on wildlife than the much sparser hunter populations of previous periods. Protection of fields from herbivores, and of stock from predators, became considerations no less important than obtaining meat. This may explain the extinction of several species within the period of the historical record (earlier extinctions might also have been caused by climatic changes).

The wild ass (*Equus asinus onager*) and the oryx (*Oryx leucoryx*) are often mentioned in the Bible (the former as a dweller of arid regions), but have been extinct in the Negev for centuries. Gazelles (*Gazella gazella*, *G. dorcas*) ibex (*Capra ibex*) and many small herbivores have survived, but probably in much smaller populations than in the Chalcolithic. Large predators were especially feared and persecuted by pastoralists: leopard, wolf and hyaena have almost, but not quite, become extinct. Populations of the smaller predators (jackal, fox, caracal wildcat) and predatory birds must have also suffered considerable losses by hunting. On the other hand, the herds of slow-moving domestic stock may have been a larger and easier source of potential food than were the wild herbivores (particularly young stragglers and stock dying in drought). For some predator populations this could have partly compensated for the increased hunting pressure.

(3) Effects on soils. The silty loess soils which cover much of the region are fairly susceptible to both wind and water erosion. The reduction of vegetative cover by ploughing and grazing should have caused an increase in erosion rates in the transition to the crop—livestock system. However, as yet no clear evidence of this change in the Negev has been detected by geomorphical methods. In the cultivated plains, the effect of reduced plant cover may have been offset by the runoff-reducing effect of the ploughed surface. On slopes with a rough stony surface, erosion is rather slow, even today. But on slopes with a smooth-surfaced layer of shallow soil overlying hard rock there are signs of fairly rapid erosion, and it is likely that in these habitats erosion rates were significantly accelerated by domestic livestock grazing and trampling.

The traditional farming—grazing system utilized a high proportion of plant production and biomass each year; 70—80% in cropland and 40—70% in grazing land. Thus the larger part of the annual nutrient (in particular nitrogen) uptake by plants was removed from the system; part of it was returned in animal excretions, but probably with a rather low recycling efficiency. The annual net input of nitrogen from natural sources (rainfall, asymbiotic and symbiotic fixation) in the semi-arid zone is only about 10 kg N ha^{-1}. In the short term, mineralization of organic nitrogen can cover deficits, but in the long term the level of organic nitrogen must come to a balance with net inputs and outputs. It can be calculated (Harpaz, 1975; Noy-Meir and Harpaz, 1977) that, as the utilization of primary production of an annual vegetation is increased from 20 to 60%, the steady-state level of soil organic nitrogen is reduced from 8000 to 1350 kg N ha^{-1}. Nitrogen becomes limiting in most years and the average annual plant biomass production that can be maintained in the long term drops from 5000 to 1700 kg ha^{-1}. It seems likely that the low levels of soil nitrogen and of primary production in most cropland and grazing land in the Negev are a result of a high level of exploitation of plant biomass and soil nutrients over thousands of years, without any actions to return nutrients to the soil. In any sites where nutrients accumulate by natural processes (around ant nests and old stock camps; in wadis and depressions where suitable obstacles cause organic debris to accumulate) the increased production by annuals and crops is very conspicuous.

RECENT CHANGES IN THE BEDOUIN SYSTEM

In our century, world-wide economic and political developments have induced changes also in the economy and the land-use system of the Negev

Bedouin. The ecological implications of these changes are probably greater than those of anything else that has happened in the region in the last 5000 years.

Early in the century, government intervention stopped tribal feuds and raids and created more peaceful and stable conditions. Motor transport replaced camel transport, depriving Bedouins of one source of income. The Bedouin responded to the new situation by reducing camel herds but increasing sheep and goat herds and by increasing the area of cultivation of wheat and barley.

Further changes began in the nineteen-thirties and increased in pace in the nineteen-forties and nineteen-fifties. Improved medical care accelerated human population growth. Increasing numbers of Bedouins found jobs outside the traditional economy; aspirations to change from tent life to a more settled and modern way of life began to spread.

In the Arab—Israeli war in 1948 some tribes left the Negev (mainly in the west) while others remained in their former territories (in the east and centre); a few were dislocated within the Negev (Marx, 1967). For those who remained, the only immediate result of the new political situation was that long-distance migrations to Sinai or Jordan (and in the nineteen-fifties also to the northern parts of Israel) were no longer allowed; thus herd movements became more restricted in range. Otherwise, the trends which had begun to transform the traditional economy in the last years of the British mandate, continued to accelerate in contact with the the rapidly expanding and modernizing Israeli economy.

The increasing cash incomes of Bedouins working for wages were partly invested in tractors, modern ploughs and harvesters. Within a few years Bedouin agriculture became almost entirely mechanized; almost all suitable and available land was brought under cultivation and sown to barley or (increasingly) wheat almost every year. To a limited extent, fertilizers (mainly N) began to be applied.

The increasing attachment to the cultivated fields and the desire for modern comforts led to widespread spontaneous settlement. A typical sequence was: mobile tent—fixed tent—tin hut—wood hut— stone house (Amiran et al., 1976); but often there were short-cuts (e.g. in the area east of Bethlehem; Shmueli, 1970). Patterns of settlement vary from one to a few houses in an isolated family cluster, to large, diffuse and sprawling settlements of several hundreds of units. More compact planned settlements have been initiated by the government, with partial success.

The effects of recent economic, social and political changes in the Bedouin livestock—grazing system have been complex and somewhat paradoxical. On the one hand, wage employment and the shift to farming and settlement have reduced the relative importance of livestock as a source of subsistence and income, and the relative attraction of stock-raising as an occupation. On the other hand, some other factors have tended to encourage an increase in the number of sheep and goats.

Veterinary care, including preventive immunization, has become readily available to Bedouin stock-owners and has reduced mortality. Restrictions on stock movement have been relaxed, allowing Bedouin to take advantage of otherwise unused pasture and stubble areas in central and northern parts of Israel. Permanent water points along pipelines in some previously waterless areas of the semi-arid zones now allow a more continuous use by stock. Most importantly, meat prices have increased steadily in the last 40 years, apparently due to the increase in population and standard of living in the country as a whole. The meat/grain price ratio, which for the 3000 years preceeding 1900 had fluctuated between 50 and 150 kg barley per sheep or full grown lamb (1.5 to 3 kg barley kg^{-1} liveweight), has increased to 400 to 600 kg wheat or barley per sheep (8—12 kg grain kg^{-1} liveweight) in the nineteen-seventies.

Thus, supplementary feeding of grain and grain-based concentrates (and certainly of any cheaper forages) to support lamb growth and to prevent mortality or reproductive failure, has suddenly become not only an economically viable, but even an attractive practice. Indeed some sheep farms in Israel now provide up to 70% of the total maintenance and production requirements of the flock from concentrated forages (partly cheap residuals, partly expensive feeds); the higher levels of meat (and milk) production achieved seem to just cover the large expenses on feed. Bedouin stock-raisers in general have been more conservative and cautious in expenditure on feeds, but most of them now give 25—30% of requirements (in some flocks up to 50%) in supplementary feeds (locally-produced barley grain, bought grain, concentrates, hays, residuals). These are given mainly to pregnant and lactating ewes in late autumn and early winter, sometimes also to weaned lambs in late spring, and to a lesser extent to goats and kids. This results in the production of larger lambs as well as in greatly reduced mortality of lambs and of adult animals. Thus, the 20% per year potential net growth rate of the female herd (which previously could be realized only in very good years, and was counterbalanced by net losses in droughts), can now be approached in almost every year.

The actual rate of increase of the total population of sheep and goats owned by Bedouin, in the last 10 years, has probably been about 5% per year. The population now [7] (200 000) is probably 2 to 2.5 times the mean level around which the stock population in the semi-arid zone fluctuated in the traditional system. The pressure on the semi-arid grazing lands has not increased by the same factor, because new feed sources (supplements, northern grazing) have simultaneously become available. Herd-owners have modified feeding and migration schedules to use these new sources. At present, there is a remarkable diversity of management strategies between individual

[7] In 1976.

owners, and a great variability in the use of supplementary feeds and in the extent of migration (Noy-Meir, 1975). Some have adopted a fairly intensive and sedentary system, with high supplement inputs (30—50%) and little or no migration away from the vicinity of the now permanent home. Others, though themselves permanently settled in the semi-arid zone, have their herds migrating (or staying) most of the time in better pastures in the non-arid regions; and they use less supplements. A few still follow the more traditional migration patterns at the margins of the desert (mostly tent-dwelling goat herders) and hardly use any supplements. Almost all possible intermediates and combinations can be found.

The overall effect, however, is that only 30% (in drought) to 40% (in a good year) of the total requirements of the livestock population owned by the Bedouin of the semi-arid region is now demanded (and indeed can be supplied) from the pasture (and stubble) resources of the region; the rest being supplied by supplementary feeds and by grazing outside the region. Therefore the effects of the drastic population increase on the semi-arid pastures have not been as severe as might otherwise be expected. However, these pastures are taking some of the pressure; even those flocks that are heavily supplemented at home in autumn and winter are still taken out to graze for at least several hours a day, though the pasture near the settlements may already be very poor. Large flocks which have come to rely almost entirely on pastures and stubble in the non-arid regions tend to move to the semi-arid zone whenever their stay in the north becomes less attractive or impossible, which may happen occasionally, for a variety of reasons.

The recent trends in the traditional semi-arid pastoral system can be summarized in a series of paradoxes.

(1) Working and walking with sheep has become socially less attractive, but owning sheep is economically still very attractive, and building up a flock has become easier.

(2) Nomadic pastoralists have become more settled, but some of their herds range further than ever.

(3) The relative contribution of the Bedouin sheep and goat herds to the livelihood of the Bedouin population itself has decreased, but their contribution to the food basket of the rest of the population in the country and to the national economy has increased.

(4) The relative importance of the pastures of the semi-arid zone to the nutrition of the Bedouin herds has decreased, but the grazing pressure on the pastures (and even more the potential danger of over-grazing) has increased.

These paradoxes are reflected in the variability and diversity of the present situation, which make it difficult to discern a stable and well-defined land-use "system". It is not easy to foresee how these paradoxes will be resolved in the future and in which direction the present transitional situation will stabilize; this depends on many ecological, economic and social factors.

Ecological effects of recent changes

The effects of the 20th century on the Negev ecosystem have been both profound and manifold. The Turkish railway to the Suez in the First World War, which wiped out the oak forests in the Sharon, probably also took its toll from the old *Pistacia* trees in the Negev (the Nizana area), which now survive only in the remotest areas. The introduction of modern firearms has greatly increased the risk to the few remaining species of large wildlife.

The main effects have resulted from the continuous increase in the density and concentration of the Bedouin population and their livestock, as well as in the area cultivated by them. The grazing, wood-gathering, and hunting pressures have increased markedly in all areas within about 5 km from the Bedouin settlements, and particularly around Beer-Sheva, where the density is highest. The resulting trends in vegetation, wildlife and soils are qualitatively similar to those described above (for the traditional system). But the extreme effects which were previously confined to small areas, have become much more widespread. In many areas of the north-eastern Negev and the Judaean Desert, not only has the shrub vegetation been reduced to a low cover of almost unpalatable species, but also the density and composition of the annual and perennial herbaceous vegetation has become very poor. In much of this area, the major deterioration seems to have occurred in the last 30—40 years. Observations on exclosures (Noy-Meir, 1975) show significant differences in composition of both annual and perennial vegetation after 2—5 years.

The largest exclosure experiment with the most dramatically demonstrable results, was a side-effect of a political situation. On satellite photographs (Otterman et al., 1975) and aerial photographs taken about 1970 (Fig. 4.4), the 1948—1967 Egyptian—Israeli border shows as a sharp line of contrast between the lighter Sinai and the darker Negev; no such line appeared on 1944 photographs (Fig. 4.3). In this period, the border was effectively closed to the Sinai Bedouin and their livestock. On the Israeli side there was little or no grazing and shrub removal. A dense shrub vegetation (mainly *Artemisia monosperma*), and litter of shrubs and annuals, have accumulated to result in the darker colour. The lower albedo in the vegetated area causes higher temperatures above it, and it has been speculated that by a convection mechanism this may increase local rainfall (Otterman, 1974). On the other side of the line, vegetative cover has been further reduced since 1944 by increased pressures of shrub removal (for firewood and the construction of brush huts), cultivation in patches (watermelon and castor bean) and livestock grazing (though *Artemisia monosperma* ranks low in the preference of most stock).

Only a few of the many new ecological processes in the region have been observed in any detail. In the early nineteen-fifties, the summer annual *Aellenia hierochuntica*, previously rare in the region, drastically increased its

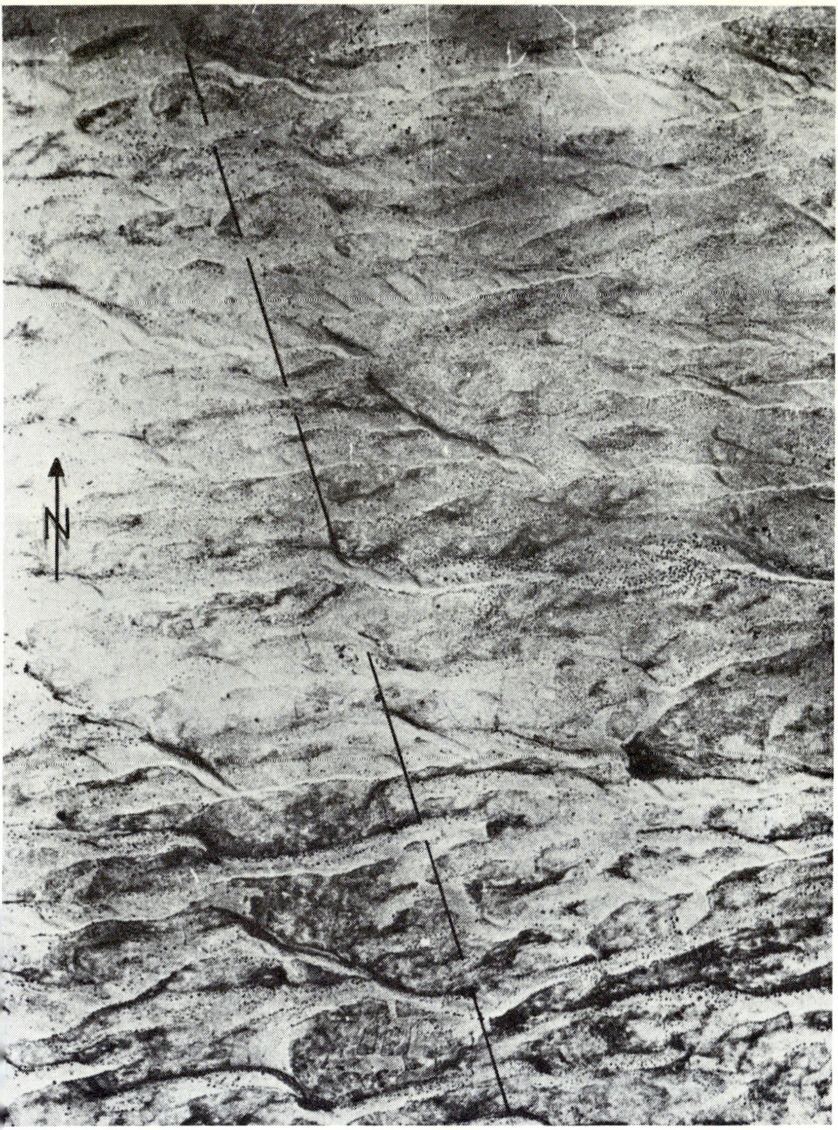

Fig. 4.3. Aerial photograph of an area of sand dunes on the border between Sinai (left) and Negev (right) in 1944. Approximate scale 1 : 8000.

populations in cultivated fields in the Beer-Sheva Plains. The population then declined, in part due to a rust disease. By 1960 the disease had disappeared, but *Aellenia* populations remained low. The probable explanation (Shimshi, 1971) is that in some fields there is a saline layer at a depth of 40—50 cm. This layer was untouched by traditional shallow ploughing, but was brought to the surface by deep ploughing which was first introduced to the area in the

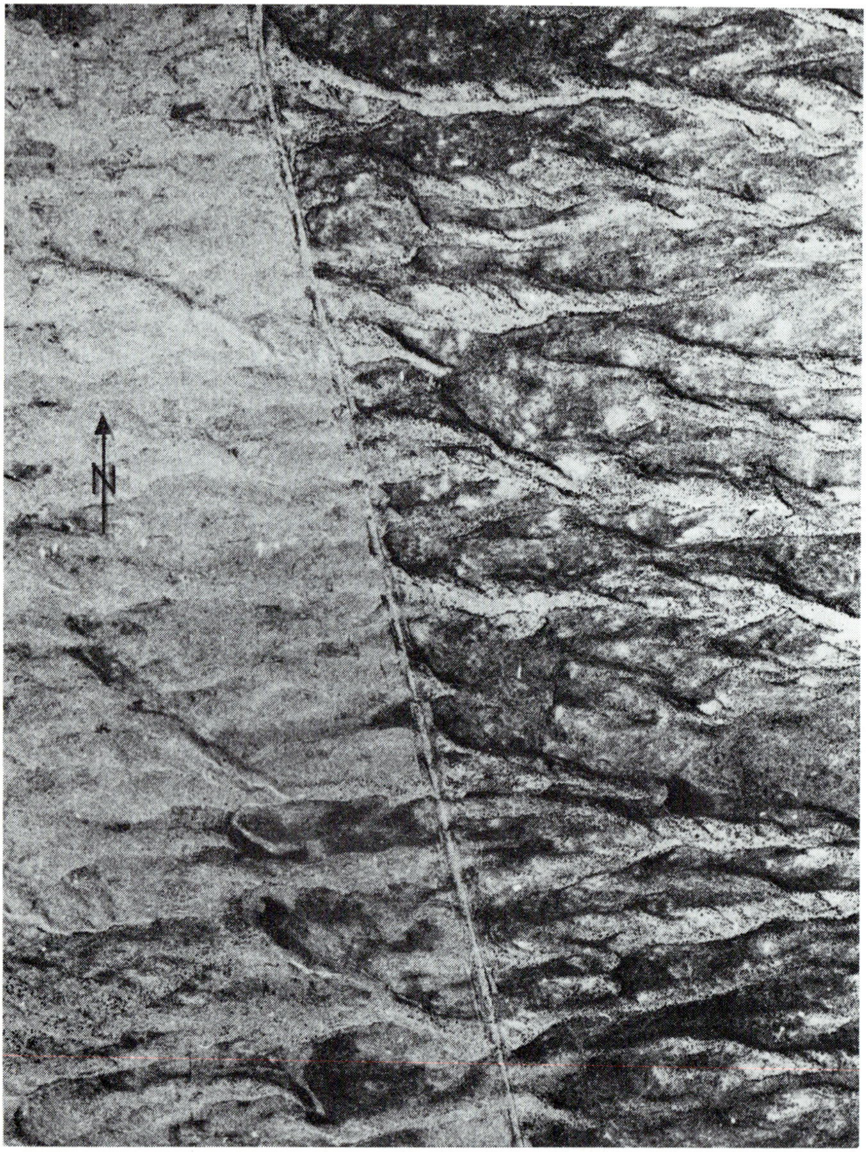

Fig. 4.4. Aerial photograph of the shown area in Fig. 4.3 in 1970.

nineteen-fifties. The new saline surface depressed growth of both the crop and winter annuals. This left sufficient moisture in the soil for *Aellenia* (which is a slow grower and a poor competitor, but salt-tolerant) to complete its growth and reproduction in summer. After several years of deep ploughing, the salt was dispersed throughout the plough layer and *Aellenia* lost its relative advantage.

The native halophytic summer annuals (*Aellenia*, *Salsola inermis*) are now most common as colonizers in all habitats where the soil has been disturbed and surface salinity increased (near roads, pipelines, construction sites, etc).

It is mainly in these modern habitats that recently introduced species have successfully spread and established (Danin et al., 1975). These are mostly halophytic chenopods from other arid zones, which in the Negev have a life-cycle similar to the native summer annuals. *Kochia indica* from north-west India was unintentionally introduced to Egypt in the Second World War, and from there to the Negev (in 1948?; Negbi, 1968). *Atriplex muelleri*, *A. semibaccata* and *A. spongiosa* were introduced from Australia (intentionally or not) in recent years. The South American shrub *Nicotiana glauca* has spread widely along roadsides and into wadis.

THE NEW CROPPING SYSTEM

Definition

In the last decades a new system of land-use has developed and stabilized in large parts of the Negev, mainly in Jewish farms in the west and north. It has the following characteristics:

(1) most arable land is cultivated to crops, predominantly wheat;

(2) most livestock on the farms is raised and fed in the shed, with little or no use of grazing land and stubble;

(3) considerable amounts of fertilizer and irrigation water (partly imported into the region) are supplied to crops;

(4) crop production per unit area is very high, and most of the products are exported from the region as cash crops.

The transformation of the north-western Negev into one of the most productive areas of the wheat belt of Israel began with the establishment of the first Jewish farms about 1940. The process gained momentum in the period 1950—1970, during which average wheat yields per ha have approximately doubled, due to the combined effects of new management practices (particularly nitrogen fertilization) and cultivars. At the same time, the supply of irrigation water through pipelines from the north allowed the creation of blocks of irrigated intensive crops in the Negev, as well as some supplementary irrigation of wheat. Sheep-raising and grazing, which originally had been part of the mixed-farming systems in most Jewish farms in the Negev, then appeared to become a relatively less efficient use of labour and capital, and decreased markedly since 1960. Most of the remaining sheep flocks are raised and fed intensively in sheds, like the more numerous dairy cattle herds.

The continuing application by farmers of high levels of inputs (fertilizer, supplementary water, pesticides) to dry-land crops, even in the uncertain semi-arid climate, is now encouraged by risk-reducing government policies: drought compensation (for crop failures above the 220 mm average rainfall line), subsidized water and guaranteed prices for products.

Land classification and tenure

Practically all land in this system is state-owned land which, in theory, is leased for limited periods, but in practice the lease is usually renewed automatically. Most of the land is held and cultivated by villages of two types: co-operatives (H: moshav) and collectives (H: kibbutz); only a small proportion of the land is in farms leased to private individuals and companies, or cultivated by government agencies. Each village of 50—100 families has about 400—800 ha comprising:

(a) irrigated land (H: shalhin), 2—3 ha per family, cultivated to vegetables and flowers (sometimes in glasshouses), fruits, forage crops and some grain crops;

(b) dry arable land (H: baal), 5—10 ha per family, cultivated to various winter and summer crops, but mainly to wheat;

(c) non-arable land (H: bur), 0—15 ha per family, which may or may not be used as natural or improved pasture.

In the kibbutz all land is held and cultivated collectively by all members. In the moshav, each family cultivates its own plot of irrigated land, while the dry land is usually not divided but is cultivated by the co-operative. The non-arable land is also common.

Each village (and each farmer in a moshav) is also given a fixed annual ration of irrigation water (about 18 000 m^3 per family); most of it is used for the small shalbin areas, but some may be used to supplement dry-land crops.

Crops and cultivars

The present distribution of the cultivated dry-land area in the new system in the Negev among crops (Table 4.5), shows an increase in wheat and a sharp decrease in barley, compared with the traditional system. New crops have appeared, such as hay crops, sunflower, safflower and cotton, but they are restricted to relatively small areas, mainly along the wetter northern margins.

The main cause of the drastic change in the wheat/barley ratio, which occurred mainly since 1960, seems to be the development of new, high-yielding and disease-resistant cultivars of wheat, which was not accompanied by a similar improvement in the barley cultivars. Before 1960, average yields (per ha) of barley in the Negev were about 40% higher than those of wheat (mainly due to better success in dry years and areas); now they are about 30% lower. The barley/wheat price ratio has increased recently from about

TABLE 4.5

Percentages of various crops in the sown area of the semi-arid zone (200—300 mm) in the traditional system (estimates) and in the modern system (D. Ariel, personal communication, 1976)

	Traditional (± 1900)	Modern (1970)
Wheat	25	74
Barley	70	9
Sorghum	2	2
Legumes (pulses)	3	
Legumes (hay)		7
Sunflower		6
Safflower		1
Oats		0.5
Cotton		0.5

0.7 (as it was in the traditional system) to nearly 1, but barley cultivation has remained limited.

The wheat revolution in the Negev (as in other parts of Israel) involved two main factors: increased fertilizer use and introduction and breeding of cultivars. From the nineteen-forties the old cultivars of *Triticum durum* (Noursi, Ettit) were replaced by cultivars of *Triticum aestivum* with higher yield potential and better baking quality. Cultivar Florence was introduced from Morocco; after 1957 it was replaced by the more rust-resistant Florence/Aurore, which in the nineteen-sixties accounted for 70% of the wheat area. In good conditions, yields up to 5000 kg ha^{-1} were obtained. During the nineteen-sixties, dwarf and semi-dwarf cultivars were first introduced from Mexico, then locally further bred and improved (Ephrat, 1968). The latter cultivars were shown to out-yield even Florence/Aurore, though in the semi-arid region their advantage is not as great as in other regions (Hofman, 1974). Since 1970 these varieties, in particular earlier maturing ones (e.g. Miriam, Lakhish, Ceeon, etc.), have almost completely replaced Florence-Aurore in the Negev. One of their advantages is that higher nitrogen levels can be applied (and higher yields obtained) without causing lodging.

Barley breeding has received less attention than wheat and the number of new cultivars (e.g. Dvir, Omer) now used is smaller.

A locally-bred variety of vetch (*Vicia sativa* cv. Assor) and oats for seeds, are grown in some of the better zones, but their areas have declined in recent years.

Sorghum is still (as in the old system) sown as an opportunistic summer crop in some areas (mainly on the north-western margins), only when and where sufficient soil moisture (200—300 mm) is stored in the soil in spring (e.g. on winter fallow or after having spring rains). Some new hybrid varieties are being introduced.

Sunflower, safflower, cotton (and some watermelons) are similarly grown as opportunistic summer crops. The area sown to these crops is small and highly variable between years.

In the nineteen-sixties, agave for fibre was planted in some areas; though establishment was successful, this effort was discontinued for economic reasons.

Crop rotation

The three-year rotation which was standard in more humid areas of Israel, wheat or barley—sorghum—hay, was at first also introduced in the northern Negev. Since 1960 attempts were made to increase the proportion of wheat in the rotation; with adequate fertilization and disease control this proved to be feasible. In the wetter part (>250 mm) of the semi-arid zone, wheat is usually grown for 3—5 years continuously, then interrupted by 1 year of summer crop (sorghum, sunflower, etc.) or a winter legume. In the drier belt (200—250 mm) other crops are seldom successful, and wheat is usually alternated with a fallow year every 2—3 years. In a regression analysis of yield data from the region (Zaban, 1977) a significant increase in wheat yield after fallow (compared with after wheat) was found, which was correlated with rainfall in the fallow year. About 50% of that rainfall was estimated to be conserved for the following crop.

Cultivation

The single shallow ploughing in the traditional system has been replaced by a much longer sequence of careful seedbed preparation: deep ploughing (40—60 cm), discing (with fertilizer application), levelling, rolling (all mechanized). These operations account for 25% of production costs (Zaban, 1977); the purpose is to create an aerated and pervious seedbed and reduce weed competition and disease (Arnon, 1972). However, some recent results (Bar-Daroma, 1966; Stoebe and Ariel, 1970) indicate only minor differences in yields between different cultivation treatments; only no ploughing at all reduced yields by 20—25%.

Special attention is given to erosion control practices (contour ploughing, road and ditch treatment). The smoother surface of the final seedbed at the beginning of the rainy season is rather susceptible to erosion.

Fertilizer application

The increased application of fertilizer (main N) has been a major factor in increasing crop yields. The annual supply of mineral nitrogen from natural sources (rain, fixation, mineralization of organic N) in an unfertilized system is about 30 kg ha^{-1} (Harpaz, 1975). This allows a maximum grain yield of 1000—1200 kg ha^{-1} which was the best obtained in the old cropping system in years with good rainfall. Yields of 500—1000 kg ha^{-1} in years with medium rainfall also reflect low soil fertility. Thus even in the semi-arid zone yields were limited by N-deficiency in nearly half of the years.

Fertilizer inputs in the new wheat farms started with 20 kg N ha^{-1} (mostly as $(NH_4)_2SO_4$) until 1950 and increased to 80—100 kg ha^{-1} in 1970. Significant responses in wheat grain yield can be observed up to 120—150 kg N ha^{-1} (when rainfall >250 mm) but the economic optimum is probably an average of 70 kg N ha^{-1} (Zaban, 1977). In drought years response is smaller, but some of the N is left unused in the soil and is to a large part available to the next crop (losses by denitrification and leaching seem to be small in dry conditions). A bio-assay method for residual available N in the soil allows an adjustment of N input each year and avoids unnecessarily high inputs (Amir and Ephrat, 1971).

Phosphate fertilizer is applied only whenever available P in the soil is less than a critical level; in practice, under continuous wheat, about 40 kg ha^{-1} are applied once in 2—4 years. No response to K fertilizer has been recorded.

Supplementary irrigation and runoff storage

Once an adequate nitrogen supply is ensured, the water supply becomes the main factor determining crop yields in the semi-arid zone in all years. The average marginal response of wheat yield to rainfall, or to irrigation water is about 1 kg m^{-3} (10 kg ha^{-1} mm^{-1}) (Lomas and Shashoua, 1973; Shimshi and Kafkafi, 1973; Zaban, 1977). The marginal response to rainfall or irrigation at critical times for the crop (particularly establishment and early growth, November—December) is often considerably higher (1.5—2 kg m^{-3}). In recent years the practice of irrigating wheat crops in November and December, and whenever water stress is observed, has become common in the north-western Negev. At present 25—35% of the area sown to wheat receives supplementary irrigation. Most of the water is pumped from the north through the national network and is applied by sprinklers. The economic viability of this practice is controversial, as the marginal return in wheat yield more than covers the subsidized cost of water to the farmer, but usually does not cover the real costs.

An alternative source of water for supplementary irrigation is local runoff (Lanir, 1976). Large amounts flow through watercourses in flash floods, one to several times each winter. In several farms in the Negev, capture and storage of such water, by damming small watercourses, has been successful. The cost per m^{-3} for irrigation of wheat is considerably lower than the real cost of transported water, and is covered by the yield gain. However, losses from the reservoir by evaporation and seepage are rapid, preventing storage for more than a few months, and necessitating early use of the water. This factor, as well as the unreliability in amount and timing of input, limit the contribution of dams to wheat yields. Therefore farmers are reluctant to invest further in this direction. Often siltation of the reservoir is also a problem, and watershed management to reduce siltation also reduces water yields.

Weed, disease and pest control

The change from a 3-year crop rotation to a wheat-after-wheat system caused an increase in weed and disease problems, which the intensified cultivation before sowing was not always sufficient to overcome; thus chemical methods were introduced. In the early nineteen-sixties the prominent weeds in wheat were annual forbs, mainly of the Cruciferae (*Sinapis alba, Erucaria boveana*); these were easily controlled by the application of 2-4D. As a result, annual grasses (*Phalaris* spp., *Avena* spp.) which are tall and prolific seeders have recently become a serious problem. These too can be controlled by new selective herbicides, but the costs are high and often suppression of one weed allows another one to take over.

The main fungal diseases affecting wheat and barley in the Negev are septoria and rusts. Though new wheat varieties are continuously being bred for disease resistance, occasional outbreaks occur and in irrigated areas are controlled by fungicides.

Insect damage has so far not been a problem in the grain crops, and insecticides are hardly used. Birds sometimes cause heavy local damage (especially in sorghum), but in general, not to the extent that special control measures are required. Rodents (the vole *Microtus guentheri*) have caused considerable yield losses in the past, particularly during population peaks which occurred once every few years (Bodenheimer, 1958). The extensive distribution of poisoned grain in vole-infested fields has reduced the population to a more or less steady low level. However, it has also severely reduced populations of other animals, including predatory birds which previously helped to control rodent populations (Mendelssohn, 1973).

Sowing

The increased yield expectation due to increased inputs into dry-land grain crop fields has also caused changes in sowing practices. Seed input has been increased to 120 kg ha^{-1}, and sowing is, of course, mechanized. Sowing is usually in November, whether or not it has rained by then. This early sowing assures maximal yield in years with favourable rainfall distribution. The risk of failure due to a dry spell after germination is taken; sometimes the crop is resown in mid-winter.

Harvest and post-harvest treatment

The sickle has been replaced by large combine harvesters, but the time of harvest of barley and wheat has not changed. The cut straw discharged from the harvester is usually later harrowed into rows and baled; the amount of straw baled per ha tends to be roughly equal to the grain yield (2000—3000 kg ha^{-1} in a fair to good year). The fields after harvest are rather variable in the amount of stubble left standing (depending on the height to which the harvester has been adjusted), and in the amount of straw, chaff and ears on the ground (depending on the efficiency of harrowing and baling). The total amount of dry matter left in the stubble fields (2000—4000 kg ha^{-1}; Noy-Meir, 1975; Zaban, 1977), is in any case much higher than in unfertilized fields in the old cropping system (500—1000 kg ha^{-1}). Many wheat growers prefer to plough the field early in summer, usually after burning the stubble and straw rests, rather than let the stubble be grazed and risk the transmission of fungal disease and weeds to the next crop. In many other farms, stubble grazing is allowed between June and October, either by sheep and cattle of the farm itself or by sheep and goats of Bedouin who rent the stubble usually for 2—3 months.

Livestock husbandry

Though most farms in the Negev have livestock there is generally little interaction between the livestock system and the dry-land (crop and range) system. Most common are intensive dairy cattle herds kept in sheds and fed largely on imported concentrate feed, supplemented by crop residues (sugar-beet tops, low-grade vegatables, peanut hay, etc.) and some locally-grown hay and irrigated forages. These herds are managed virtually independently of the surrounding semi-arid ecosystem, except for its climate. Adaptation to the latter is attained by the local breed (Israeli-Friesian) and by suitable buildings.

Intensive sheep flocks are less common now, and are of two types: (a) locally-bred improved Awassi, mainly for milk production for the cheese industry, (b) German-Merino for meat production. Both types rely mainly on imported concentrates and on crop residues. The milk sheep use green spring pasture when available within 2—3 km from the shed.

Some farms with large areas of non-arable land in the northern margin of the semi-arid zone (250—300 mm), have semi-extensive beef herds of mixed breed. The mother herd and the young calves graze on rangeland and stubble and the weaned calves are fattened on concentrates in a feedlot. In summer and fall when dry pasture is abundant but low in protein, the mother herd is given a nitrogen supplement in the form of either oil-cake protein or chicken droppings. In late fall and early winter, when the pasture is poorest, supplementation is increased by feeding barley or concentrates.

About 15% of the total yearly nutritional requirements of the mother herd (and the unweaned calves) comes from the supplements, 35% from stubble and 50% from rangeland pasture. The stocking density on the hill range is 10—20 cows km^{-2}. The herd is managed in a loose and flexible rotation between paddocks of 1—3 km^2.

Range improvement is largely limited to fencing and provision of water points. After 5—10 years of grazing in this system, the annual pasture has a higher density and larger proportion of tall early-growing grasses than adjacent unfenced areas, which are grazed by Bedouin flocks without any control on density or season of grazing. Range improvement by fertilizer application (mainly N) and by planting fodder shrubs (mainly *Atriplex halimus*) has been attempted in experimental areas. Though these attempts were successful, the magnitude of the response has been variable and has not encouraged farmers to invest in wider application. Small areas of cultivable land have been sown to annual legumes (*Medicago truncatula*, *M. polymorpha*, etc.) to provide winter grazing for the cattle; often the legume population has successfully re-established from seed for several years, sometimes even after a wheat year (Y. Ofer, personal communication, 1976).

Ecological evaluation

(1) Production, human population and welfare. Under the new cropping system, grain yields per area of arable land in a good year have gone up from 1.2 tons ha^{-1} in the traditional system to 3.0 tons ha^{-1} in recent years. In drought years, yields have remained low, 0.3—0.5 tons ha^{-1}. The 55 000 ha in this system in the Negev represent about half of the dry-land farming area in the country. Depending on the year, grain production from this area is 1/6 to 1/2 of the total production in Israel, which has been 200—250 thousand tons per year in recent years (Anonymous, 1977).

It is difficult to estimate the human population density supported by this system locally, since most farmers in the region derive part of their income from other productive systems which are based on imported water (irrigated crops) and feed (intensive livestock). A rough estimation of the relative proportions of income from various sources indicates that the density of the population sustained now by dry-land cropping is probably of a similar order of magnitude to that which has been sustained by the traditional cropping—grazing system. However, the standard of living of the rural population now is much higher than in the old system.

(2) Stability. The transition to the new high-input system has only just occurred and in some ways it is still developing. An early indication of its stability may be the fact that some severe droughts (1958/1959, 1962/1963, 1965/1966, 1968/1969—1970/1971, 1975/1976), far from causing

a critical setback, have hardly slowed down the transition. A major reason for that may be the fact that farmers have been protected against major losses of their investment in the fields by the government drought compensation scheme (which covers the area with more than 220 mm average rainfall). However, it has been estimated that the gain from increased grain production in the Negev in good years more than offsets the cost of the compensation scheme to the national economy, and that the scheme might well be extended to an even drier zone (Zaban, 1977). The economic viability of other forms of government support, such as highly subsidized water which may be partly used for wheat irrigation, is being questioned. The obvious justification was in getting the new rural settlements smoothly established. In the long run (as water becomes more scarce and costly in Israel generally), the new agriculture in the Negev may have to minimize its dependence on supplementary irrigation. There are some alternative solutions that could increase water availability, in particular more efficient use of local rainfall and runoff; but these could hardly compensate for the present use of imported water. Apart from this, the long-term stability of the system could be threatened from two quarters: (a) an exogenous increase in the input costs relative to the price of grain, and (b) reduced productivity or increased costs induced by accumulative ecological processes. Such processes have not yet been studied in depth and can only be described in outline here.

(3) Ecological effects. The deeper ploughing eradicates from the fields all shrubs and some weed species. Many *Acacia* trees still stand within fields, but of course no seedling establishment is possible (Halevy and Orshan, 1972). Populations of many annual weeds survive the ploughing as does the perennial *Cynodon dactylon* in some areas. The smooth surface prepared for sowing increases the risk of erosion by early rains; the actual damage varies but usually erosion appears to be contained at a local scale (Tamir, 1970).

The high annual inputs of nitrogenous fertilizer are most probably causing a gradual build-up of the organic nitrogen in store in the soil which in the long-term is likely to be useful. The loss of a recycling function, by the avoidance of stubble grazing in part of the fields, does not seem to be very significant by comparison. The high concentrations of mineral nitrogen, and irrigation of summer crops where practised, may have significant effects on the soil microflora. The continuous wheat system may cause exhaustion of minerals other than the supplied N and P; in some cases, unexplained yield reductions after several years of wheat have been reported, but so far no specific deficiency has been identified.

Nitrogen enrichment, new cultivation techniques and use of herbicides have dramatically altered the composition of the weed flora but have apparently not reduced the severity of the weed problem. The weeds dominant at present are tall, nitrogen-responsive grasses similar to wheat itself, and their control is rather expensive.

Vole control by poisoning has prevented major outbreaks, but apparently hardly reduced the mean level of rodent populations; it has greatly reduced populations of predatory mammals and birds. Some wild species have taken advantage of the greatly increased availability of plant food and water in the new agro-system (and of reduced predation). For instance, partridge populations in and near the farming areas in the Negev have already increased in recent years, and their control is now demanded. Porcupine, hare and possibly gazelle populations may be responding similarly.

NEW INTEGRATED CROPPING GRAZING SYSTEMS?

Are they necessary?

In the present agricultural systems in the Negev (both Bedouin and Jewish), productivity of grain (per unit area) and of milk and meat (per head) is considerably higher than it was in the traditional cropping—grazing system. This is mainly due to considerable inputs of fertilizers (and other chemicals), concentrated feed, and (in some areas) water. At the same time the close integration between cropping and livestock-raising, which characterized the old system, has been weakened. Grazing livestock and grazing land, which had a central role in the old system, have lost much of their importance. The present use of grazing land is far from optimal; some land is heavily over-grazed, while some land is under-utilized or not used at all.

Is this because, in a modern economy, use of grazing land and crop—livestock integration in the semi-arid zone are no longer economically useful? (or perhaps even wasteful).

It is true that with the present structure of prices and subsidies, wheat-growing and livestock-raising in feedlots can each separately give a fairly stable level of profits; thus grazing and integration are no longer essential for economic viability. However, this viability might be rather sensitive to market price ratios and to government policies. Better integration of the crop and livestock components, at both regional and farm level, and better use of grazing resources, should make the agriculture of the Negev more robust in the long term. They would almost certainly also have some immediate economic benefits for both the individual farmer and the national economy, e.g. by reducing expenses on imported feeds. These seem to be mainly social reasons why these potential benefits are not realized, in both the Jewish and the Bedouin agricultural systems.

Social factors

In the Bedouin system, the supply of shepherding labour has gone down (as the job lost in status) and its cost has gone up. But this factor alone has not, as yet, been sufficient to stop herds from increasing, nor to reduce significantly their use of grazing land. However, the relative contribution of the already poor (and probably still deteriorating) grazing lands in the Bedouin areas to nutrition and production in the herds is decreasing. If this trend continues, most Bedouin herds in the Negev and Judaean Desert will eventually be kept in feedlots and fed on concentrates and residues, in an intensive system similar to that in Jewish farms. Stubble grazing will probably continue, but grazing the adjacent rangeland will become virtually useless.

This could happen even though much of the land is capable, with proper control of grazing (and perhaps some range improvement), of contributing substantially to sheep and goat production. However, in the present system of land-use and tenure (similar to the traditional system) the grazing land in the Bedouin areas is used by everybody but owned by none (now nominally by the government). Thus there is no incentive to any stock-owner to limit voluntarily the number of stock or the grazing season in order to maintain pasture productivity in any area, let alone invest in improving it. There are no signs of co-operation between individual owners in this matter. The present competitive exploitation of the common grazing land could be stopped only by fencing it in and either sub-dividing it into individual leases, or having it managed and controlled by a government agency. There are some social and political obstacles to any such reorganization. Moreover, it would have to be shown that the expected productivity of the rangeland under the new system justifies the investment required.

In the Jewish agricultural systems, land-use organization should not be a problem, since grazing land can be leased in blocks by villages already organized on a co-operative (moshav) or collective (kibbutz) basis. However, the typical structure of a Negev moshav (an equally small irrigated block to each farmer, and a common dry cropland) has no formal provisions for raising livestock on grazing land, either by individuals or in common. In the kibbutz, each branch of the farm (e.g. dry-land crops, dairy cattle, sheep— is managed more or less independently by a group of members, and accounted for separately. This is conducive to efficiency in each branch, but not always to activities which depend on close integration between branches. Livestock branches involving grazing and shepherding have tended to be dropped due to scarcity of labour, or relatively low returns, over a few years; their re-establishment requires a major capital investment.

Experiments to increase pasture production

These social and organizational problems could probably be overcome, if it were clearly demonstrable that under conditions of modern farming in the Negev, a system including grazing livestock has a considerable economic advantage over the present (cropping only) system. Such a comparison has been difficult due to lack of information about the productivity of semi-arid pastures in conditions of careful management and high inputs, similar to those already used in crops.

Recent experiments are providing such information. The Tadmor Experimental Farm at Migda (20 km north-west of Beer-Sheva; a loess plain with 250 mm mean annual rainfall) was established in 1962. A 5-year grazing experiment (1967—1972) showed that the natural annual grass—forb pasture (without any treatment except fencing-in 1—12 ha plots) could maintain continuously sheep densities well above the previous estimate of one sheep ha^{-1}, with little or no need for supplementary feeding (Tadmor et al., 1974).

The maximum average lamb production (40 kg liveweight ha^{-1}) was attained at a density of 2.5 sheep ha^{-1}. Sowing with annual legumes further increased primary and secondary production by about 50%, even though after 5 years the legumes had been replaced by grasses (Eyal et al., 1975).

Preliminary fertilization trials showed a good response. In the second phase of experiments (1971—1975), primary production of ungrazed native pasture with high inputs of N and P was investigated at Migda by a Dutch—Israeli team (Van Keulen, 1975). In years with medium to good rainfall, primary production in N—P-fertilized pastures (150—200 kg N ha^{-1} + 50 kg P ha^{-1}) was 6—8 t dry matter ha^{-1} (predominantly annual grasses: *Phalaris minor*, *Hordeum murinum*) compared with about 3 t ha^{-1} in unfertilized pastures in Migda and 2 t ha^{-1} in similar pastures outside the farm fence. In a dry year the absolute response was smaller, but the ratio similar. In pastures given P only, annual legumes (*Trigonella arabica*) which became dominant in good years produced almost the same dry-matter yields as the N-grass pastures, but failed in a dry year.

The primary production of the native annual pasture is thus similar to that of wheat, at similarly high soil fertility. Since 1975, experiments have been aimed at finding the best way of translating this into a high secondary (meat) production. Preliminary results indicate that in a good or medium year, the high nitrogen pasture can maintain 10—15 sheep ha^{-1} at a nutrition level adequate for near-optimal lamb growth during the green season, provided that the pasture is deferred until a critical green biomass of 500—600 kg ha^{-1} has accumulated (Benjamin et al., 1977). The consequences for long-term production over a period which includes also dry years, and in a system which includes both crop and pasture areas remain to be examined. An option being investigated is increased sheep fecundity (by choice of twinning breeds and by hormonal treatment); this would require higher inputs of supplementary feed but could still be economically advantageous (Eyal and Benjamin, 1977).

Successful experiments in establishing annual legume pastures (*Medicago* spp.) for periods of 2—4 years suggest that they could have a role in an intensified cropping—grazing system, perhaps in rotation with wheat.

There have been several experiments with fodder shrubs. The native *Atriplex halimus* established rapidly upon planting and formed dense stands though sheep could survive on them (at about one sheep ha^{-1}) intake and lamb production were much lower than in native annual pasture (Eyal et al., 1975). Several introduced shrubs (*Atriplex*, *Kochia*, *Cassia* spp.) have been found to be more palatable to sheep, and should provide better nutrition (but are also more susceptible to grazing) (Forti, 1971, 1977). In any case, shrubs are likely to have a useful role only as part of an integrated grazing system, mainly as a reserve fodder in seasons when the annual pasture is poorest (and possibly on sites which are unsuitable for annuals anyway). Shrub planting is rather expensive, but a cheaper method for establishing *Atriplex halimus* by direct seeding has been proposed (Orev, 1970, 1972).

Prospects of wider application

The most successful results from pasture improvement (nitrogen, legumes, shrubs) in the semi-arid zone have been obtained at Migda, at 250 mm mean rainfall and on soils suitable for wheat. To what extent can these results be extrapolated to areas with lower rainfall or less favourable soils (more shallow, stony, saline or sandy), where the cropping alternative is hardly or not at all feasible and grazing is the only possible land-use? Scattered observations from other places indicate that production of annuals responds drastically to nitrogen in loessial wadis in areas with annual rainfall down to 75 mm, and that there is still a moderate response on loessial plains at 150 mm and on stony slopes at 250 mm. Fodder shrubs (particularly *Atriplex halimus*) have been succesfully established on stony slopes around Beer-Sheva (200 mm) (Orev, 1970). The native vegetation (both annual and perennial, in various habitats) seems to be capable of marked spontaneous improvement in density and composition after several years of exclosure. Increased production is also observable upon any micro-topographical change that detains or concentrates runoff.

Further experiments and assessments of economic viability of improvements are desirable. But there is already strong evidence that in most parts of the semi-arid region, the potential pasture productivity is considerably higher (by a factor of 2—5) than the present productivity, and that much of this potential could be realized by controlled grazing and by relatively cheap pasture improvements (fencing is probably by far the largest expense). This knowledge can be the basis for the development of new production systems, combining and integrating crops with grazing livestock. These systems can provide a higher level of food production for the country, and a higher income for both Jewish and Bedouin farmers, than both previous and present systems.

OTHER LAND-USES

Throughout this chapter, attention has centred on the agricultural—pastoral use of the semi-arid region, which has been, and probably will be, the main use for most of the area. In concluding, other land-uses must be at least briefly considered, in particular afforestation, industry and nature conservation.

Abraham had planted a tamarisk at the well of Beer-Sheva (Genesis 21, 33). For a long period after that, only a few trees (other than fruit trees) were planted in the Negev, usually in similar situations. Trees can hardly establish in this climate without being watered for at least the first summer or two. In the last 35 years there have been many attempts at establishing tree plantations in various parts of the semi-arid region. The most successful have been: *Tamarix* spp. on deep sands and in wadis, *Eucalyptus* spp. in depressions and dammed wadis, *Pinus halepensis* (which occurs naturally in the Mediterranean region in Israel) on hills in the >230 mm rainfall zone. In these cases a substantial proportion of trees have reached maturity, but no spontaneous regeneration has been observed as yet. The areas concerned are small (except one large pine forest at 250—270 mm), and their value is mainly for recreation and landscaping rather than for commercial wood production. It is likely that future afforestation efforts in the Negev will concentrate on limited areas, in direct relation to the need of the growing urban population.

Mineral resources and mining are insignificant in the semi-arid region, but are important just beyond its arid margins (phosphate, Dead Sea potash and bromide). Chemical industries based on these deposits have developed around Beer-Sheva, and are projected to expand considerably in areas further away from the city. These industries involve some noxious pollutants. Unless utter caution is exercised in their planning and operation they could eventually have a critical effect on the suitability of substantial areas in the semi-arid Negev for either human habitation or agricultural use.

The industrial and urban expansion, as well as the prospects for intensified agro-pastoral systems, raise the question of conservation. So far the semi-arid zone has been neglected in this respect. It has neither the attractive evergeen scrub and woodland of the Mediterranean zone, nor the scenic topography and unusual plant and animal life of the desert. As yet, only two or three nature reserves have been established in the area, and none of them is larger than a few km^2. However, the semi-arid region is a bio-geographical transition and meeting zone, with a diverse flora and fauna (including several endemic species) which is of unique scientific interest. The landscape of open plains, low chains of hills and meandering wadis has its own attractions. The fact that the typical semi-arid ecosystem is one that has been influenced and modified by Man and domestic animals over a long history does not detract from the importance of preserving something of it. This could be done by establishing more reserves, but also by giving more attention to ecological and landscape aspects in the overall planning and development of the region.

ACKNOWLEDGEMENTS

We are most grateful to Hagit Levi and to Haim Zaban for allowing us to quote from their respective theses while still in preparation. We also wish to thank Dorothy Lee and Linda Warren for typing the manuscript.

REFERENCES

Aharoni, Y. (Editor), 1974. Beer-Sheba. 1. Excavation at Tel Beer-Sheba 1969—1971, seasons. Tel Aviv University.

Amir, J. and Ephrat, J., 1971. A biological method for evaluating soil nitrogen availability and forecasting nitrogen fertilizer needs of wheat. Agric. J. 63: 385—388.

Amiran, D.H.F., Ben-David, J. and Shinar, A., 1976. The process of sedentarization and settlement among the Bedouin of the Negev. Research Report to Ford Foundation. Dep. Geography, Hebrew University, Jerusalem.

Anonymous, 1977. Internal Reports, 1964—1975. Central Office for Statistics, Prime Ministers Office, Tel-Aviv.

Aref, A. al, 1934. The History of Beer-Sheva and Its Tribes. Jerusalem (in Arabic; Hebrew translation, M. Kapeliuk, Tel-Aviv, 1937).

Arnon, I., 1972. Crop Production In Dry Regions, Vols. I and II. Leonard Hill, London, 650 and 693 pp.

Bailey, C. and Danin, A., 1975. Desert Plants in Bedouin life. Notes on the Bedouin, 5: 1—48. Midreshet Sde-Boqer, Field Studies School (in Hebrew).

Bar-Daroma, M., 1966. Comparison of cultivation methods in the Negev. Hassadeh, 46: 1069, 1187 (in Hebrew).

Ben-David, A., 1974. Talmudische Ökonomie. I. Georg Olms, Hildesheim, New York.

Benjamin, R.W., Noy-Meir, I. and Seligman, N.G., 1977. Effect of grazing deferment at the beginning of the growing season on pasture growth, feed intake by sheep and liveweight increase of lambs. Hassadeh, 57: 1868—1873 (in Hebrew).

Bodenheimer, F.S., 1958. Animal ecology today. Monogr. Biol. 6. Junk, The Hague.

Boyko, H., 1949. On the climax vegetation of the Negev with special reference to pasture problems. Palest. J. Bot., Rehovot Ser., 7: 17—35.

Dan, Y., Yaalon, D.H., Koyumdjinsky, H. and Raz, Z., 1962. The Soils and Soil Association Map of Israel. Israel Ministry of Agriculture and Hebrew University of Jerusalem.

Danin, A. and Orshan, G., 1970. Distribution of indigenous trees in the Northern and Central Negev Highlands. La Yaaran, 20: 115—119. (Hebrew with English summary).

Danin, A., Orshan, G. and Zohary, M., 1975. The vegetation of the Northern Negev and the Judaean Desert of Israel. Isr. J. Bot., 24: 118—172.

Ephrat, J., 1968. Thirty years of wheat breeding in Israel. Israel Agriculture. Ministry of Agriculture, Tel-Aviv (in Hebrew).

Epstein, E., 1939. Bedouin of the Negev. Palest. Explor. Q., pp. 59—73.

Evenari, M., Shanan, L. and Tadmor, N.H., 1971. The Negev: The Challenge Of A Desert. Harvard University Press, Cambridge, Mass., 345 pp.

Eyal, E. and Benjamin, R.W., 1977. Fecundity of Merino and Awassi sheep in extensive and intensive management systems. Manuscript.

Eyal, E., Benjamin, R.W. and Tadmor, N.H., 1975. Sheep production on seeded legumes, planted shrubs and dryland grain in a semi-arid region of Israel. J. Range Manage., 28: 100—107.

Felix, Y., 1963. Agriculture in Palestine in the Period of the Mishna and the Talmud. Magnes Press, Jerusalem (in Hebrew), 368 pp.

Forti, M., 1971. Introduction of fodder shrubs and their evaluation for use in semi-arid zones of the North-western Negev. The Negev Institute for Arid Zone Research, Beer-Sheva (mimeographed).

Forti, M., 1977. Plants for arid zones. In: E. Zohar (Editor), The Desert, Past, Present, Future. Reshafim Publishing House, Tel-Aviv, pp. 273—279 (in Hebrew).

Ganor, A., Markewitch, R., Kessler, Y. and Rosnan, N., 1973. Climate of Sinai. Ktavim Meteorologiim, No. 22. Meteorological Service, Bet Dagan, Israel.

Halevy, G. and Orshan, G., 1972. Ecological studies on *Acacia* species in the Negev and Sinai. I. Distribution of *Acacia raddiana, A. tortilis* and *A. gerrardii* ssp. *negevensis* as related to environmental factors. Isr. J. Bot., 21: 197—208.

Harpaz, Y., 1975. Simulation of the nitrogen balance in semi-arid regions. Ph.D. Thesis, Hebrew University, Jerusalem, 134 pp.

Hofman, M., 1974. Technological progress and economic returns from wheat research. M.Sc. Agric., Thesis. Faculty of Agriculture, Hebrew University of Jerusalem, Rehovot (in Hebrew with English summary).
Holling, C.S., 1973. Resilience and stability of ecological systems Ann. Rev. Ecol. Syst., 4: 1—24.
Josien, T., 1955. La faune chalcolithique des gisements palestiniens de Bir es-Safadi et Bir Abou Matar. Isr. Explor. J., 5: 246—256.
Katznelson, J., 1956. Variability of rain in Palestine and statistical methods for its measurement. Ktavim Meteorol. Ser. E, No. 4 Meteorological Service, Bet Dagan, Israel (in Hebrew).
Katznelson, V., 1959. The climate of the Negev. Meteorol. Stud., Ser. E, No. 10. Israel Meteorological Service, Bet Dagan (in Hebrew).
Lanir, V., 1976. Multi-purpose small storage dams for utilization of marginal water resources. Hassadeh, 56: 889—901 (in Hebrew).
Lomas, J. and Shashoua, Y., 1973. The effect of rainfall on wheat yields in arid regions. UNESCO Ecol. Conserv., 5: 531—537 (Proc. Uppsala Symp. Plant Response to Climatic Factors).
Marx, E., 1967. Bedouin of the Negev. Manchester University Press.
Mendelssohn, H., 1973. Ecological effects of chemical control of rodents and jackals in Israel. In: M.T. Farvar and J.P. Milton (Editors), The Careless Technology: Ecology and International Development. Natural History Press, Garden City, N.Y., pp. 527—544.
Musil, A., 1908. Arabia Petraea. III. Ethnologischer Reisebericht. Hofler, Wien.
Negbi, M., 1968. The status of summer annuals in Palestine. Isr. J. Bot., 17: 217—221.
Negev, A. (Editor), 1972. Archaeological Encyclopedia of the Holy Land. Weidenfeld and Nicholson, Jerusalem.
Noy-Meir, I., 1975. Primary and secondary production in sedentary and nomadic grazing systems in the semi-arid region: analysis and modelling. Res. Rep. Ford Foundation, Dep. Bot., Hebrew University, Jerusalem.
Noy-Meir, I. and Harpaz, Y., 1977. Agro-ecosystems in Israel. Agro-Ecosystems, 4(1/2): 143—167.
Orev, Y., 1970. Browse shrubs for range improvement. World Crops, 22: 410—411.
Orev, Y., 1972. Demonstration of a new shrub seeding method. In: Wildland Shrubs, Their Biology and Utilization. USDA Forest Service General Technical Report INT-1, pp. 405—406.
Otterman, J., 1974. Baring high-albedo soils by overgrazing: a hypothesized desertification mechanism. Science, 186: 531—533.
Otterman, J., Waisel, Y. and Rosenberg, E., 1975. Western Negev and Sinai eco-systems: comparative study of vegetation, albedo, and temperatures. Agro-ecosystems, 2: 47—60.
Perrot, J., 1955. The excavations of Tel Abu-Matar, near Beersheba. II. Isr. Explor. J., 5: 73—84.
Seligman, N.G., Tadmor, N.H. and Raz, Z., 1962. Range survey of the central Negev. Bull. No. 67. The National and University Institute of Agriculture, Bet Dagan, Israel.
Shimshi, D., 1971. Population dynamics of *Aellenia hierochuntica* (Bornm) Aellen. Isr. J. Bot., 20: 44—47.
Shimshi, D. and Kafkafi, U., 1973. Irrigation and nitrogen fertilizer interaction semi-dwarf wheat. Report to Ministry of Agriculture, Tel-Aviv, mimeographed (Hebrew).
Shkolnik, A., Borut, A. and Choshniak, J., 1972. Water economy of the bedouin goat. Symp. Zool. Soc. London, 31: 229—242.
Shmueli, A., 1970. The settlement of the Bedouin of the Judaean Desert. Gome, Tel-Aviv, Israel Housing Ministry (in Hebrew).
Stoebe, E. and Ariel, D., 1970. The effect of preparatory cultivation methods on the yields of dry-land crops in rotation. Internal report, Israel Ministry of Agriculture. mimeographed (in Hebrew).
Tadmor, N.H., Eyal, E. and Benjamin, R.W., 1974. Plant and sheep production on semi-arid annual grassland in Israel. J. Range Manage., 27: 427—432.
Tamir, R., 1970. Erosion, run-off and yields of winter grain from land in the south (of Israel) with different seed-bed preparation. 3rd Ann. Symp. on Erosion problems. Div. of Soil Conservation, Ministry of Agriculture, Tel-Aviv, pp. 18—20. (in Hebrew).
Van Keulen, H., 1975. Simulation of Water Use and Herbage Growth in Arid Regions. Pudoc, Wageningen, 176 pp.
Zaban, H., 1977. Land use systems in the semi-arid zone of Israel. Ph.D. thesis, University of Reading.
Zohary, M., 1962. Plant Life of Palestine: Israel and Jordan. Ronald Press Co., New York, N.Y., 262 pp.
Zohary, M., 1976. A New Analytical Flora of Israel. Am Oved, Tel-Aviv (in Hebrew), 540 pp.

Chapter 5

UTILIZATION OF GRASS- AND SHRUBLANDS OF THE SOUTH-WESTERN UNITED STATES *

CARLTON H. HERBEL

Range Scientist, Jornada Experimental Range, Science and Education Administration—Agricultural Research, U.S. Department of Agriculture, Las Cruces, New Mexico (U.S.A.)

INTRODUCTION

The rangelands of the south-western United States may seem limitless, but as a renewable source for livestock grazing, they are fragile, easily damaged by abuse, and subject to droughts. The original vegetation on large parts of these lands has been depleted by a combination of these factors, and a rapid increase of unwanted shrubs. Table 5.1 shows the area in rangeland and the livestock numbers in Table 5.1 are the totals for each state, the rangeland areas considered in this chapter are roughly the western half of Texas and Oklahoma, most of New Mexico, and all but the south-western quarter of Arizona.

CLIMATE

Annual precipitation ranges from less than 250 mm in the lower elevations of New Mexico, Arizona and western Texas to 750 mm in the eastern portions of the plains grasslands in central Texas and Oklahoma (see Fig. 5.1

TABLE 5.1

Rangelands in thousands of ha and livestock numbers in thousands in states of the south-western United States

	Texas	Oklahoma	New Mexico	Arizona
Rangeland [b]	47 355	9205	26 488	23 522
Cattle, beef [c]	16 552	6401	1855	1181
Sheep [d]	2688	89	575	510
Goats [d]	1150	[a]	[a]	[a]

[a] Small numbers, not estimated.
[b] Wooten et al. (1962).
[c] U.S. Department of Agriculture (1975a).
[d] U.S. Department of Agriculture (1975b).

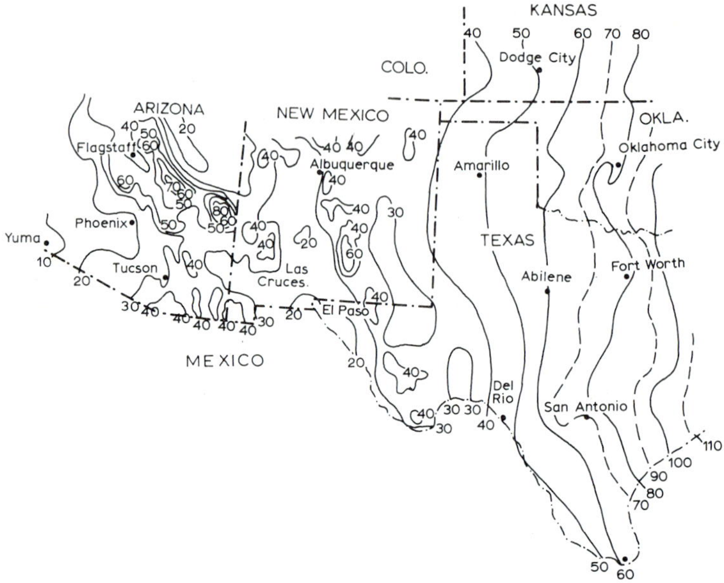

Fig. 5.1. Average annual precipitation (cm) in the south-western United States (from U.S. Department of Commerce, 1975).

and Table 5.2). The precipitation not only varies greatly within and among seasons and years, but also among locations separated by only short distances. About 70% of the mean annual precipitation occurs during the spring—summer period in the plains grasslands of Texas, Oklahoma and eastern New Mexico. In western New Mexico and southern Arizona, the growing season precipitation occurs during the summer, and the spring period is normally very dry. The latter areas have a slight increase in winter rainfall. Spring—summer precipitation for the entire area is generally due to convectional thunderstorms from the Gulf of Mexico, whereas the winter precipitation is due to broad fronts sweeping southward across the plains or moving westward from the Pacific. The entire region is frequently plagued by drought and during a prolonged drought may have a desert-like appearance. It may also have high winds during some periods, which result in considerable erosion when coinciding with a reduction in vegetational cover due to drought or over-grazing. The mean annual evaporation ranges from 2160 mm at Oklahoma City to 3050 mm at some of the lower elevations in southwestern New Mexico and southern Arizona. The frost-free period (consecutive days when the minimum temperature is above 0°C) averages about 180 days in the north-eastern part of the region to 259 days at Tucson.

The major vegetational types of the region are determined primarily by precipitation. Arizona and the western third of New Mexico have been included by Trewartha (1961) in the Intermontane Precipitation Region (2a

TABLE 5.2

Average climate features (U.S. Department of Commerce, 1975)

	Jan.	Feb.	March	April	May	June	July	Aug.	Sep.	Oct.	Nov.	Dec.	Ann.
Amarillo, Texas													
Precipitation (mm)	17	16	21	34	86	73	59	66	48	45	17	20	500
Temperature (°C)	3	5	9	14	19	25	27	26	22	16	8	4	15
Del Rio, Texas													
Precipitation (mm)	23	22	21	35	69	58	33	39	66	50	16	21	453
Temperature (°C)	11	13	17	22	26	29	30	30	27	22	15	11	21
Oklahoma City, Oklahoma													
Precipitation (mm)	33	35	50	79	132	114	60	64	77	64	40	36	783
Temperature (°C)	3	5	9	16	20	26	28	28	23	17	9	5	16
Las Cruces, New Mexico													
Precipitation (mm)	12	13	8	4	8	13	33	43	31	19	8	12	203
Temperature (°C)	5	7	11	15	20	25	26	25	22	16	9	6	16
Tucson, Arizona													
Precipitation (mm)	21	21	13	7	3	7	52	73	25	16	16	23	279
Temperature (°C)	10	12	14	19	23	28	30	28	27	21	14	11	20
Flagstaff, Arizona													
Precipitation (mm)	46	45	37	30	13	18	58	72	40	39	25	42	465
Temperature (°C)	−3	−1	2	6	10	15	19	18	15	8	2	−1	8

(Fig. 5.2), bounded on the east by the Rocky Mountains and on the west by the Pacific Coast Ranges. A unique feature of this area is the existence of two rainy seasons, one maximum occurring in winter and the other in summer, making it transitional between the winter maximum found on the Pacific Coast and the summer maximum typical of the Great Plains. In general, the winter maximum is strongest on the western edge of the region, with the summer rains increasing in importance towards the east. Winter storms are mainly the result of the Pacific high pressure areas shifting southward, allowing lows to move inland, bringing with them masses of moist maritime air (Trewartha, 1961). The July—August rainfall is derived from air masses moving in from the Gulf of Mexico (Jurwitz, 1953). Most of these storms are local, convectional thunderstorms and do not result in the widespread activity characteristic of the winter period (Bryson, 1957).

The Great Plains has been termed the Interior Region by Trewartha (1961). It has a typical continental climate characterized by a predominance of warm season precipitation. Here, for the most part, winter is the season of lowest precipitation. Trewartha (1961) considers this a region of Type 3 precipitation. Sub-type 3a, primarily centering in New Mexico, receives its maximum moisture in July and August. The flow of maritime air into the region makes New Mexico a major centre of thunderstorm activity. No trace of a secondary winter maximum exists. Instead, the cold season is very dry due to the continental influence of the prevailing westerly winds.

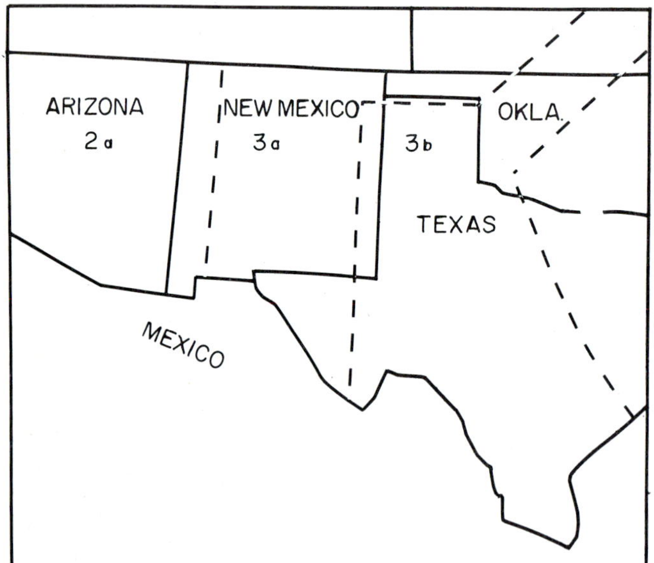

Fig. 5.2. Precipitation regions of the south-western United States (from Trewartha, 1961).

A unique precipitation pattern termed 3b by Trewartha (1961), exists in the southern part of the Great Plains, and includes major portions of Texas and Oklahoma. Here, little moisture is received from November through March. Instead, a double maximum of precipitation occurs in the warm season. One peak is reached in May, resulting from the flow of humid air from the Gulf of Mexico into the area where it is subsequently warmed by contact with the heated land surface. This is followed by a decrease in rainfall in late June, July and August which apparently results from a tongue of dry northerly air associated with a large low pressure cell positioned in the mid-troposphere over the southern Great Plains (Wexler and Namias, 1938). Movement of the cell in September results in a second maximum period of precipitation in late summer.

LAND, PLANT AND ANIMAL RESOURCES

Four land resource regions have been identified in the portions of Texas, Oklahoma, New Mexico and Arizona considered in this chapter (Austin, 1972) (Fig. 5.3). An outline of the land and plant resources discussed in the following pages is shown in Table 5.3.

South-Western Plateaus and Plains Range and Cotton Region (Fig. 5.3.I)

The land resource areas within this region are in southern Texas: Rio Grande Plain, Texas Central Basin and Edwards Plateau. The Rio Grande Plain (Fig. 5.3, 83), totalling 82 600 km^2, is about 80% rangeland. The mean annual precipitation within this area ranges from 500 to 870 mm. Mean annual temperature is 21°C and the average frost-free period ranges from 260 to 320 days. The elevation ranges from sea-level to 300 m. This plain is nearly level to gently undulating. Grumusols and reddish chestnut soils are extensive. Lithosols are confined to low, narrow ridges and low plateau-like areas. Sandy Regosols occur along the coast of the Gulf of Mexico, along the Rio Grande River and in the north-eastern part of the area. A clayey Regosol is found in the western part of the area. Calcisols are in sandy and silty old alluvium on stream terraces and deltas in the southern and western part of the area and in alluvial soils on narrow floodplains and deltas (Austin, 1972).

The vegetation of the Rio Grande Plain is dominated by the *Prosopis—Acacia—Schizachyrium—Setaria* (Fig. 5.4) and the *Prosopis—Schizachyrium* associations. The dominant species of the former associations are: *Prosopis glandulosa* [1], *Acacia rigidula*, *Schizachyrium scoparium* var. *littoralis* and *Setaria macrostachya* (Fig. 5.4).

[1] Plant names follow Correll and Johnston, 1970.

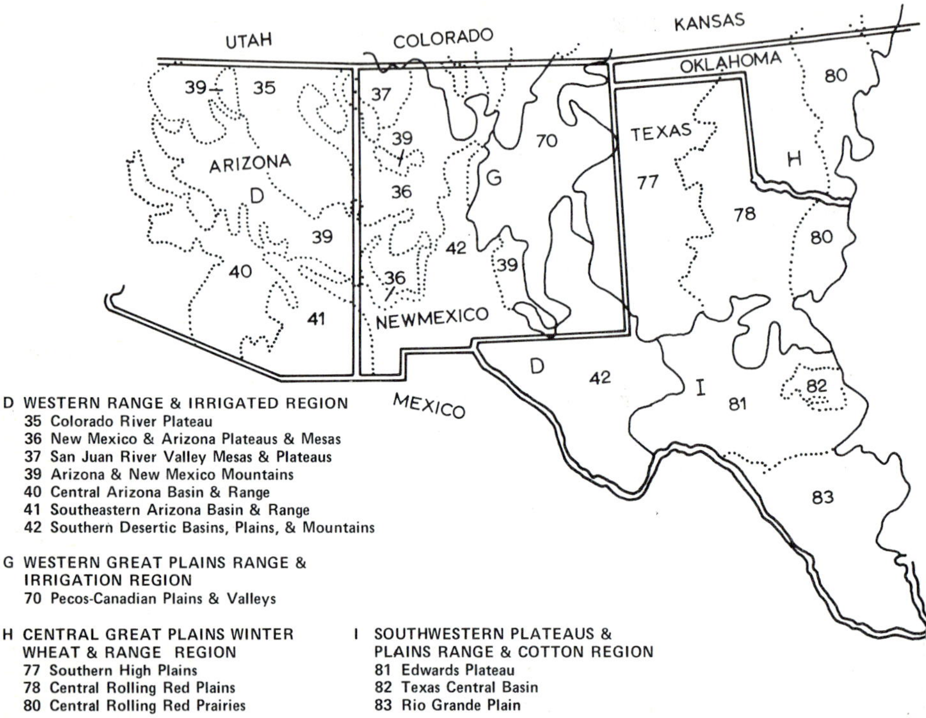

Fig. 5.3. Land resources of the south-western United States (from Austin, 1972).

Other major components of the association are: *Acacia berlandieri*, *A. farnesiana*, *Bothriochloa saccharoides*, *Cenchrus myosuroides*, *Chloris ciliata*, *C. verticillata*, *Opuntia leptocaulis*, *O. lindheimeri*, *Pappophorum bicolor* and *Trichloris pluriflora*. The dominants of the *Prosopis—Quercus—Schizachyrium* association are *Proposis glandulosa*, *Quercus virginiana* and *Schizachyrium scoparium* var. *littoralis*. Other major components are: *Brachiaria ciliatissima*, *Panicum virgatum*, *Paspalum hartwegianum*, *P. setaceum*, *Sorghastrum avenaceum* and *Trachypogon secundus* (Kuchler, 1964, Gould, 1975).

The Texas Central Basin (Fig. 5.3, 82), 7800 km^2, is about 90% rangeland and is located in the eastern part of the Edwards Plateau. The mean annual precipitation within this area ranges from 625 to 750 mm. The mean annual temperature is 18°C and the frost-free period averages 240 days. The elevation ranges from 240 to 390 m. The rolling to hilly uplands are crossed by a few smooth valleys and many steep hills and ridges. Reddish Prairie soils occupy the smoother slopes of uplands and valleys, Lithosols occupy the steep hill slopes and rough stoney land occupies the steepest slopes, cliffs of limestone and granite ridges (Austin, 1972).

The vegetation of the Texas Central Basin is dominated by the *Prosopis*—

TABLE 5.3

Land and plant resources of the south-western United States (Kuchler, 1964; Austin, 1972)

I. South-western Plateaus and Plains Range and Cotton Region (Fig. 5.3, I)
- A. Rio Grande Plain (Fig. 5.3, 83)
  - (1) *Prosopis—Acacia—Schizachyrium—Setaria* association (Fig. 5.4)
  - (2) *Proposis—Quercus—Schizachyrium* association
- B. Texas Central Basin (Fig. 5.3, 82)
  - (1) *Proposis—Quercus—Schizachyrium* savanna
- C. Edwards Plateau (Fig. 5.3, 81)
  - (1) *Juniperus—Quercus—Schizachyrium* association (Fig. 5.5)

II. Central Great Plains Winter Wheat and Range Region (Fig. 5.3, H)
- A. Central Rolling Red Prairies (Fig. 5.3, 80)
  - (1) *Schizachyrium—Stipa* association
  - (2) *Quercus—Schizachyrium* cross timbers
- B. Central Rolling Red Plains (Fig. 5.3, 78)
  - (1) *Proposis—Buchloe* association (Fig. 5.6)
- C. Southern High Plains (Fig. 5.3, 77)
  - (1) *Bouteloua—Buchloe* association (Fig. 5.7)
  - (2) *Quercus—Schizachyrium* association

III. Western Great Plains Range and Irrigated Region (Fig. 5.3, G)
- A. Pecos. Canadian Plains and Valleys (Fig. 5.3, 70)
  - (1) *Bouteloua—Hilaria* steppe
  - (2) *Bouteloua—Hilaria—Larrea* shrubsteppe
  - (3) *Bouteloua—Buchloe* association
  - (4) *Juniperus—Pinus* woodland

IV. Western Range and Irrigated Region (Fig. 5.3, D)
- A. Southern Desertic Basins, Plains and Mountains (Fig. 5.3, 42)
  - (1) *Flourensia—Larrea* shrub savanna
  - (2) *Bouteloua—Hilaria* prairie (Fig. 5.8)
  - (3) *Bouteloua—Hilaria—Larrea* shrubsteppe
- B. Arizona and New Mexico Mountains (Fig. 5.3, 39)
  - (1) *Juniperus—Pinus* woodland
  - (2) Transition between *Quercus—Juniperus* woodland and *Cercocarpus—Quercus* association
  - (3) *Pinus* forest (Fig. 5.9)
  - (4) *Pinus—Pseudotsuga* forest
- C. South-eastern Arizona Basin and Range (Fig. 5.3, 41)
  - (1) *Bouteloua—Hilaria—Larrea* shrubsteppe
  - (2) *Larrea—Flourensia* association
- D. New Mexico and Arizona Plateaus and Mesas (Fig. 5.3, 36)
  - (1) *Juniperus—Pinus* woodland
  - (2) *Bouteloua—Hilaria* steppe
- E. Central Arizona Basin and Range (Fig. 5.3, 40)
  - (1) *Larrea—Franseria* association
- F. San Juan River Valley Mesas and Plateaus (Fig. 5.3, 37)
  - (1) *Atriplex—Sarcobatus* association
- G. Colorado River Plateau (Fig. 5.3, 35)
  - (1) *Artemisia* association
  - (2) *Juniperus—Pinus* woodland
  - (3) *Bouteloua—Hilaria* steppe

Fig. 5.4. The *Prosopis—Acacia—Schizachyrium—Setaria* association in the Rio Grande Plain near Uvalde, Texas (photo: G.B. Donart).

Quercus—Schizachyrium savanna. The dominant species are *Prosopis glandulosa*, *Quercus* spp. and *Schizachyrium scoparium*. Other major components of the vegetation are: *Andropogon gerardi*, *Bothriochloa barbinodis*, *Bouteloua curtipendula*, *Buchloe dactyloides*, *Diospyros texana*, *Juniperus ashei*, *Quercus marilandica* and *Q. stellata* (Kuchler, 1964).

The Edwards Plateau (Fig. 5.3, 81) has about 85 700 km^2, and roughly 95% is in rangeland. The mean annual precipitation within this area ranges from 375 to 875 mm. The mean annual temperature ranges from 16 to 21°C, and the frost-free period averages 220 to 240 days. The elevation ranges from 300 to 1200 m. This broad dissected limestone plateau has gently undulating divides and broad valleys. Valley sides are steeply sloping to rolling or hilly. Shallow Lithosols, underlain by limestone, occupy the steeper slopes of the Edwards Plateau. Grumusols occur in valleys and on nearly level uplands: Calcisols occupy better drained level to gentle slopes where soil materials are less clayey, and alluvial soils occur on narrow flood plains (Austin, 1972).

The vegetation of the Edwards Plateau is dominated by the *Juniperus—Quercus—Schizachyrium* association (Fig. 5.5). The plant dominants are *Juniperus ashei*, *Quercus virginiana* and *Schizachyrium scoparium*. Other major components are: *Andropogon gerardi*, *Aristida glauca*, *A. purpurea*

Fig. 5.5. The *Juniperus—Quercus—Schizachyrium* association in the Edwards Plateau near Sonora, Texas (photo: G.B. Donart).

Bouteloua curtipendula, *Buchloe dactyloides*, *Hilaria belangeri*, *Prosopis glandulosa* and *Sporobolus cryptandrus* (Kuchler, 1964; Gould, 1975).

The discussion of animal species found in the region follows Shelford (1963). This includes a treatment of presently dominant species and those that may have been important in the past. The following animals are common in the South-western Plateaus and Plains Range and Cotton Region; subspecies of the pygmy mouse (*Baiomys taylori*), harvest mouse (*Reithrodontomys fulvescens*), black-tailed jack rabbit (*Lepus californicus*), fox squirrel (*Sciurus niger*), badger (*Taxidea taxus*), spotted skunk (*Spilogala gracilis*), pocket gopher (*Geomys bursarius*), red wolf (*Canis niger*), coyote (*C. latrans*), white-tailed deer (*Odocoileus virginianus*), turkey (*Meleagris gallopavo*), scaled quail (*Callipepla squamata*) and mourning dove (*Lenaidura macroura*). Mammals that were dominant over the area in former times were the bison (*Bison bison*) and the pronghorn antelope (*Antilocapra americana*). Other birds characteristic of the area are the horned lark (*Eremophila alpestris*) and the boat-tailed grackle (*Cassadix mexicanus*). The common orthopterans are: *Melanoplus differentialis*, *Chortophaga viridifasciata*, *Ophulella palidna*, *O. speriosa*, *Orchelimum concinnum*, *Scudderia texensis*, *Spharagemon bolli* and *Psoloessa texana*. A hemnipteron present is the stink bug (*Mormidea lugens*). The chief coreid is *Mecidea longula*. The most common ant is *Crematogaster laeviuscula.* Two species of snails *Bulimulus alternatus* and *Helicina orbiculata*, occur in the trees and shrubs (Shelford, 1963).

Central Great Plains Winter Wheat and Range Region (Fig. 5.3, H)

The land resource areas within this region, mainly in Texas and Oklahoma, are: Central Rolling Red Prairies, Central Rolling Red Plains and Southern High Plains (Austin, 1972). The mean annual precipitation in the Central Rolling Red Prairies (Fig. 5.3, 80) ranges from 580 to 890 mm, and the mean annual temperature from 16 to 18°C, and the average frost-free period 200 to 240 days. The elevation ranges from 300 to 460 m. On this dissected plain, the divides are undulating to gently rolling, and the valley sides are steep. Reddish Prairie soils occupy the smoother upland slopes in most of this area, but Reddish Chestnut soils are most extensive on the slopes in the south-western part of the area. Lithosols are dominant on steep slopes and Regosols in deep sands. Alluvial soils are on floodplains and low terraces along the rivers. About half of the area is in rangeland and about half in cropland (Austin, 1972).

The vegetation of the Central Rolling Red Prairies is dominated by the *Schizachyrium—Stipa* association in the south and the *Quercus—Schizachyrium* cross timbers in the north (Kuchler, 1964). The dominant species in the *Schizachyrium—Stipa* association are *Schizachyrium scoparium* and *Stipa leucotricha*. Other major species are: *Andropogon gerardi, Bouteloua curtipendula, Panicum virgatum, Sorghastrum avenaceum* and *Sporobolus asper*. The dominants of the *Quercus—Schizachyrium* cross timbers are: *Schizachyrium scoparium, Quercus marilandica* and *Q. stellata.* Other major components are: *Andropogon gerardi, Bouteloua curtipendula, Sorghastrum avenaceum* and *Stipa leucotricha* (Kuchler, 1964).

The mean annual precipitation in the Central Rolling Red Plains (Fig. 5.3, 78) ranges from 500 to 750 mm (Austin, 1972), the mean annual temperature from 14 to 18°C, the mean frost-free period from 185 to 230 days. The elevation ranges from 450 to 900 m, increasing gradually from east to west. The broad divides are nearly level to gently sloping, and the valleys have short steep slopes on this dissected plain. Reddish Chestnut soils and Chestnut soils occupy most of the smoother uplands. In about half the area, consisting of more hilly lands, the soils are thin or have weakly expressed profiles — Regosols in deep sandy materials, Lithosols on slopes underlain by consolidated rocks, and Calcisols in areas with a strong horizon of calcium carbonate. About half of this area is rangeland and the other half is cropland (Austin, 1972).

The vegetation of the Central Rolling Red Plains is dominated by the *Prosopis—Buchloe* association (Kuchler, 1964). The dominant vegetation is *Buchloe dactyloides* and *Prosopis glandulosa* (Fig. 5.6). Other important

Fig. 5.6. The *Prosopis—Buchloe* association in the Central Rolling Red plains near Dickens, Texas.

plants are: *Bouteloua gracilis*, *Juniperus pinchotii*, *J. virginiana*, *Schedonnardus paniculatus* and *Yucca angustifolia* (Kuchler, 1964).

The Southern High Plains (Fig. 5.3, 77) has about 130 300 km^2, roughly 40% of which is in rangeland (Austin, 1972). The mean annual precipitation within this area ranges from 380 to 585 mm, the mean annual temperature from 13 to 16°C, and the frost-free period from 180 to 220 days. The elevation ranges from 760 to 1520 m. These smooth, high plains have gentle slopes except for the very steeply-sloping breaks along the major rivers. The deep sands in the south-western portion of this area have irregular dune topography. The principal soils of the uplands are Chestnut and Reddish Chestnut soils in loamy mantles. Calcisols are dominant on the more sloping parts and in shallow valleys, Lithosols on the steeper slopes and breaks, and Regosols and Reddish-Brown soils in the deep sands (Austin, 1972).

The vegetation of the Southern High Plains is generally dominated by the *Bouteloua—Buchloe* association, with the *Quercus—Schizachyrium* association dominating the more sandy sites. The dominant plants of the *Bouteloua—Buchloe* association (Fig. 5.7) are: *Bouteloua gracilis* and *Buchloe dactyloides*. Other major species are: *Bouteloua curtipendula*, *Machaeranthera pinnatifida*, *Opuntia polyacantha*, *Sporobolus cryptandrus* and *Yucca angustifolia*. The dominants of the *Quercus—Schizachyrium* associ-

Fig. 5.7. The *Bouteloua–Buchloe* association in the Southern High plains near Bronco, Texas.

ation are *Quercus mohriana* and *Schizachyrium scoparium*. Other major species are *Andropogon hallii*, *Artemisia filifolia*, *Bouteloua gracilis*, *Buchloe dactyloides*, *Eriogonum annuum*, *Juniperus pinchotii*, *Prosopis glandulosa*, *Quercus havardii*, *Sporobolus cryptandrus* and *Yucca angustifolia* (Kuchler, 1964).

The major animals in the Central Great Plains Winter Wheat and Range are: pronghorn antelope, black-tailed jack rabbit, desert cottontail (*Sylvilagus audubonii*), plains harvest mouse (*Reithrodontomys montanus*), grasshopper mouse (*Onychomys* spp.), deer mouse (*Peromyscus maniculatus*), prairie vole (*Microtus ochrogaster*), thirteen-lined ground squirrel (*Citellus tridecemlineatus*), spotted ground squirrel (*C. spilosoma*), short-tailed shrew (*Blarina brevicauda*), least shrew (*Cryptotis parva*), coyote, badger, Swainson's hawk (*Buteo swainsoni*), rough-legged hawk (*B. lagopus*), greater prairie chicken (*Tympanuchus cupido*), western meadowlark (*Sturnella neglecta*), horned lark, dickcissel (*Spiza americana*), McCown's longspur (*Rhynchophanes mccownii*), lark bunting (*Calamospiza melanocorys*), garter snake (*Thamnophis radix*), western diamondback rattlesnake (*Crotalus atrox*), bullsnake (*Pituophis catenifer*) and blue racer (*Coluber constrictor*). One of the abundant insect groups is the orthoptera. Some of the important ones are: *Encoptolophus sordidus costalis*, *Melanoplus keeleri luridus*,

Conocephalus saltans, *Acrolophitus hirtipes*, *Amphitornus coloradus*, *Psoloessa delicatula*, *Aulocara elliotti* and *Metator pardalinus*. Other common insects are: the lygaeid (*Ligyrocarsis diffusus*), plant bug (*Lugus oblineatus*), predatory bug (*Orius insidiosis*), damsel bug (*Nabis ferus*), stink bug (*Euschistus variolarius*), chinch bug (*Blissus leucopterus*), leafhopper (*Agallia constriata*), twelve spotted cucumber beetle (*Diabrotica twelve-punctata*), scarab (*Aphodius distinctus*), lady beetle (*Hippodamia convergens*), robber fly (*Promachus* spp.) and the wheat stem sawfly (*Cephus cinctus*) (Shelford, 1963).

Western Great Plains Range and Irrigated Region (Fig. 5.3, G)

West of the Southern High Plains is the Pecos-Canadian Plains and Valleys area in eastern New Mexico. This area totals 74 900 km^2 with more than three-quarters in rangeland. About 10% is controlled by various agencies of the Federal Government. The mean annual precipitation in this area varies from 300 to 400 mm, the mean annual temperature from 10 to 16°C, and the frost-free period from 160 to 200 days. The elevation ranges from 1200 to 1950 m. Most of the slopes of these dissected high plains are gentle to rolling, but bands of steep slopes and rough broken land border the stream valleys. Shallow soils and soils having weakly expressed profiles are widespread — Lithosols, rough broken land, and rockland on gentle to moderate slopes of uplands, and Regosols in deep sands. Reddish-Brown soils occur on the smoother uplands throughout most of the area, but Reddish Chestnut soils and Chestnut soils occur in the deeper materials in the northern part of the Pecos-Canadian Plains and Valleys (Austin, 1972).

The vegetation of the Pecos-Canadian Plains and Valleys is dominated by the *Bouteloua—Hilaria* steppe association, the *Bouteloua—Hilaria—Larrea* shrubsteppe association, the *Bouteloua—Buchloe* grass association and the *Juniperus—Pinus* woodland association. The vegetation of the *Bouteloua—Buchloe* association was previously described in the section on the Southern High Plains. The dominant plants of the *Bouteloua—Hilaria* association are *Bouteloua gracilis* and *Hilaria jamesii*. Other major species are: *Andropogon hallii*, *Artemisia tridentata*, *Bouleloua curtipendula*, *Schizachyrium scoparium* and *Yucca angustifolia*. The *Bouteloua—Hilaria—Larrea* shrubsteppe is dominated by *Bouteloua eriopoda*, *Hilaria mutica* and *Larrea tridentata*. Other major plants are *Acacia constricta*, *Aristida divaricata*, *A. longiseta*, *Bouteloua curtipendula*, *B. gracilis*, *Opuntia* spp., *Prosopis glandulosa*, *Sporobolus airoides*, *S. cryptandrus* and *Xanthocephalum sarothrae*. The dominant plants of the *Juniperus—Pinus* association are *Juniperus monosperma*, *J. osteosperma* [2], *Quercus emoryi*, *Q. undulata* and *Sporobolus cryptandrus* (Kuchler, 1964).

[2] Plant names followed by [2] are from Kearney and Peebles (1969). All other nomenclature follows Correll and Johnston (1970).

The vertebrate influents and dominants of the Pecos-Canadian Plains and Valleys originally included the bison, pronghorn antelope and possibly deer. Less prominent animals, the conspicuous box turtle (*Terepene ornata*), kangaroo rat (*Dipodomys* spp.), bullsnake, garter snake, rattlesnake, lark bunting, scaled quail (*Callipepla squamata*), and the western meadowlark, were present. Grasshoppers of several species also exert considerable influence on the community. There are numerous burrowing insects present, including mutillids, digger wasps (*Ammophila procera*) and tiger beetles (*Cicindela* spp.). Coyotes and cottontails are also numerous (Shelford, 1963).

Western Range and Irrigated Region (Fig. 5.3, D)

This is a diverse region of plateaus, plains, basins and many isolated mountain ranges. The land resource areas in western Texas, southern and western New Mexico and all but the desertic portion of Arizona, are: Southern Desertic Basins, Plains and Mountains; Arizona and New Mexico Mountains; New Mexico and Arizona Plateaus and Mesas; South-eastern Arizona Basin and Range; Central Arizona Basin and Range; San Juan River Valley Mesas and Plateaus and Colorado River Plateau (Austin, 1972). The Southern Desertic Basins, Plains and Mountains of western Texas and southern New Mexico (Fig. 5.3, 42) with a total of 160 000 km^2, are about 95% rangeland. About a third of the land is under the jurisdiction of the Federal Government. The mean annual precipitation ranges from 200 to 400 mm, the mean annual temperature from 10 to 26°C, and the frost-free period from 200 to 240 days, over most of the area. The elevation ranges from 750 to 1500 m in basins and valleys and more than 2500 m in the mountains. Broad desert basins and valleys are bordered by gently to strongly sloping fans and terraces. Steep mountain ranges and many small mesas are common throughout the area. Red Desert soils and Lithosols are dominant in the area. Others are Calcisols in highly calcareous materials and Regosols in deep sands (Austin, 1972).

The major vegetation types of the Southern Desertic Basins, Plains and Mountains are the *Flourensia—Larrea* shrub savanna, the *Bouteloua—Hilaria* prairie and the *Bouteloua—Hilaria—Larrea* shrubsteppe. The vegetation of the *Bouteloua—Hilaria—Larrea* shrubsteppe was previously described in the section describing the vegetation of the Pecos-Canadian Plains and Valleys. The dominant plants of the *Flourensia—Larrea* savanna are *Flourensia cernua* and *Larrea tridentata*. Other major plant components are: *Acacia constricta*, *Aristida divaricata*, *A. longiseta*, *Opuntia imbricata*, *Prosopis glandulosa*, *Scleropogon brevifolius* and *Yucca elata*. The dominants of the *Bouteloua—Hilaria* prairie are *Bouteloua gracilis* and *Hilaria mutica* (Fig. 5.8). Other major components of the vegetation are *Bouteloua curtipendula*, *B. eriopoda*, *Opuntia imbricata* and *Xanthocephalum sarothrae* (Kuchler, 1964; Gould, 1975).

Fig. 5.8. *Hilaria mutica* type in the *Bouteloua—Hilaria* Prairie association in the Southern Desertic Basins, Plains and Mountains near Las Cruces, New Mexico.

The Arizona and New Mexico Mountains (Fig. 5.3, 39) total 90 900 km^2. About 60% is controlled by the Federal Government, and an additional 10% is in Indian reservation; about 40% is rangeland. The elevation ranges from 1350 to 3000 m, with some mountain crests at 3750 m. Much of the area is so rough and steep that it has little economic use, but it includes some deeply dissected plateaus. The mean annual precipitation varies from 250 to 875 mm, increasing with elevation. The mean annual temperature ranges from 4°C at high elevations to 13°C at the lower elevations. Brown, Reddish-Brown and Reddish Chestnut are the principal soils in the foothills and valleys, but Lithosols, stony land and rock outcrops occupy large areas on both mountains and foothills (Austin, 1972).

The major vegetation types of the Arizona and New Mexico Mountains area are the *Juniperus—Pinus* woodland and a transition between the *Quercus—Juniperus* woodland and *Cercocarpus—Quercus* association; at higher elevations, the *Pinus* forest and the *Pinus—Pseudotsuga* forest. The vegetation of the *Juniperus—Pinus* woodland was previously described under the section on the Pecos-Canadian Plains and Valleys. The major plants of the *Quercus—Juniperus* woodland and the *Cercocarpus—Quercus* association are: *Juniperus deppeana*, *J. monosperma*, *Quercus emoryi*, *Q. gambelii*, *Q. grisea* and *Cercocarpus ledifolius* [2]. Some of the other species are: *Arcto-*

staphylos pungens, *Bouteloua curtipendula*, *B. gracilis*, *Fallugia paradoxa*, *Nolina microcarpa*, *Pinus cembroides*, *Purshia tridentata*, *Quercus havardii*, *Q. turbinella*, *Q. undulata* and *Rhus microphylla*. The dominant plant of the *Pinus* forest is *Pinus ponderosa* (Fig. 5.9). The other major components of the vegetation are: *Blepharoneuron tricholepis*, *Ceanothus fendleri*, *Festuca arizonica*, *Muhlenbergia montana*, *Pinus cembroides*, *Poa fendleriana*, *Pseudotsuga menziesii*, *Quercus gambelii* and *Stipa* spp. The dominant species of the *Pinus—Pseudotsuga* forest are *Pinus ponderosa* and *Pseudotsuga menziesii*. The other major plant species are: *Blepharoneuron tricholepis*, *Ceanothus fendleri*, *Festuca arizonica*, *Holodiscus discolor* and *Salix* spp. (Kuchler, 1964).

South-eastern Arizona Basin and Range (Fig. 5.3, 41) totals about 45 800 km^2, of which 95% is rangeland. About 30% of the area is controlled by the Federal Government. The elevation ranges from 750 to 1800 m. There are some basins and dry lake beds in this nearly level broad plain. The mean annual precipitation ranges from 200 to 500 mm, the mean annual temperature from 13 to 21°C, and the frost-free period from 180 to 240 days. The principal soils are Red Desert, Brown, Reddish-Brown and Calcisols. Alluvial soils are extensive on floodplains and on the lower younger alluvial fans (Austin, 1972).

The major vegetation types of the South-eastern Arizona Basin and Range are the *Bouteloua—Hilaria—Larrea* shrubsteppe and the *Larrea—Flourensia* association (Kuchler, 1964). The *Bouteloua—Hilaria—Larrea* shrubsteppe association was previously described in the section on the Pecos-Canadian Plains and Valleys. The dominants of the *Larrea—Flourensia* association are *Flourensia cernua* and *Larrea tridentata*. Other major components of the vegetation are: *Acacia constricta*, *A. greggii*, *Agave lecheguilla*, *Bouteloua breviseta*, *Dasyliron* spp., *Opuntia* spp. and *Prosopis glandulosa* (Kuchler, 1964).

About 98% of the New Mexico and Arizona Plateaus and Mesas (Fig. 5.3, 36), which totals 67 900 km^2, is rangeland. About 20% of the area is controlled by the Federal Government and another 20% is Indian reservation. The mean annual precipitation ranges from 250 to 325 mm, the mean annual temperature is about 10°C and the average frost-free period ranges from 120 to 180 days. The elevation ranges from 1500 to 2100 m. These plateaus and mesas generally have gentle slopes. Sierozems are extensive in the drier portions of the area. Regosols in deep sands and alluvial soils are important on floodplains throughout the area (Austin, 1972).

The major vegetation types of the New Mexico and Arizona Plateaus and Mesas are the *Juniperus—Pinus* woodland and the *Bouteloua—Hilaria* steppe, both of which were previously described in the section on the Pecos-Canadian Plains and Valleys.

The Central Arizona Basin and Range (Fig. 5.3, 40) totals 49 200 km^2 85% of which is used for range. About 50% of the land is controlled by the

Fig. 5.9. The *Pinus* forest in the Arizona and New Mexico Mountains north of Globe, Arizona.

Federal Government and 10% is Indian reservation. The mean annual precipitation is about 250 mm. The mean annual temperature ranges from 16 to 23°C, and the frost-free period from 225 to 300 days. The elevation ranges from 300 to 750 m. The low basins are bordered by gently sloping alluvial fans and are separated by mountain ranges. The most extensive soils are Red Desert soils on the older alluvial fans and terraces, alluvial soils on floodplains and the younger alluvial fans, Calcisols in calcareous materials and Lithosols on valley walls and mountain slopes (Austin, 1972).

The vegetation of the Central Arizona Basin and Range is primarily of the *Larrea—Franseria* association. The dominants are *Franseria dumosa* [2] and *Larrea tridentata*. The other major components of the vegetation are: *Acacia greggii*, *Cercidium floridum* [2], *Dalea spinosa* [2], *Encelia farinosa* [2], *Franseria deltoidea* [2], *Hilaria rigida* [2], *Opuntia echinocarpa* [2] and *Prosopis juliflora* var. *velutina* [2] (Kuchler, 1964).

The San Juan River Valley Mesas and Plateaus (Fig. 5.3, 37) occupy 12 200 km^2 in north-western New Mexico. About 90% is rangeland, about 30% is controlled by the Federal Government and much of the remainder is in an Indian reservation. The mean annual precipitation ranges from 200 to 250 mm. The mean annual temperature is 10°C and the frost-free period is about 150 days. The elevation is 1500 to 1800 m. The gently sloping broad valleys and plains are bordered by deeply dissected bands of steep slopes and sharp local relief. The dominant soils are Sierozems and Desert soils. Alluvial soils occur on the narrow floodplains of major streams and Regosols in windblown soils in the same position. Lithosols are extensive and there are large areas of badlands in the dissected belts adjacent to the stream valleys (Austin, 1972).

The vegetation of the San Juan River Valley Mesas and Plateaus is dominated by the *Atriplex—Sarcobatus* association. The dominant species are *Atriplex confertifolia* and *Sarcobatus vermiculatus*. Other major components of the vegetation are: *Eurotia lanata*, *Grayia spinosa*, *Kochia americana* and *Suaeda torreyana* (Kuchler, 1964).

About two-thirds of the Colorado River Plateau (Fig. 5.3, 35) is either controlled by the Federal Government or in Indian reservations. The mean annual precipitation varies from 200 to 400 mm, the mean temperature from 7 to 13°C and the frost-free period from 150 to 220 days. The elevation ranges from 1500 to 2100 m. Gently sloping plateaus and mesas are deeply and abruptly cut by canyons. Sierozems and Calcisols occupy the plateau tops, older terraces and alluvial fans. Lithosols and badlands are also extensive (Austin, 1972).

The vegetation of the Colorado River Plateau is dominated by the *Artemisia* association, *Juniperus—Pinus* woodland and the *Bouteloua—Hilaria* steppe. The vegetation of the latter two associations was previous described in the section on the Pecos-Canadian Plains and Valleys. The *Artemisia* association is dominated by *Artemisia tridentata*. Other plant components are

Agropyron smithii, Artemisia nova [2], *Atriplex confertifolia, Chrysothamnus* spp. and *Tetradymia canescens* [2] (Kuchler, 1964).

Only a few species of animals occur through all the various resource areas in the Western Range and Irrigated Region. A generally distributed permeant is the pronghorn antelope. The bison has never been prominent in this region. Mule deer (*Odocoileus hemionus*) and the black-tailed jack rabbit occur throughout the region. The wolf (*Canis lupus*) has dens, mostly at the higher elevations. The coyote and bobcat (*Lynx rufus*) help in reducing populations of rabbits and wood rats (*Neotoma albigula*). The badger is generally distributed throughout. The banner-tailed kangaroo rat (*Dipodomys spectabilis*) is one of the most characteristic rodents of the region. The antelope jack rabbit (*Lepus alleni*) and black-tailed jack rabbit are common influents in the Southern Desertic Basins, Plains and Mountains and the South-eastern Arizona Basin and Range. The enemies of the jack rabbits include the golden eagle (*Aquila chrysaetos*), great horned owl (*Bubo virginianus*), red-tailed hawk (*Buteo jamaicensis*) and the ferruginous hawk (*B. regalis*). Other rodents, besides those mentioned above, often encountered in these latter resource areas are: desert cottontail, grasshopper mouse, Merriam's kangaroo rat (*Dipodomys merriami*) and the Bailey's pocket mouse (*Perognathus baileyi*).

Small birds common to the area are: horned larks, black-throated sparrows (*Amphispiza bilineata*), lark sparrows (*Chondestes grammacus*), grasshopper sparrows (*Ammodramus savannarum*), loggerhead shrike (*Lanius ludovicianus*) and nighthawks (*Chordeiles* spp.). Some of the common reptiles are: earless lizards (*Holbrookia maculata approximans* and *H. texana*), Devil's River whiptail (*Cnemidophorus perplexus*), large kingsnake (*Lampropeltus getulus splendida*), rattlesnake and bullsnake. Some of the common insects are: locust (*Ageneotettix deorum*), squash bugs (*Mecidea longula, Liorhyssus hyalinus* and *Arhyssus lateralis*), false chinch bug (*Nysius ericiae*), damsel bug (*Nabis alternatus*), stilt bug (*Jalysus wickhami*), lantern bug (*Oliarus pima*), leafhopper (*Neokolla curcubita* and *Exitianus obscurinervis*), leaf beetles (*Promecosoma virida* and *Pachybrachis nigrofasciatus*), snout beetles (*Centrinaspis* spp., *Pantomorus albosignatus* and *Mitostylus setosus*) and ants (*Pogonomyrmex barbatus, P. californicus, Dorymyrmex pyramicus* and *Myrmecocystus melliger*) (Shelford, 1963).

Some of the common mammals of the Central Arizona Basin and Range are: rock pocket mouse (*Perognathus intermedius*), Bailey's pocket mouse, rock squirrel (*Citellus variegatus*), hispid cotton rat (*Sigmodon hispidus*), spotted skunk, kit fox (*Vulpes macrotis*), desert cottontail, white-throated wood rat, peccary (*Pecari tajacu*) and white-tailed deer. Some of the common birds are: Gila woodpecker (*Centurus uropygialis*), ash-throated flycatcher (*Myiarchus cinerascens*), purple martin (*Progne subis*), roadrunner (*Geococcyx californianus*), Gambel's quail (*Lophortyx gambelii* and the cactus wren (*Camphylorhynchus brunneicapillum*). Common reptiles are:

tiger rattlesnake (*Crotalus tigris*), black-tailed rattlesnake (*C. molossus*), Mojave rattlesnake (*C. scutulatus*), bullsnake, Gila monster (*Heloderma suspectum*), collared lizard (*Crotaphytus collaris*), regal horned lizard (*Phrynosoma solare*), uta lizard (*Uta* spp.), whiptail lizard (*Eumeces skiltonianus*), desert spiny lizard (*Sceloporus magister*), zebra-tailed lizard (*Callisaurus draconoides*) and the desert tortoise (*Gopherus agassizi*). Some of the common invertebrates are: crab spider (*Misumenops celer*), grasshopper (*Trimerotropis pallidipennis* and *Platylactista azteca*), bruchid (*Acanthoscelides amicus*), snout beetle (*Apion ventriocosum*) and the ants (*Novomessor cockerelli* and *Pogonomyrmex barbatus nigrescens*) (Shelford, 1963).

Common animals in the Colorado River Plateau and the Arizona and New Mexico Mountains are: mule deer, mountain lion (*Felis concolor*), coyote, black bear (*Euarctos americanus*), porcupine (*Erethizon dorsatum*), least chipmunk (*Eutamias minimus*), deer mouse (*Peromyscus maniculatus*) and bushy-tailed wood rat (*Neotoma cinerea*). The major arthropods are: Black Hills beetle (*Dendroctonus ponderosa*), bark beetle (*Ips integer*), shot borer (*Orthotomicus ornatus*) and carpenter ant (*Camponotus herculeanus modoc*). Some of the major birds are: pigmy nuthatch (*Sitta pygmaea*), white-breasted nuthatch (*S. carolinensis*), Mexican junco (*Junco phaeonotus*) and Steller's jay (*Cyanocitta stelleri*) (Shelford, 1963).

The most common animals in the San Juan River Valley Mesas and Plateaus are: Ord's kangaroo rat (*Dipodomys ordii*), sagebrush chipmunk (*Eutamias minimus*), little pocket mouse (*Perognathus longimembris*), Great Basin pocket mouse (*P. parvus*), black-tailed jack rabbit and the pronghorn antelope. The common birds are: horned lark, mourning dove, common nighthawk (*Chordeiles minor*), vesper sparrow (*Posocetes gramineus*), western kingbird (*Tyrannus verticalis*) and loggerhead shrike (*Lanius ludovicianus*). Common invertebrates are: spider (*Metepeira foxi*), leafhoppers (*Eutettix insanus* and *Aceratagallia cinerea*) and grasshoppers (*Trimerotropis pallidipennis* and *Melanoplus occidentalis*) (Shelford, 1963). The dominant mammals and birds given above generally occur at the same elevations throughout the region.

PAST LAND-USE

In 1600 there were probably 45 million bison occurring in all parts of the grassland of North America, except California. The size of herds in the Pecos-Canadian Plains and Valley area in eastern New Mexico was smaller than those in the Central Great Plains. In the latter region, extensive grazing by bison and prairie dogs (*Cynomys* spp.) tended to change the mid grasses dominated by *Schizachyrium scoparium* to short grasses of the *Bouteloua—Buchloe* association. The pronghorn antelope had a total population similar to that of the bison, but they tended to occupy drier portions of the grassland than did the bison (Shelford, 1963).

The range cattle industry of the United States started in the South-western Plateaus and Plains Region in South Texas early in the nineteenth century (Fugate, 1961). Most of the early cattle and cowboys came from Spain by way of Mexico. From about 1840 to 1865, the cattle industry developed as stock farmers and others learned the new business and developed new techniques. The range cattle industry began to expand in 1866 after the Civil War. Destitute soldiers were returning from the war; hide hunters were killing the bison, the United States Army was pushing Indians onto reservations and railroads were connecting the West with the more populous eastern United States. All these factors helped settle the West and created markets for cattle. Meanwhile, Texas ranges were stocked with multitudes of cattle, virtually worthless where they stood. During the period between 1866 and 1880, the cattle industry spread over most of the West. The trails going north from Texas were not single trails but merely a general direction of travel made up of numerous small trails converging at river fords and mountain passes.

Although sheep were among the animals first introduced from Europe into North America, they were relatively unimportant on the western rangelands before 1865 (Stoddart et al., 1975). Increases in range-sheep production were minimal until there were railroads to carry the wool crop. After this development, sheepmen found much range that was well suited to their flocks and were attracted by the profits to be made from the grazing of free lands. From 1866 to 1901 there was a period of trail-herding in the sheep industry similar to that in the range-cattle industry. Sheep were trailed from western rangelands to fattening and marketing points in the mid-west. The pioneer sheepmen found they could increase their grazing land by seasonal migrations, sometimes covering vast areas. They grazed their animals yearlong, using the high, cool mountains during the summer months and the lower, warmer areas during the winter.

The range-livestock industry boomed in the inflationary period after the Civil War, and except for a brief period of panic in 1873, phenomenal increases in numbers occurred until 1885 (Stoddart et al., 1975). The grass resource was considered to be limitless by many livestock owners, but the winter of 1885—1886, a winter of unprecedented severity, caused the livestock industry to collapse because there was no unused forage and no emergency feed. It has been estimated that this winter left 85% of the animals dead over wide areas on the range. As a consequence of this disaster, much of the speculative money supporting the boom was withdrawn. However, the remaining individuals had a long-term interest in raising livestock and were the real pioneers of the present range-livestock industry.

In southern New Mexico, livestock numbers increased rapidly in the eighteen-eighties, especially near watering points. Many of the water wells were established between 1900 and 1910 so that the area was fully stocked by 1910 (Buffington and Herbel, 1965). Wooton (1908) reported that the condition of New Mexico range was deteriorating and that it was not as productive as it was earlier. Cattle numbers also increased rapidly in Arizona in the eighteen-eighties. There were only 5000 cattle in the Arizona territory in 1870 (Martin, 1975). Rapid movements of cattle from Texas and Mexico increased the number to 1 095 000 in 1890. Griffiths (1901) found the range in southern Arizona to be seriously depleted as a result of prolonged over-stocking.

Land policies greatly affected the settlement and course of range management in the United States. At the time of its inception, the Federal Government had no land. However, shortly afterwards, several states turned over their unowned land to the Government and this became the first public domain. Though these original lands were the basis for the development of land policy, later acquisitions hold special interest to readers of this chapter. Most of the Western lands were acquired for indirect considerations rather than for agricultural uses. For instance, the South-west Territory was obtained from Mexico in 1846 to provide good harbours on the West Coast. Since no great value was attached to the land involved in these acquisitions, there was little concern for its judicious administration (Stoddart et al., 1975).

The land-disposal measures enacted by the United States Government changed with time, as new circumstances arose, and as different economic segments were able to influece Congress. Zimmet (1966) identified five objectives of land-disposal policies: (1) sale of land to produce revenue for the Government; (2) facilitate the settlement and growth of new communities; (3) reward veterans of wars; (4) promote education and the construction of internal improvements, such as railroads, by grants of land, and (5) retention for reservations, national forests and public domain.

One result of the Government's failure to regulate use of the public land was serious damage to range. During the period 1866 to 1934, when the Taylor Grazing Act was passed, little provision was made for administering grazing on public land. The result of this general lack of supervision was an intense competition among users to secure as much grazing as possible from the public lands. Such conditions resulted in extensive over-grazing on the range resource. Likewise, permitting land to be farmed in arid and semi-arid areas led to much damage because it was unsuited for crop production. In most cases the farmers were from more humid climes and they discovered that the land was not sufficiently productive for cropping. During that time, the forage was destroyed and the land was subjected to erosion, which was sometimes so severe as to interfere seriously with revegetation after abandonment of farming practices (Stoddart et al., 1975).

PRESENT CONDITIONS

The vegetation on some rangelands has improved since the nineteen-thirties. Moderate stocking is more common, and many ranchers have improved their ranges by seeding, controlling unwanted plants, or periodic resting of the vegetation from grazing. The portions of Texas and Oklahoma discussed in this chapter have little land owned by the Federal Government. However, 34% of the land within New Mexico is controlled by an agency of the Federal Government, and 45% of Arizona is federally controlled (Stoddart et al., 1975). Most of this land is leased to ranchers for livestock grazing. These rangelands are no longer the almost exclusive domain of ranchers and public land administrators. Absentee owners, speculators, miners, sportsmen and other recreationists have an increased interest in land practices. Public concern about aesthetics, habitat destruction, or the balance of nature has at times been great enough to halt range improvement projects.

Widespread concern about how public lands are managed has greatly increased planning costs of administration of public lands. To some extent, this concern also limits management options for private landholders (Martin, 1975). As Byerly (1970) indicated, we must now solve the problems of agricultural production with methods that are socially and economically acceptable. Environmental impact statements must now be prepared before major projects are initiated on public lands. Maximum sustained production of forage and livestock is no longer the only consideration in operations on public lands. However, with the increasing demand for food by the rapidly growing population of the world, increasing effort is placed on production of animal protein from lands not suited for cultivation. The challenge to the range manager is to bring a proper balance between the biological realities of the site and the demands of the consumer (Stoddart et al., 1975).

In general, ranges have relatively low biological productivity (Clawson, 1972). They are sometimes termed economic residuals after more productive sites are converted to higher economic uses such as improved pastures or croplands (Paulsen, 1975). While limited plant growth is typical of most ranges, rough topography, remoteness, shallow soil, low rainfall and severe temperatures are also characteristic of some rangelands (Thomas and Ronningen, 1965). In addition, socio-economic factors further complicate the producer's efforts.

Most of the ranges described here are grazed primarily by cattle. However, in the Edwards Plateau and Texas Central Basin, sheep and/or goats are sometimes mixed with cattle to better utilize all classes of forage. Sheep graze many of the forbs and goats browse on the *Quercus* and other palatable shrubs (Merrill, 1954). Vegetation control exerted by sheep and goats is a fringe benefit besides the income from animal products. There is little evidence, however, that browsing by sheep or goats will reduce the growth and spread of *Prosopis*, *Juniperus*, *Acacia* or *Larrea* (Martin, 1975).

Efforts to improve productivity of rangeland must often be restricted because of economic considerations. Sometimes only relatively inexpensive means can be employed, and these must be amortized over long periods. In many cases, however, herbage and browse on rangeland could contribute far more feed for livestock and wildlife than it now provides (Paulsen, 1975). One of the least expensive practices on rangeland is to manipulate grazing, but the results are inconclusive (Herbel, 1971).

Original plant productivity has been reduced over large areas by past grazing abuses, brush invasion, droughts and past attempts to cultivate non-arable land. In many areas, the degradation is so complete that seeding is required to restore at least a portion of the former productivity (Herbel, 1973b). In some situations seeding increases productivity above pristine levels.

Large parts of the south-western United States have been invaded by unwanted shrubs. There are 37 million ha of *Prosopis* spp. throughout the area (Platt, 1939). *Artemisia filifolia* and *Quercus havardii* are problems on sandy soils of the Central Great Plains Winter Wheat and Range Region. *Juniperus* spp. are dominant in the Edwards Plateau and parts of the Pecos-Canadian Plains and Valleys, and the mid-elevations in the Western Range and Irrigated Region. *Larrea tridentata* and *Flourensia cernua* occupy about 20 million ha at lower elevations of the Western Range and Irrigated Region. *Opuntia* spp. are a serious problem in parts of the South-western Plateaus and Plains Range and Cotton Region, the Pecos-Canadian Plains and Valleys and the Western Range and Irrigated Region.

The density of brush on rangelands of the south-western United States has increased rapidly since 1900. A survey by the U.S. Soil Conservation Service found 36 million ha, or 82% of Texas grasslands were infested with one or more low-value woody plants (Smith and Rechenthin, 1964). Smith and Rechenthin reported that 22 million ha were covered with dense stands of brush requiring control measures before any other range improvement could be made. Yet in spite of efforts to control brush by chemical and mechanical methods, its density is increasing. The survey showed that the area of *Prosopis* alone has increased by half a million ha during the period of 1938–1963. The increase in brush in South Texas was primarily a local expansion and not a general expansion of the range of the species (Johnston, 1963). In most instances, species have moved from gulleys and watercourses onto upland sites.

Livestock grazing began about 1900 on the upland areas of southern New Mexico. Notes from the land survey in 1858 enabled Buffington and Herbel (1965) to reconstruct vegetation types from that time to 1963 on the Jornada Experimental Range near Las Cruces, New Mexico. *Prosopis* dominated only 5% of the area in 1858, but this had increased to 50% in 1963. Most of the stands of *Prosopis* in 1858 were associated with Indian activities. The rapid increase in recent years is due to dispersal of *Prosopis* seed by livestock. As *Prosopis* begins to dominate a sandy site, low dunes form, and grass cover is greatly reduced.

Larrea dominated 0.4% of the study area on the Jornada Experimental Range in 1858 and 14.2% of the area in 1963 (Buffington and Herbel, 1965). *Flourensia* dominated 0.4% of the study area in 1858 and 8.6% of the area in 1963. It originally grew on the slopes next to mountains, but it has moved down the slopes and is now most prevalent on heavier soils, where it competes with *Hilaria mutica* and *Scleropogon brevifolius*. As *Prosopis* and *Flourensia* began to dominate the slopes next to mountains, the original grass stands became less dense. Eventually *Larrea* moved onto those sites and gained dominance over the *Prosopis*, *Flourensia* and the residual grass stand. 10 to 15 cm of topsoil has been lost from slopes now dominated by *Larrea*, leaving the larger pebbles and stones to form an erosion pavement. *Larrea* will also invade sites where the original grass stand has been depleted.

RECOMMENDED LAND-USE PRACTICES

Each ranch has different characteristics and objectives, and must be managed accordingly. Ranches differ in the amount of improvements (fencing, water developments, equipment), the proportion of various soil and vegetation types, wildlife species, recreational opportunities and livestock characteristics (kind, breed and class). Often the degree of management is determined by economic conditions and the willingness of an operator to undertake practices involving some financial risk. Generally, the less costly practices will also reap less benefits. An example of low intensity management is to build fences, increase the watering points and initiate a grazing system. The changes in productivity are very slow, and thus the return on investment is very slow (Herbel, 1973a). An example of a much more intensive system would be pastures seeded to *Eragrostis curvula* and grazed in conjunction with wheat and *Sorghum sudanense*. This system was described by McIlvain (1976) and discussed in detail later. The costs are much greater with this system, but the potential benefits are much higher.

Each land manager should assess the potential economic benefits to be obtained from each major vegetation and soil type. He must consider values obtained from livestock, wildlife, recreation and water; then he can determine which treatment or combination of treatments to use. Possibilities include grazing systems, seeding, control of unwanted plants, burning, fertilization, water-spreading, fencing and increasing the watering points. After improvements have been completed, or at any point in the process, the land manager may modify his plans to maximize profits while maintaining the resource. Similarly, with changing technology or improving economic conditions, the manager may decide to intensify his improvement efforts (Herbel et al., 1974).

Grazing management

Merrill (1954) compared continuous year-long grazing at three intensities with deferred-rotation grazing at a moderate rate on the Edward Plateau near Sonora, Texas. There were three herds in the four-unit rotation system; each unit was grazed 12 months, then rested 4 months. Thus, during a 4-year cycle, each unit was deferred once during each of the 4-month periods. Stocking was with a combination of cattle, sheep and goats. The study was initiated in 1949. After 11 years the stocking rate of the units in the deferred-rotation system was increased 33% from 12.4 animal units km^{-2} to 16.6 animal units km^{-2} (Merrill, 1969). These units carried the increased grazing pressure and at the same time made greater range improvement than any of the units grazed continuously. Average annual net returns for 1959—1965 were \$ 1.78, \$ 2.91 and \$ 1.63 ha^{-1} with continuous stocking at the rate of 6.2, 12.4 and 18.6 animal units km^{-2}. The average net return for the same period on the rotation units was \$ 4.15 ha^{-1} (Merrill, 1969).

A two-unit, one-herd rotation; a four-unit, three-herd rotation; and year-long continuous grazing were compared on the Edwards Plateau near Barnhart, Texas (Huss and Allen, 1969). The four-unit rotation was the same as that described by Merrill (1954). In the two-unit rotation, the units were alternately grazed and deferred for 3- and 6-month periods (e.g. one unit was deferred 1 March to 1 June, grazed 1 June to 1 December, and deferred 1 December to 1 March). Thus, during a 2-year period, each unit was deferred 12 months with deferment during each season. All treatments were stocked with cattle and sheep at 10 animal units km^{-2}. During 1959—1965 the average annual net returns were \$ 3.06, \$ 3.90 and \$ 4.17 ha^{-1} for continuous, four-unit rotation, and two-unit rotation grazing, respectively (Huss and Allen, 1969). Huss and Allen found that combination use of cattle and sheep was more profitable than grazing either class alone. Merrill (1969) also found that combination grazing with cattle, sheep and goats was more profitable at Sonora than using sheep or cattle alone.

Fisher and Marion (1951) compared rotation and continuous grazing at a moderate rate on the Central Rolling Red Plains near Spur, Texas. The grazing season was about 1 May to 1 October. The rotation system consisted of grazing each of three units for 1 month and deferring it for 2 months for the 1942—1949 period. They concluded that: (1) rotational grazing did not improve the vegetational composition from 1942 to 1947; (2) rotational grazing increased differential use of the major grass species as the season progressed or during drought, and in some instances resulted in less moisture penetration on sites occupied by the more desirable species, and (3) gains of yearling steers grazing on the rotation units were slightly lower than those of yearling steers grazing on the continuous units.

Various grazing systems were compared at the Texas Experimental Ranch in the Central Rolling Red Prairies near Throckmorton during 1960—1968

(Kothmann et al., 1970). A moderate stocking rate with cows and supplemental feeding level of 0.7 kg day^{-1} of cottonseed cake during winter were used in comparing grazing systems. The three systems were year-long-continuous, a two-unit rotation similar to that studied at Barnhart, Texas, by Huss and Allen (1969), and a four-unit rotation similar to that studied at Barnhart by Huss and Allen and at Sonora, Texas, by Merrill (1954). Calf production per animal unit averaged 200, 208 and 221 kg for the moderate year-long continuous, two-unit rotation and four-unit rotation, respectively, for the 8 years.

Since 1969, 300 ranchers in Texas have initiated a high intensity—low frequency grazing system (Leithead, 1974). It provides for one herd of cattle to graze five to nine units for 15 to 40 days each, depending on the amount of forage available. When the forage in each unit is fully used, the herd is moved to another unit and is not returned to that unit for 4 to 12 months. Leithead (1974) found that the shorter grazing periods, 15 to 20 days, resulted in better livestock distribution and less selective grazing. In this system, units are not grazed during the same calendar period in successive years. Some conclusions by Roberts (1967) in studies of this system in South Africa are pertinent to this discussion. He indicated that livestock should not be moved at any set time nor should the units be stocked in a set sequence. At the time of year when plants are growing rapidly, the livestock should be moved frequently, perhaps as often as every 5 days, to prevent injury to plants. When the plants are dormant, livestock movements may be determined by nutrient requirements of livestock.

McIlvain and Shoop (1969) compared several grazing systems on native range in the Central Rolling Red Plains on the Southern Great Plains Field Station near Woodward, Oklahoma. They concluded that the following grazing systems have not proven superior to continuous year-long grazing at the same stocking rates: (1) summer and winter grazing; (2) alternate-year grazing; (3) three-unit rotations with rotations at 2-months, 1-month, 15-day and 10-day intervals, and (4) two-unit, 6-week, one-herd rotation. They cited some major reasons for the success of continuous year-long grazing in that area, as (1) forage production primarily depends on summer rainfall — and monthly forage production during the summer can vary from 25—700 kg ha^{-1}; (2) most species are grazed by cattle at one time or another; (3) many of the "increaser" species are excellent grazing plants and they may be very productive under certain conditions; (4) cattle use the forage before other losses occur; (5) young shoots and regrowth forage is more palatable and more nutritious than more mature forage; (6) grazed plants save moisture for later green growth, and (7) favourable growing seasons combined with proper management allow ranges to recover a desirable species composition. Some additional reasons for the success of year-long continuous grazing may be (1) use is light during the growing season under year-long continuous grazing, and (2) lighter stocking per unit area means less compaction of the soil by livestock when the soil is wet (Herbel, 1973a).

More recently McIlvain (1976) has compared native range with three complementary pasture systems: (1) native range plus *Eragrostis curvula*; (2) native range plus wheat—*Sorghum sudanense*, and (3) *E. curvula* plus wheat—*Sorghum.* The control treatment was continuous, moderate, year-long stocking of native range at 3.6 ha per steer. This latter system produced more beef at a lower cost while maintaining vigour and condition of the range, than in studies at Woodward (McIlvain and Shoop, 1965). In the native range plus *Eragrostis curvula* system, a total of 2.0 ha was required per steer; 90% of the total land was native range and 10% was *Eragrostis*. This proportion was used to avoid more intensive practices, such as hay-making, if a larger proportion of the land was seeded to *Eragrostis.* The approximate chronological order of grazing was:

Late October and November	— native range
December to March—April	— *Eragrostis* aftermath
March—April to early May	— native range
Early May to early September	— *Eragrostis*
Early September to early October	— native range

In the native range plus wheat—*Sorghum* system, a total of 1.8 ha was required per steer; 75% of the total land was native range, and 25% was double-cropped by planting wheat each October and *Sorghum sudanese* each June. The approximate chronological order of grazing was:

Late October to mid-March	— native range
Mid-March to early June	— wheat
Early June to early August	— native range
Early August to October	— *Sorghum*

The *Eragrostis* plus wheat—*Sorghum* system required 0.8 ha per steer; 50% of the total land was planted to *Eragrostis* and 50% was double-cropped with wheat and *Sorghum.* The approximate chronological order of grazing was:

Late October to mid-March	— *Eragrostis*
Mid-March to early June	— wheat
Early June to early August	— *Eragrostis*
Early August to October	— *Sorghum*

Some of the results from the various schemes are shown in Table 5.4. Beef production on these rangelands was doubled and quadrupled by using complementary pastures that required progressively higher levels of management and economic inputs per hectare. The major value of *Eragrostis curvula* as tame pasture is quantity. The essence of wheat—*Sorghum* as farmed forage for graze-out is quality. The native range provided stability and flexibility to complementary pasture systems. The philosophy of complementing low-producing rangelands with high-producing tame pastures or farmed forages, or both, makes it possible to: (1) use green forages in dry periods; (2) graze and rest each forage resource for its proper growth and development; (3) avoid grazing areas when and where poisonous plants are a problem; (4) use

TABLE 5.4

Generalized year-long beef production data, Central Rolling Red Plains, Wodward, Oklahoma (from McIlvain, 1976)

Grazing scheme	ha/Steer	Gain/Steer	Gain/ha
Native range (NR)	3.6	177	49
NR + *Eragrostis*	2.0 [a]	163	81
NR + wheat—*Sorghum*	1.8 [b]	191	104
Eragrostis + wheat-*Sorghum*	0.8 [c]	168	207

[a] 1.8 ha NR + 0.2 ha *Eragrostis.*
[b] 1.4 ha NR + 0.4 ha wheat—*Sorghum.*
[c] 0.4 ha *Eragrostis* + 0.4 wheat—*Sorghum.*

pastures for breeding cows; (5) use quality forages as green creeps for calves or steers needing a rapid gain and following them with cows or younger cattle to clean up the remaining forage; (6) improve the economics for practices such as brush control, insect control, fencing or fertilization, and (7) exploit the proven potential of higher milking dams, faster-growing calves, hybrid vigour and multiple births (McIlvain, 1976).

Working in the Southern Desertic Basins, Plains and Mountains, Herbel and Nelson (1969) developed the Best Pasture Grazing System. The system consists of establishing an objective for each range unit and stocking accordingly. The system is opportunistic, in that the use of forbs and short-lived grasses is maximized. They are of little value to the permanent range resource, but contribute much to livestock nutrition. No set stocking plan is established for a specific time period because of considerable variation in weather conditions that affect plant growth. In this system, the livestock are moved when the vegetation on another unit can be grazed to better advantage for both plants and animals than the unit being grazed. In the large range units in parts of the West, periodic opening and closing of watering places can be used to rotate grazing pressure to different areas within a range unit (Martin and Ward, 1970).

In range areas in the western United States, stocking must be adjusted to compensate for a highly variable forage crop. Flexible herd management has been suggested by several workers as the best method for maximizing livestock production without damaging the range resource during dry periods (Ares, 1952; Reynolds, 1954; Stubblefield, 1956; Paulsen and Ares, 1962; Boykin, 1967). Most of the severe damage to rangelands by grazing animals occurs during droughts. With flexible herd management, the herd is made up of not more than 55—60% breeding animals during average years. The remainder of the herd is composed of yearlings and replacement heifers. In years of low forage production, adjustments in the size and composition of the herd are planned to bring the herd within the carrying capacity of the range. Readily saleable animals such as weaners and yearlings are marketed. This is also a good time to cull the breeding herd of old animals and those with poor production. In the years of above-average precipitation, part or all of the natural increase from the breeding herd can be held over until the spring or fall, depending on conditions.

Hyder and Bement (1977) listed three requirements for a grazing system: (1) stocking the ranges to achieve not more than moderate use of the forage in the growing season every year, and to maintain a satisfactory botanical composition and productivity of herbage and livestock; (2) grazing should be deferred during the growing season once every few years to renew the vigour and productivity of preferred plants, and (3) include a period of heavy grazing in the dormant season once every few years to reduce unpalatable or ungrazed plants, equalize plant composition and promote more uniform grazing. There is no apparent justification for leaving a range unit ungrazed for an entire year. Some authors (e.g. Hormay and Talbot, 1961) justify rotation plans in terms of seed production, seed planting by trampling and seedling establishment. On most range ecosystems the climax species are long-lived perennials. Often these plants are poor seed producers and do not reproduce readily from seed. If the desirable species are depleted by drought or over-grazing, it is often difficult, if not impossible, to obtain recovery by manipulating the grazing. Therefore, it is extremely hazardous to deliberately over-stock an area during the growing season (Herbel, 1973a). The primary purpose of grazing management is to eliminate excessive grazing, especially in the growing season, in order to increase the vigour and productivity of existing plants (Hyder and Bement, 1977).

A grazing system must be highly flexible. Plant and animal requirements must be considered. For example, some of the range units in a ranch operation may be manipulated to furnish highly nutritious forage during the time of the year when livestock need a higher plane of nutrition. This may be done at a sacrifice of some of the "highly desirable" range species on these units. Further, it should be recognized that the critical growth stage of plants varies from year to year because of weather conditions. Because of grazing history and weather conditions, it may be more important to defer grazing in some years than others. Range units should be grazed when the key species are damaged least by grazing and when forage best meets the nutritional requirements of the livestock. This often means grazing on no pre-determined sequence (Herbel, 1973a).

Grazing systems should also be tailored to fit a variety of vegetation types, soil types and herd management plans. This means that there may be considerable variation in specific details from one ranch operation to the next. In some areas, continuous grazing may be the most profitable system. In some instances, it may be desirable to use an intensive system to attain a certain measure of improvement and then change to a different system for maximum net returns while maintaining the resource. The vagaries of weather, particularly drought, are common problems on rangelands and considerable flexibility must be built into any grazing plan, or for that matter, any activity on rangeland (Herbel, 1973a).

Seeding

Seeding semi-arid rangelands is generally a difficult undertaking because of limiting climatic, soil and/or topographic features. The good sites with a favourable climate are in cultivated crops. Therefore the task is to seed and grow range plants where even the hardiest crop plants are not productive and are difficult or impossible to establish.

Most range grasses should not be seeded deeper than 2 cm. Establishing seedlings is often difficult because of an adverse micro-environment (rapid drying, unfavourable temperatures and crusting of the soil surface). Harsh environmental conditions in the surface soil often prevent successful seedling establishment. Army and Hudspeth (1960) and Hudspeth and Taylor (1961) reported that sufficient moisture for seedling emergence could not be maintained on bare surface soil except under extremely favourable weather conditions in the Central Great Plains Winter Wheat and Range Region. Drier regions would have more intensive problems. The major objectives of preparing seedbeds for range seeding are: (1) to prepare a favourable micro-environment for seedling establishment; (2) remove or substantially reduce competing vegetation, and (3) if possible, leave litter on the surface of the soil to reduce erosion hazards and to improve the micro-climate (Herbel, 1972). Only a limited seeding success was obtained after pitting with a pitting disc and ripping (Thomas and Young, 1956; Dortignac and Hickey, 1963; Dudley and Hudspeth, 1964). Ripping lines often seal-over in a relatively short time. Narrow pits can fill with soil rather rapidly on some sites.

The broad shallow pits made with the basin-forming machine developed by Frost and Hamilton (1965) made a good seedbed and lasted longer than conventional pits. Over a 4-year period, average production of seeded *Cenchrus ciliaris* was 773 kg ha^{-1} on broad pits and 283 kg ha^{-1} on conventional pits on a site in the South-eastern Arizona Basin and Range near Tucson (Slayback and Cable, 1970). Pitting is generally most successful on medium to heavy-textured soils on flat or gently sloping sites.

Contour furrows form good seedbeds on medium to heavy-textured soils. It is desirable to use interrupted furrows to prevent a larger water loss if a furrow wall breaks and to preclude the necessity of furrowing exactly on the contour (Herbel, 1972). Firming the soil beneath the seed while planting was more successful than firming the soil surface following seeding. Firming the soil after seeding has the following disadvantages: (1) it may push the seed deeper than desired; (2) it may restrict aeration in the seed zone, and (3) it enhances crusting on some soils (Herbel, 1972).

A method of seeding arid and semi-arid areas infested with brush has been discussed by Abernathy and Herbel (1973). The brush and competing vegetation are controlled by a root-plough. Basin pits are formed, and about 40% of the area is seeded with a press-wheel seeder. An attempt is made to concentrate the brush and water on the seeded area. Brush control, pitting, seeding and brush placement are accomplished with one pass over the land (Fig. 5.10).

Concentrating water, as with various land-forming procedures, does not always ensure seedling establishment. The surface soil still dries rapidly, particularly in hot, arid and semi-arid areas. This rapid drying may lead to the formation of a heavy crust on medium- to heavy-textured soils. In those instances, if the surface could be shaded to reduce evaporation, seedling emergence and establishment would be greatly enhanced (Herbel, 1972).

Plant species used for range seeding vary with climatic and site conditions and management of a specific range unit. Improved eco-types of many grass species have been selected for: superior seedling vigour; drought, disease and insect tolerance; forage and seed production and the ability to reproduce vegetatively. Considerable use is made of seed harvests of native species. It is important to choose native eco-types of local origin, generally within 300 km north and 450 km south of the area to be seeded. Some of the major native and introduced species used for seeding in the South-west follow.

Fig. 5.10. An area dominated by *Larrea tridentata* (right) was root-ploughed and successfully seeded to *Eragrostis lehmanniana* and *Bouteloua eriopoda* (left).

South-western Plateaus and Plains Region: *Cenchrus ciliaris*, *Chloris gayana*, *Bothriochloa ischaemum*, *B. barbinodis*, *Bouteloua curtipendula*, *Schizachyrium scoparium*, *Panicum coloratum*, *P. antidotale*, *Leptochloa dubia* and *Stipa leucotricha.* Central Great Plains Winter Wheat and Range Region: *Andropogon gerardi*, *A. hallii*, *Schizachyrium scoparium*, *Bothriochloa caucasica*, *Buchloe dactyloides*, *Bouteloua gracilis*, *B. curtipendula*, *Eragrostis curvula*, *E. trichodes*, *Sporobolus airoides*, *Panicum obtusum*, *P. virgatum*, *Agropyron smithii* and *Stipa leucotricha.* Pecos-Canadian Plains and Valleys: *Agropyron smithii*, *Sporobolus airoides*, *Bouteloua gracilis*, *B. curtipendula*, *B. gracilis*, *Eragrostis chloromelas*, *E. lehmanniana*, *Agropyron desertorum*, *Atriplex canescens*, *Sporobolus airoides* and *Oryzopsis hymenoides.* Higher elevations of the Western Range and Irrigated Region, particularly following fires in the *Pinus* associations: *Poa ampla*, *Festuca arizonica*, *F. ovina* var. *duriuscula*, *Agropyron intermedium*, *Alopecurus pratensis*, *Bromus inermis*, *Dactylis glomerate*, *Lolium perenne* and *Phleum pratense.*

Control of unwanted plants

Undesirable plants can be controlled and ranges can be revegetated with forage species, but it takes judicious use of control methods and sound grazing practices. Woody plants such as *Prosopis*, *Larrea*, *Flourensia*, *Quercus* and *Juniperus* cannot be eliminated by good grazing practices alone. The brush has to be controlled before the range can benefit from other practices such as grazing management, seeding or water-spreading. The most effective method for control of woody plants depends on the site, the species and the degree of infestation. Any control of brush requires considerable attention to detail to obtain maximum benefits. Control of unwanted plants is generally less costly when invasion is just beginning and the plants are small and scattered. In this situation, a method is selected that will not destroy the residual forage plants. On sandy soils heavily infested with brush, a broadcast chemical method will control the undesirable plants and result in an increase of forage plants. Mechanical control methods are generally avoided on sandy soils because of the wind-erosion hazard if a good plant cover is not maintained. Fortunately, natural re-vegetation is often quite rapid on sandy soils after chemical control of the brush. A mechanical method of control accompanied by seeding may be required on soils with a medium to heavy texture, a heavy infestation of brush and a poor stand of desirable plants. Natural re-vegetation after brush control is often very slow on medium to heavy-textured soils (Herbel et al., 1974).

Controlling stands of mixed brush species with a single spray application is often difficult because species vary in degree and time of susceptibility to herbicides. To be effective, foliage applications of herbicides must be applied at the proper stage of growth. Treating *Prosopis* even 1 week too early drastically reduces the effectiveness of spraying with 2,4,5-T (Valentine and Norris, 1960). To achieve adequate initial control of many woody plants, two or more herbicidal applications are necessary. Two aerial spray applications of 0.6 kg ha^{-1} of 2,4,5-T, 1 to 3 years apart killed 23—64% of the *Prosopis* on sand dunes in southern New Mexico (Herbel and Gould, 1970). Perennial grass yields on areas sprayed twice in 1958—1961 averaged 234 kg ha^{-1} during 1963—1968, compared to 39 kg ha^{-1} on an adjacent unsprayed area. The sand dunes have levelled and there has been less wind erosion on the sprayed areas.

Herbicides such as 2,4-D, 2,4,5-T, silvex, dicamba and picloram control many plant species. The phenoxy herbicides are effectively used to control *Quercus* and *Artemisia* growing in the Central Great Plains Winter Wheat and Range Region. Mixtures of picloram and 2,4,5-T or dicamba and 2,4,5-T are sometimes used to control *Prosopis* in Texas, particularly where it is growing in association with other unwanted plants.

Individual plant treatments of dry herbicides are an effective and economic method of controlling sparse stands of brush. Monuron, picloram, dicamba and other herbicides are applied as powder, granules or pellets around the base of target plants (National Research Council, 1968).

Bulldozing and mechanical grubbing, root-ploughing, discing and cabling or chaining are the major mechanical methods of brush control on rangeland. Bulldozing is effective on sparse stands of many species. Bulldozer blades or front-end loaders may be fitted with a stinger blade which is pushed under the crown of the plant to ensure uprooting of the bud zone. Experienced operators can lift and push over a shrub or small tree in one operation (National Research Council, 1968). Bulldozing is used extensively to control *Juniperus* stands in the South-west. Grubbing with a stinger blade is used to control *Prosopis* growing on medium-textured soils in the Pecos-Canadian Plains and Valleys and the Southern High Plains.

A root-plough is a horizontal blade attached to a track-type tractor. Root-ploughing cuts off the shrub or small tree at depths of approximately 40 cm for *Prosopis* or other re-sprouting species and 15—20 cm for non-sprouting species. Root-ploughing kills 90% or more of all the vegetation growing on the area. The method is best adapted to dense brush areas having little or no residual grass and where seeding of desirable species is possible. It is used to control *Prosopis* in Texas (Rechenthin et al., 1964), *Larrea* and *Flourensia* in New Mexico (Abernathy and Herbel, 1973) and chaparral in central Arizona (Cable, 1975).

In discing, the brush is uprooted with a large disc plough or tandem disc. It is limited to small shallow-rooted plants like *Artemisia*, *Flourensia* and *Larrea*. It also destroys grasses growing on the area, so, like root-ploughing, it should only be used in areas where desirable plants can be established (National Research Council, 1968).

Chaining and cabling involve the dragging of an anchor chain or heavy duty cable, 90—120 m long, behind two track-type tractors (Fisher et al., 1959). The method is effective in controlling non-sprouting species like some *Juniperus* spp. Cabling is used to control *Opuntia imbricata* in New Mexico. It is useful in knocking down *Prosopis* trees previously killed by aerial spraying, thereby reducing the cost of working livestock. Trees left standing prevent the operator from seeing his livestock.

There is some research on biological control of unwanted plants in the area, but results to date have not justified field-scale applications in the South-west.

Burning as a tool

Where prescribed burning is an applicable tool, many objectives can be achieved simultaneously. Increased herbage yields, increased utilization, increased availability of forage, improved wildlife habitat (more food with unburned patches for cover), control of undesirable shrubs and control of various insects and diseases can all be achieved with one burn (Wright, 1974). However, much of the area discussed in this chapter rarely has enough fuel for widespread fires.

Box (1967) burned dense stands of *Hilaria mutica* in the Central Rolling Red Plains in the spring just when *Prosopis glandulosa* had begun growth. All the *Prosopis* were top-killed and about 10% were root-killed. Most of the trees re-sprouted from the base, but insects attacked the new sprouts and kept them eaten to the base. No insects were found on trees in the unburned plots. Similar results were obtained on *Condalia*. Box (1967) reported that a wildfire in the same area killed 26% of the *Bothriochloa saccharoides*, 84% of the *Aristida longiseta*, 32% of the *Buchloe dactyloides*, 48% of the *Panicum obtusum*, 40% of the *Setaria leucopila* and 72% of the *Sporobolus cryptandrus*. Although all plants were damaged, those high in the successional scale, *Buchloe* and *Bothriochloa*, were harmed least. Because of the variety of seasons and weather conditions, one must use care in interpreting the results of a burning study. If the data are from a wildfire, the fire probably occurred during a dry period. If the data are from a prescribed burn, the fire probably occurred when the weather and plant conditions were optimal. Fires during dry periods are harmful because they magnify drought stress on plants, whereas fires during wet periods are generally beneficial because moisture is not limiting and fires increase soil temperature and stimulate nitrification (Wright, 1974). As an example, *Hilaria mutica* produced 3151 kg ha^{-1} after burning during a wet year and only 700 kg ha^{-1} after burning during a dry year. The unburned controls produced 1263 and 1068 kg ha^{-1}, respectively (Wright, 1972).

In the Southern Desertic Basins, Plains and Mountains in southern New Mexico, Herbel and Nelson (1974) conducted a study to determine if *Hilaria mutica* could be grazed during the winter—spring when it is mostly dormant.

The old growth on *Hilaria* was burned in the summer after the first storm for 3 years. However, calf performance the following winter—spring was not improved by burning the previous summer.

Prescribed burning can be recommended for controlling *Aplopappus tenuisectus* in southern Arizona (Tschirley and Martin, 1961). However, Martin (1975) indicated that populations of *Aplopappus* tend to fluctuate without any control measures and therefore treatment could not be recommended. Martin further stated that prescribed burning cannot be generally recommended for the Central and South-eastern Arizona Basin and Range or the Southern Desertic Basins, Plains and Mountains. Fires that kill shrubs usually kill grasses too, and fires that spare the grasses may not kill the shrubs. Burning in that area usually favours annual grasses and pioneering perennial grasses at the expense of the climax perennial grasses.

Burning must be combined with other methods to control *Quercus* spp. in order to improve forage production in the chaparral portions of the Arizona Mountains (Cable, 1975). Five successive annual burnings were required to reduce the number of shrub *Quercus* sprouts below pre-burn numbers (Pond and Cable, 1960). Burning a former chaparral area in winter, 7 years after root-ploughing and seeding, effectively top-killed shrub sprouts but did not harm the grass (Pase, 1971). This type of burning, repeated periodically, might offer a method of suppressing shrub growth on areas where the chaparral has been controlled.

Burning of live, non-sprouting *Juniperus* trees is an effective, economical method of control in Arizona and New Mexico (Arnold et al., 1964). Where *Juniperus* is invading grasslands, trees up to 1 m high are easily killed and the herbaceous understory will carry a fire. Burning is also used after chaining, bulldozing or piling to get rid of unwanted slash and to kill missed plants. *Juniperus* trees in open stands can be individually ignited and burned by the use of propane or oil burners (Vallentine, 1971). The canopy is enveloped in flame by igniting the trees at the base. All the non-sprouting *Juniperus* trees were killed by scorching 60% of the crown.

The benefits of ground fires in the *Pinus* forests, dominated by *Pinus ponderosa*, in the Arizona and New Mexico Mountains are: (1) increased water production; (2) increased forage production; (3) providing space for hunting and camping; (4) preventing or reducing destruction of the harvest by wildfire, and (5) improving the aesthetic values (Vallentine, 1971). Humphrey (1962) concluded that the moderately open savanna stands of *Pinus*, essentially free of understory trees and shrubs, and supporting a good stand of grasses and forbs, has been maintained by lightning-caused fires. These recurrent natural fires prevented dense thickets of young *Pinus* and understory brush and kept fires from developing in the tree crowns by preventing large accumulations of fuel.

Management after a burn is essential (Wright, 1974). Grazing animals will frequently concentrate on a burn because the feed is more palatable, nutritious and available. Therefore, burning must be done on a manageable unit basis.

Fertilization

In some parts of the region, low amounts of available nitrogen (N) and phosphorus (P) in the soil limit plant growth. Fertilizing with N and P is economical only where there is adequate moisture and plant species that respond to the added nutrients.

In the Central Rolling Red Plains, forage yields of *Eragrostis curvula* increased about 50% with 34 kg N ha^{-1} (McIlvain and Shoop, 1970). A single application in April increased gains of yearling steers by 10%, carrying capacity by 25%, beef production per unit area by 31%, and profit per ha by 36%. *Eragrostis* fertilized with N was more palatable, stayed green longer at the beginning of droughts, made more rapid regrowth after grazing and produced more seed.

On rangeland dominated by *Bouteloua gracilis* in the Pecos-Canadian Plains and Valleys, 45 kg N ha^{-1} increased herbage production from 974 kg ha^{-1} on the control to 1487 kg ha^{-1} and 67 kg N ha^{-1} increased production from 974 to 1801 kg ha^{-1}. This same range fertilized with 45 kg N ha^{-1} produced steer gains of 54 kg ha^{-1}, whereas unfertilized range produced gains of only 26 kg ha^{-1} (Dwyer, 1971). The increase was due mainly to the greater number of yearling steers that could be grazed on the fertilized range.

On the Rio Grande Plain, P fertilization increased herbage yields and provided needed P to the grazing livestock (Vallentine, 1971). Calving was increased from 69 to 98% and the weight of calf produced per ha was approximately doubled.

On flood plains dominated by *Hilaria mutica* in the Southern Desertic Basins, Plains and Mountains, it was concluded that fertilization with N and P was uneconomical in all but the best moisture conditions (Herbel, 1963). In a year when favourable soil moisture conditions existed over a continuous 60-day growing period, 101 kg N ha^{-1} increased herbage yield from 3126 kg ha^{-1} on the control to 7790 kg ha^{-1}, but the benefits of annual treatment were small in two other years with above average rainfall. Some possibilities exist for using moderate rates of fertilizer to increase forage production on flood plains where this will permit longer growing season deferment of adjacent upland pastures.

On an area in the South-eastern Arizona Basin and Range dominated by *Eragrostis lehmanniana* and *Aristida glabrata*, check plots yielded an average of 2772 kg ha^{-1} (Holt and Wilson, 1961). Fertilization with 28, 56 and 112 kg N ha^{-1} increased yields 66 to 158%. Cattle preferred the fertilized forage and exhibited no preferences for grass species. On unfertilized areas cattle showed marked species preferences. The fertilizer extended the green-feed period up to 6 weeks.

Water-spreading

Water-spreading has two main functions: (1) increasing forage production by spreading of floodwater and the storage of water in the soil profile, and (2) reducing gully erosion and downstream flooding and sedimentation (Monson and Quesenberry, 1958). Stream channels, dry most of the time, that flow for only short periods after heavy rain storms, generally provide the water supply for water-spreading. The watershed area above the water-spreading site should provide at least one flooding over the site per year for satisfactory forage production and additional floodings each year are advantageous (Miller et al., 1969).

Frequent and heavy deposits of sediment interfere with the effectiveness of the spreader system. Such deposits retard plant growth and often kill younger plants (Stokes et al., 1954). Hubbell and Gardner (1950) found that sediments carried by floodwater had an adverse effect on the yield of all grasses in the study area in the San Juan River Valley Mesas and Plains, except *Agropyron smithii.* Of the other grasses studied by them, *Sporobolus airoides* was only slightly affected by sedimentation, but *Hilaria jamesii* and *Bouteloua gracilis* were readily killed by sedimentation. The additional water provided by the water-spreading substantially increased production of all the above grasses where sedimentation did not pose a problem to the particular species.

Game habitat

Properly planned and executed game range improvement projects may increase game production through improved quantity and/or quality of the feed supply. Thus, most range improvement aids in increasing game animals. Range improvement also aids where game animals have increased in numbers to the point where the forage supply is being exhausted or the range is in a deteriorating conditon. In the latter, an alternative would be to reduce the game herd to the carrying capacity indicated by the range condition (Lamb and Pieper, 1971).

Hamilton (1976) increased gross income from hunting leases by 300% from 1965—1975 on the Chaparrosa Ranch in the Rio Grande Plain. He found that spraying *Prosopis* in strip patterns maintained populations of deer while improving livestock production. Spraying an entire 1600 ha pasture drastically reduced the deer herd in that pasture.

The mixed conifer forest in the Arizona and New Mexico Mountains is used by elk (*Cervus canadensis*), deer (*Odocoileus hemionus*), turkey (*Meleagris gallopavo*), grouse (*Dendrogapus obscurus fuliginosus*), bighorn sheep (*Ovis canadensis*) and beaver (*Castor canadensis*) (Lamb and Pieper, 1971). Forage yields are highest in forest openings and these are preferred by the grazing animals (Reynolds, 1962). Intensive livestock grazing can cause deer and elk to avoid an area. The presence of some slash remaining from harvesting timber may tend to favour deer over cattle and therefore reduce direct competition (Reynolds, 1966). Light-to-moderate grazing by livestock apparently has little effect on big-game animals (Clary, 1975). Reductions of the timber overstory, whether to improve yields of water or timber, will virtually always increase herbage. Big-game numbers usually increase with the improved forage supply. Seeding after wildfire improves the habitat for both deer and elk (Kruse, 1972).

The *Pinus—Juniperus* zone is of critical importance to deer and is used to some extent by turkeys and elk (Lamb and Pieper, 1971). Range improvement work has included cabling, chaining, bulldozing, broadcast burning, single tree cutting, tree crushing and various combinations of these treatments. Any of these methods will improve a game range if properly designed. Cleared areas should be kept narrow so that deer using the openings are always close to cover. Ideally, the openings should not be over 400 m wide. The steep slopes and the rocky ridges with shallow soil should be left untreated. Thinning *Juniperus* trees where there are more than 370 ha^{-1} also improves game habitat. The control method chosen to improve forage for livestock should cause minimal damage to browse plants for wildlife (Paulsen, 1975).

The major game species in the Arizona chaparral is deer. Deer populations are relatively low where the brush is dense and herbaceous understory is sparse (Cable, 1975). Urness (1974) found that deer spent a quarter to half as much time on chaparral areas cleared by root-ploughing as in untreated brush. However, he suggested that the deer probably received much more benefit per unit time on the cleared areas because of the relatively high volumes of high quality forage and because deer spent their time on these areas exclusively for feeding while resting and ruminating on untreated areas. Management plans for deer on chaparral must provide browse and cover.

The principal game species on the South-eastern Arizona Basin and Range and the Southern Desertic Basin, Plains and Mountains are antelope, deer, quail and doves. Habitats for these animals can often be improved without reducing, or even while increasing, the livestock capacity of a range (Herbel et al., 1974). Where a mixture of shrubs and grasses grow, an occasional fire seems to improve the browse. Antelope and game birds prefer mixed grass—forb vegetation to brush types. On areas infested with *Larrea*, any treatment that increases herbaceous vegetation will help wildlife. Increasing the number of watering points where they are far apart will also benefit wildlife (Lamb and Pieper, 1971).

The major objective for game-range improvement is to break-up homogeneous vegetation patterns into heterogeneous vegetation patterns. This gives the variety of feeding, nesting and resting cover that is vital for wildlife. In some areas, heterogeneous vegetation patterns can be obtained by removing brush from the best sites and leaving the areas of poorest soils untreated.

In general, good range management practices and good wildlife conditions are highly compatible, but some modifications in some range practices may improve conditions for game (Lamb and Pieper, 1971).

Some examples of optimizing ranch production follow. In the Rio Grande Plain, *Prosopis* would be sprayed in strips and *Cenchrus ciliaris* seeded on the more productive soils. In the Edwards Plateau and the Texas Central Basin, the stands of *Juniperus* would be thinned by chaining, *Bouteloua curtipendula* would be seeded on selected sites and a grazing system would be initiated. On the Central Rolling Red Plains, practices would include spraying *Artemisia filifolia* and seeding about 10% of the ranch unit with *Eragrostis curvula*. On the Southern High Plains, *Prosopis* would be sprayed and *Hilaria mutica* would be burned. The *Juniperus* would be controlled on selected sites on the Pecos-Canadian Plains and Valleys, the Arizona and New Mexico Mountains, the New Mexico and Arizona Plateaus and Mesas and Colorado River Plateau and some areas would be seeded with *Agropyron smithii*, *Bouteloua gracilis* and *B. curtipendula*. Some of the practices in the Southeastern Arizona Basin and Range, and the Southern Desertic Basins, Plains and Mountains would include spraying the *Prosopis* and root-ploughing the *Larrea* and seeding with *Eragrostis lehmanniana* and *E. chloromelas*. The land manager should select those practices that increase productivity of the land while considering the total environment.

REFERENCES

Abernathy, G.H. and Herbel, C.H., 1973. Brush eradicating, basin pitting and seeding machine for arid to semi-arid rangeland. J. Range Manage., 26: 189—192.

Ares, F.N., 1952. Size and composition of the herd. Am. Cattle Prod., 34(7): 14, 16, 18.

Army, T.J. and Hudspeth, E.B., Jr, 1960. Alteration of the micro-climate of the seed zone. Agron. J., 52: 17—22.

Arnold, J.F., Jameson, D.A. and Reid, E.H., 1964. The pinyon-juniper type of Arizona: effects of grazing, fire and tree control. U.S. Dep. Agric. Prod. Res. Rep. No. 84.

Austin, M.E., 1972. Land resource regions and major land resource areas of the United States. U.S. Dep. Agric. Handb., 296.

Box, T.W., 1967. Brush, fire and West Texas rangeland. Proc. Tall Timbers Fire Ecol. Conf., 6: 7—19.

Boykin, C.C., 1967. Profitability and flexibility of two range cattle systems in the Rolling Red Plains of Texas. J. Range Manage., 20: 375—379.

Bryson, R.A., 1957. The annual march of precipitation in Arizona, New Mexico and north-western Mexico. Inst. Atmospheric Physics, Univ. Arizona, Tucson.

Buffington, L.C. and Herbel, C.H., 1965. Vegetational changes on semi-desert grassland range from 1858 to 1963. Ecol. Monogr., 35: 139—164.

Byerly, T.C., 1970. The environment and agriculture: issues and answers. Agric. Sci. Rev., 8(1): 1—8.

Cable, D.R., 1975. Range management in the chaparral type and its ecological basis: The status of our knowledge. U.S. Dep. Agric., For. Serv. Res. Paper RM-155.

Clary, W.P., 1975. Range management and its ecological basis in the ponderosa pine type of Arizona: the status of our knowledge. U.S. Dep. Agric., For. Serv. Res. Paper RM-158.

Clawson, M., 1972. Range management in the United States for the next one to three generations. J. Range Manage., 25: 328—332.

Correll, D.S. and Johnston, M.C., 1970. Manual of the Vascular Plants of Texas. Texas Research Foundation, Renner, 1881 pp.

Dortignac, E.J. and Hickey, W.C., Jr., 1963. Surface runoff and erosion as affected by soil ripping. U.S. Dep. Agric., For. Serv. (unserialized publication).

Dudley, R.F. and Hudspeth, E.B., Jr., 1964. Pitting and listing treatments on native shortgrass rangeland. Tex. Agric. Exp. Stn, Prog. Rep. 2313.
Dwyer, D.D., 1971. Nitrogen fertilization of the blue grama range in the foothills of south-central New Mexico. N. Mex. Agric. Exp. Stn., Bull. 585.
Fisher, C.E. and Marion, P.T., 1951. Continuous and rotation grazing on buffalo and tobosa grassland. J. Range Manage., 4: 48—51.
Fisher, C.E., Meadors, C.H., Behrens, R., Robinson, E.D., Marion, P.T. and Morton, H.L., 1959. Control of mesquite on grazing lands. Tex. Agric. Exp. Stn, Bull. 935.
Frost, K.R. and Hamilton, L., 1965. Basin forming and reseeding of rangeland. Am. Soc. Agric. Eng. Trans., 8: 202, 203, 207.
Fugate, F.L., 1961. Origins of the range era in South Texas. Agric. History, 35: 155—158.
Gould, F.W., 1975. Texas plants, a checklist and ecological summary. Tex. Agric. Exp. Stn MP-585.
Griffiths, D., 1901. Range improvements in Arizona. U.S. Dep. Agric. Bur. Plant Ind. Bull. 4.
Hamilton, W., 1976. The Chaparrosa — a study in resource integration. Proc. Symp. on Integration of Resources for Beef Cattle Production. Soc. Range Manage., Denver, CO, pp. 71—75.
Herbel, C.H., 1963. Fertilizing tobosa on flood plains in the semi-desert grassland. J. Range Manage., 16: 133—138.
Herbel, C.H., 1971. A review of research related to development of grazing systems on native ranges of the western United States. U.S. Dep. Agric., Jornada Exp. Range Rep. No. 3, 32 pp.
Herbel, C.H., 1972. Using mechanical equipment to modify the seedling environment. Wildland shrubs — their biology and utilization. U.S. Dep. Agric., Gen. Tech. Rep. INT-1, pp. 369—381.
Herbel, C.H., 1973a. Grazing systems on native range. Great Plains Agric. Coun. Publ. No. 63, Univ. of Nebr., Lincoln, pp. K-1—K-19.
Herbel, C.H., 1973b. Some developments related to seeding western rangelands. Range Research and Range Problems, Crop Sci. Soc. Am. Spec. Publ. No. 3, Madison, Wisc., pp. 75—80.
Herbel, C.H. and Gould, W.L., 1970. Control of mesquite, creosote bush and tarbush on arid rangelands of the South-western United States. Proc. 11th Intl. Grassl. Congr., Surfers Paradise, Queensl. Aust., pp. 38—44.
Herbel, C.H. and Nelson, A.B., 1969. Grazing Management on semi-desert ranges in southern New Mexico. U.S. Dep. Agric., Jornada Exp. Range Rep. No. 1.
Herbel, C.H. and Nelson, A.B., 1974. Utilizing tobosa (*Hilaria mutica* (Buckl.) Benth.) during the winter and spring. Proc., XII Int. Grassl. Congr. Grassl. Utilization, Part 1, pp. 254—259.
Herbel, C.H., Steger, R. and Gould, W.L., 1974. Managing semi-desert ranges of the South-west. N. Mex. Agric. Circ. 456.
Holt, G.A. and Wilson, D.G., 1961. The effect of commerical fertilizers on forage production and utilization on a desert grassland site. J. Range Manage., 14: 252—256.
Hormay, A.L. and Talbot, M.W., 1961. Rest-rotation grazing — a new management system for perennial bunchgrass ranges. U.S. Dep. Agric. Prod. Res. Rep. 51.
Hubbell, D.S. and Gardner, J.L., 1950. Effects of diverting sediment-laden runoff from arroyos to range and crop lands. U.S. Dep. Agric. Tech. Bull. 1012.
Hudspeth, E.B., Jr. and Taylor, H.M., 1961. Factors affecting seedling emergence of Blackwell switchgrass. Agron. J., 53: 331—335.
Humphrey, R.R., 1962. Range Ecology. Ronald Press, New York. N.Y., 234 pp.
Huss, D.L. and Allen, J.V., 1969. Livestock production and profitability comparisons of various grazing systems; Texas Range Station, Tex. Agric. Exp. Stn. Bull. B-1089.
Hyder, D.N. and Bement, R.E., 1977. The status of seasons of grazing and rest in grazing management. Second U.S. - Australian Range Workshop, Adelaide, pp. 73—82.
Johnston, M.C., 1963. Past and present grasslands of South Texas and north-eastern Mexico. Ecology, 44: 456—465.
Jurwitz, L.R., 1953. Arizona two-season rainfall. Weatherwise, 6: 96—99.
Kearney, T.H. and Peebles, R.H., 1969. Arizona Flora. University of California Press, Berkeley, 1085 pp.
Kothmann, M.M., Mathis, G.W., Marion, P.T. and Waldrip, W.J., 1970. Livestock production and economic returns from grazing treatments on the Texas Experimental Ranch. Tex. Agric. Exp. Stn, Bull. B-1100.
Kruse, W.H., 1972. Effects of wildfire on elk and deer use of a ponderosa pine forest. U.S. Dep. Agric. Rocky Mt. For. and Range Exp. Stn. Res. Note RM-226.
Kuchler, A.W., 1964. Potential natural vegetation of the conterminous United States. Amer. Geog. Soc. Spec. Publ., 36.
Lamb, S.H. and Pieper, R., 1971. Game range improvement in New Mexico. N. Mex. Inter-Agency Range Comm. Rep. 9, Las Cruces.
Leithead, H.L., 1974. High intensity—low frequency grazing increases livestock production. Proc., XII Int. Grassl. Congr., Grassland Utilization, Part I, pp. 355—361.
Martin, S.C., 1975. Ecology and management of south-western semi-desert grass—shrub ranges: the status of our knowledge. U.S. Dep. Agric. For. Serv. Res. Paper RM-156.

Martin, S.C. and Ward., D.E., 1970. Rotating access to water to improve semi-desert cattle range near water. J. Range Manage., 23: 22—26.
McIlvain, E.H., 1976. Seeded grasses and temporary pastures as a complement to native rangeland for beef cattle production. Proc., Sym. on Integration of Resources for Beef Cattle Production, Soc. Range Manage., Denver, CO, pp. 20—31.
McIlvain, E.H. and Shoop, M.C., 1965. Forage, cattle and soil responses to stocking rates and grazing systems on sandy rangeland in the Southern Plains. Abstr. 18th Annu. Meeting Am. Soc. Range Manage., pp. 31—34.
McIlvain, E.H. and Shoop, M.C., 1969. Grazing systems in the Southern Great Plains. Abstr. 22nd Annu. Meeting, Am. Soc. Range Manage., p. 21.
McIlvain, E.H. and Shoop, M.C., 1970. Fertilizing weeping lovegrass in western Oklahoma. Proc. First Weeping Lovegrass Symp., Noble Foundation, Ardmore, Okla., pp. 61—70.
Merrill, L.B., 1954. A variation of deferred rotation grazing for use under South-west range conditions. J. Range Manage., 7: 152—154.
Merrill, L.B., 1969. Grazing systems in the Edwards Plateau of Texas. Abstr. 22nd Annu. Meeting, Am. Soc. Range Manage., pp. 22—23.
Miller, R.F., McQueen, I.S., Branson, F.A., Shown, L.M. and Buller, W., 1969. An evaluation of range floodwater spreaders. J. Range Manage., 22: 246—257.
Monson, O.W. and Quesenberry, J.R., 1958. Putting flood waters to work on rangelands. Mont. Agric. Exp. Stn Bull. 543.
National Research Council, 1968. Weed Control. Sub-committee on Weeds, Natl. Acad. Sci. Publ. 1597, 471 pp.
Pase, C.P., 1971. Effect of a February burn on Lehmann lovegrass. J. Range Manage., 24: 454—456.
Paulsen, H.A., Jr., 1975. Range Management in the central and southern Rocky Mountains: a summary of the state of our knowledge by range ecosystems. U.S. Dep. Agric., Forest Serv. Res. Paper RM-154.
Paulsen, H.A., Jr. and Ares, F.N., 1962. Grazing values and management of black grama and tobosa grasslands and associated shrub ranges of the Southwest. U.S. Dep. Agric. Tech. Bull. 1270.
Platt, K.B., 1959. Plant control — some possibilities and limitations, I. The challenge to management. J. Range Manage., 12: 64—68.
Pond, F.W. and Cable, D.R., 1960. Effect of heat treatment on sprout production of some shrubs of the chaparral in central Arizona. J. Range Manage., 13: 313—317.
Rechenthin, C.A., Bell, H.M. Pederson, R.J. and Polk, D.B., 1964. Grassland restoration. II, Brush control. U.S. Dep. Agric. Soil Conserv. Serv., Temple, Tex.
Reynolds, H.G., 1954. Meeting drought on southern Arizona rangelands. J. Range Manage., 7: 33—40.
Reynolds, H.G., 1962. Use of natural openings in a ponderosa pine forest of Arizona by deer, elk and cattle. U.S. Dep. Agric., Rocky Mt. For. Range Exp. Stn Res. Note 78.
Reynolds, H.G., 1966. Slash clean-up in a ponderosa pine forest affects use by deer and cattle. U.S. Dep. Agric. Rocky Mt. For. Range Exp. Stn Res. Note RM-64.
Roberts, B.R., 1967. Non-selective grazing — an evaluation. Farmers Weekly (S. Africa): July 19, pp. 40—49; July 26 pp. 34—39; August 2, pp. 34—37.
Shelford, V.E., 1963. The Ecology of North America. University of Illinois Press, Urbana, 610 pp.
Slayback, R.D. and Cable, D.R., 1970. Larger pits aid reseeding of semi-desert rangeland. J. Range Manage., 23: 333—335.
Smith, H.N. and Rechenthin, C.A., 1964. Grassland restoration — the Texas brush problem. U.S. Dep. Agric., Soil Conserv. Serv. Temple, Tex.,
Stoddart, L.A., Smith, A.D. and Box, T.W., 1975. Range Management. McGraw-Hill, New York, N.Y., 532 pp.
Stokes, C.M., Larson, F.D. and Pearse, C.K., 1954. Range improvement through waterspreading. Foreign Operations Adm., Washington, D.C.
Stubblefield, T.M., 1956. Greater returns from cow-yearling operations on South-west ranges. J. Range Manage., 9: 8—10.
Thomas, G.W. and Young, V.A., 1956. Range pitting and reseeding trials on the Texas Range Station near Barnhart, 1950—1955. Tex. Agric. Exp. Stn Prog. Rep. 1882.
Thomas, G.W. and Ronningen, T.S., 1965. Rangelands: our billion acre resource. Agric. Sci. Rev., 3(4): 10—17.
Trewartha, G., 1961. The Earth's Problem Climates. University of Wisconsin Press, Madison, 344 pp.
Tschirley, F.H. and Martin, S.C., 1961. Burroweed on southern Arizona rangelands. Ariz. Agric. Exp. Stn Tech. Bull. 146.
Urness, P.J., 1974. Deer use changes after root ploughing in Arizona chaparral. U.S. Dep. Agric., Rocky Mt. For. Range Exp. Stn Res. Note RM-255.
U.S. Department of Agriculture, 1975a. Cattle. Crop Rep. Board, Stat, Rep. Serv. Washington, D.C.
U.S. Department of Agriculture, 1975b. Sheep and Goats. Crop Rep. Board, Stat, Rep. Serv., Washington, D.C.
U.S. Department of Commerce, 1975. Climates of the States, Vol. 2, Western States. Water Information Centre, Inc., Port Washington, N.Y., 975 pp.

Valentine, K.A. and Norris, J.J., 1960. Mesquite control with 2,4,5-T ground spray application. N. M. Agric. Exp. Stn Bull. 451.
Vallentine, J.F., 1971. Range Development and Improvements. Brigham Young University Press. Provo, Ut., 516 pp.
Wexler, H. and Namias, J., 1938. Mean monthly isentropic charts and their relation to departures of summer rainfall. Am. Geophys. Union Trans., 19: 164—170.
Wooten, H.H., Gertel, K. and Pendleton, W.C., 1962. Major use of land and water in the United States. U.S. Dep. Agric. Econ. Rep. 13.
Wooton, E.O., 1908. The range problem in New Mexico. N. M. Agric. Exp. Stn Bull. 66.
Wright, H.A., 1972. Fire as a tool to manage tobosa grasslands. Proc., Tall Timbers Fire Ecol. Conf., 12: 153—167.
Wright, H.A. 1974. Range burning. J. Range Manage., 27: 5—11.
Zimmet, S., 1966. The Public Domain. U.S. Dep. Interior.

Chapter 6

LAND-USE IN THE SOUTHERN MARGINS OF THE SAHARA

G.E. WICKENS * and L.P. WHITE **

* *Royal Botanic Gardens, Kew, Richmond, Surrey (Great Britain)*
** *General Technology Systems Ltd., Brentford, Middlesex (Great Britain)*

INTRODUCTION

Land-use along the southern margins of the Sahara is entirely dependent upon the vagaries of the climate. A good rainy season will result in some productivity in what might otherwise be regarded as uninhabitable desert. A poor rainy season will force the nomadic peoples of the region to seek their livelihood further to the south. It is a way of life adapted to the environment, and occurs along the entire southern margin of the Sahara, from the Atlantic Ocean in the west to the Red Sea in the east (Fig. 6.1). As already stated, its northern limit cannot be precisely defined since it is dependent upon the chance rainfall that makes survival possible. For convenience the southern limit is given as the 450 mm isohyet, which also approximates to the southward extent of the Pleistocene sandsheet and the Sahel Domain of the phytogeographers. This Sahel Domain, which is characterized by thorn scrub and thorn savanna formations dominated by *Acacia* spp., must not be confused with the rather vague and loose use of the term "Sahel" by many geographers.

Nomadic tribes, in order to survive, accept neither political, geophysical

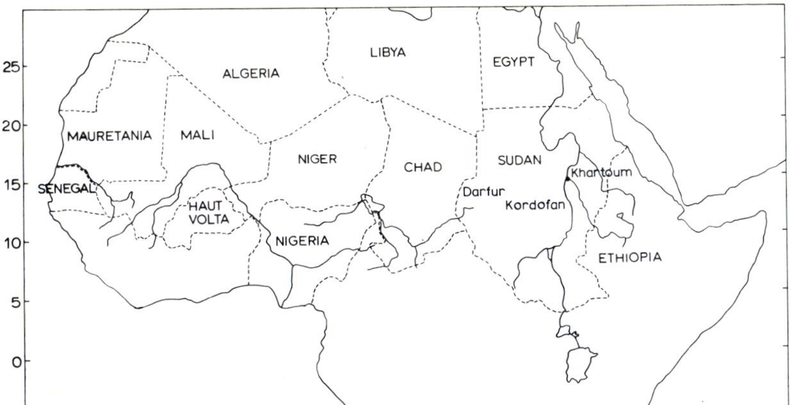

Fig. 6.1. The sub-Sahara political divisions.

nor meteorological boundaries and may habitually spend months of the dry season much further south; the arbitrary 450 mm isohyet boundary is purely for the convenience of the authors. Within the limits imposed by a single chapter it is only possible to give a general outline of the area and its utilization, and of necessity, much will have to be omitted.

ENVIRONMENT

Climate

Apart from some minor altitudinal effects, the climate in our zone of interest is very uniform for the full width of the African continent. It is dominated by the Inter-Tropical Convergence, where moist air from the Equator meets the dry air of the permanent high pressure zone over the Sahara. It is the seasonal shift of the convergence zone towards the north in the summer that brings what rain there is to the area. The overall quantity of rain from year to year depends on the latitude which the convergence reaches — the further north its penetration, the higher the rainfall in the zone as a whole. Unseasonal rainfall is rare (Fig. 6.2).

During the winter the region is dominated by hot, dry, north-easterly trade winds which often pick up large quantities of dust in the area itself and from the Sahara proper, carrying them on occasions as far as the Equator, and out into the Atlantic where it is detectable in ocean sediments as far as the coast of Brazil. (Rapp, 1974). These dust-laden winds, which may persist for weeks at a time, are known in West Africa as the "Harmattan". In addition to these very widespread movements of dust, the area also experiences more intensive sandstorms of shorter duration, particularly in the

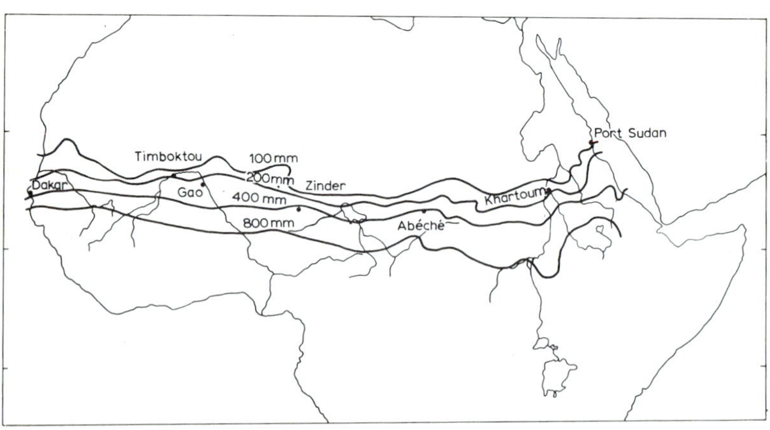

Fig. 6.2. Rainfall.

spring. These are known as "Haboobs" in the Sudan Republic.

The rains penetrating the area are the extremities of the south-westerly monsoons, arising in the Gulf of Guinea and the Congo Basin and extending as far east as the Ethiopian Highlands. They are extremely variable. Though the ultimate northerly position of the Inter-Tropical Convergence affects the overall rainfall in any season, there are still considerable local variations, particularly in the lower rainfall areas. In the wetter lands to the south there is a double peak of rainfall as the front associated with the convergence first moves north then returns southward at the end of the rainy season. In the north these peaks merge into one, providing a short rainy season — a mere 2 months at Timboktou.

Rainfall figures have little meaning on an annual basis. A year of almost no rain may be followed by a wet season during which there is the equivalent of several years of "average rainfall". Rain, when it comes, is usually in the form of brief and torrential convectional precipitation from line squalls and thunderstorms (Table 6.1).

Temperatures range from extremely hot in the summer to winter mildness, with no frost except at the very highest altitudes on the Aïr Plateau, the Ennedi, Tibesti and Jebel Marra massifs. The diurnal range is very wide, particularly in the summer months, where it is as much as 38°C in many places. Soil surface temperatures as high as 81°C have been recorded (Kendrew, 1961) but they fall off sharply with depth (Table 6.2).

The main effect of the climate on plants is one of desiccation. Most of the natural vegetation has no marked xerophytic adaptation and must rapidly use what moisture is available in order to mature (in the case of annuals) and set seed within a short space of time. Because in addition to being in short supply, the rain occurs in heavy, intermittent showers, where the soil surface is not mobile and is bare of vegetation, surface crusts commonly form. This in turn leads to high runoff, reducing the amount of water available for use by the plants in many situations.

The resulting soil erosion also seriously affects both natural vegetation and crops. Other adverse effects of the climate are deflation of loose surfaces by wind erosion and burial of plants by wind-blown sand and dust.

Apart from annual variations the area has historically been prone to

TABLE 6.1

Mean annual rainfall (mm)

Dakar, Senegal	516
Timboktou, Mali	285
Zinder, Niger	576
Abéché, Chad	488
Khartoum, Sudan	163
Port Sudan, Sudan	109

TABLE 6.2

Extreme temperatures (°C)

	Max.	Min.
Dakar, Senegal	42.7	11.6
Gao, Mali	47.0	9.0
Port Sudan, Sudan	47.0	10.0

drought and recovery cycles (Winstanley, 1976). The Sahel drought of the early nineteen-seventies thus appears to have been a single episode in the cycle rather than the manifestation of a long-term trend of climatic deterioration. It was exacerbated, however, by population and grazing pressures which, if they persist, will hold back the recovery of the vegetation. The relationships between the short-term cycles, measured in decades, and long-term arid to pluvial trends are now being investigated (Grove and Warren, 1968; Wickens, 1975). The effects are related to global weather patterns and the implications of the possibility of drought prediction to the management of the fragile ecosystems of the sub-Sahara fringe are considerable. A number of research programmes on climatology relevant to this problem are underway, promoted by international and major national agencies.

Geomorphology

Parts of five major geographic units occur within the sub-Sahara zone; the basins of the Senegal and Niger Rivers, the Chad basin, the Nile basin and the Red Sea littoral. All four of the basins extend well into the high rainfall lands to the south and there are considerable inflows of water into the area along the major rivers and their tributaries. However, with the exception of the Chad catchment, which is a closed drainage basin, the rivers flow through and out of the zone, loosing large proportions of their flow by evaporation and diversions for irrigation. On the Red Sea littoral the numerous short drainage systems are only active during the summer rains.

The present geomorphology of the zone has been largely shaped by alternations between extremely arid and pluvial conditions during the Pleistocene. At various times active dune fields have extended as far south as the present 1000 mm isohyet and have even blocked-off the courses of the Niger and Senegal rivers. Indications are that the present 150 mm isohyet has been 500 km further south than at present (Grove and Warren, 1968; Wickens, 1975). On the other hand, during pluvial periods, Lake Chad has been many times its present size, extending to include the present-day Bodélé depression to the north. Much of the now inactive or seasonal drainage network in the area has probably been perennial at some times in the past.

Geologically the whole area is underlain by the African Shield, composed of Pre-Cambrian Basement Complex rocks, mainly of granitic gneiss. Locally the shield is raised up into a number of highland blocks, which in the case of the Tibesti and Jebel Marra massifs incorporate large areas of more recent volcanic rocks. Post-Basement sedimentary rocks ranging from Paleozoic to Tertiary in age have developed to considerable thicknesses in down-warps of the shield. The Tertiary Continental Terminal Sandstones in Niger and the Mesozoic Nubian Sandstones and the Tertiary Um Ruwaba beds of the Sudan, in particular, are important as aquifers, holding very large quantities of exploitable groundwater.

The erosion products of the predominantly siliceous rocks give rise to generally infertile sandy soils.

Most of the land surface has been strongly influenced in one way or another by the previous, even more arid, climate of 22 000 to 12 000 years ago. Where water erosion has occurred it has been typically drastic, giving rise to deeply incised channels with unstable beds. Furthermore, the whole area has been overstepped a number of times by dune fields extending out of the Sahara. The most extensive of these incursions deposited a sand mantle from the Atlantic to the Nile and as far south in places as latitude 10°N. Though this sand sheet is now largely stabilized and colonized by vegetation, active dune forms are clearly recognizable in it. The present limit of the active sand sheet of the Sahara proper corresponds approximately with the 150 mm isohyet, though south of this, in many places, the stabilized sand

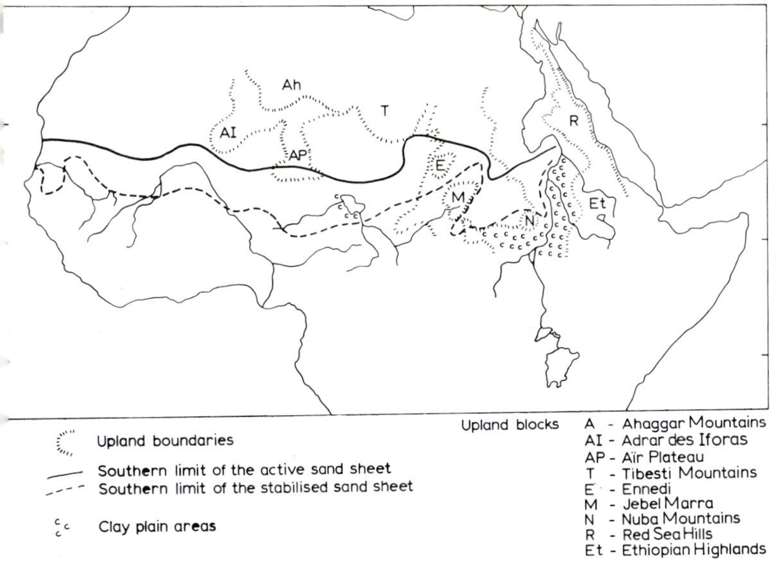

Fig. 6.3. Geomorphology.

sheet has been locally activated by pressure of grazing.

The approximate limits of the sand sheet are shown on Fig. 6.3. Of particular interest is the wind-shadow effect behind the Jebel Marra massif, which apparently obstructed the transport of sand by the wind from the north-east.

Within the limits marked, though the sand formations are very extensive, they are by no means continuous and large areas of older drift formations and rock may be exposed at the surface. South of the Atbara junction with the Nile and east of the White Nile, there is a very distinctive area of clay plains, where a thick mantle of dark clay originating from erosion of the predominantly volcanic Ethiopian Highlands has been spread over older formations. Extensive clay plains are also found to the west of Lake Chad, but on the eastern shore these have been largely over-stepped by a dune field, now stabilized, which also extends out into the present lake.

Active or recent alluvial formations are not extensive. For the most part the flood plains of even major rivers like the Nile and the Niger are relatively narrow, the clay plains pre-dating the present regime of the Nile. The situation with rivers rising in high rainfall zones and then passing into arid plains has in some places resulted in the splitting-up of the main river channels to form inland deltas. This is the case with the Niger below Segou in Mali, and the Gash which disperses itself almost completely in the region of Kassala in the eastern Sudan.

Soils

The characteristic features of soil formation are mechanical weathering and, owing to the high evaporation : precipitation ratio, the tendency for soluble salts to accumulate in the soil profile. Where mantle materials have been stable long enough for soil profiles to develop they are exclusively of the pedocal type. Leaching is practically non-existent except in periodically flooded soils. Although these soils tend to accumulate calcium carbonate and gypsum in their profiles, the development of massive and cemented calcrete horizons, common in the northern sub-Sahara zone where winter rather than summer rainfall is the rule, does not generally take place. The greatest concentrations of soluble salts occur in fine-textured soils in situations with poor drainage. These can strongly affect the composition of the natural vegetation and severely inhibit crop growth.

Two other factors typical of the soils are (1) a very low organic matter content, due both to low primary productivity and to rapid oxidation of residues, and (2) the formation of surface crusts (mentioned earlier) due to raindrop impact followed by baking.

There are four main types of soil in the zone.

(1) Aeolian sands are found extensively both as stabilized and still active sand sheets. They consist of more than 90% silica sand and show practically

no profile development apart from slight accumulations of organic matter in the A1 horizon.

(2) Residual skeletal soils occur on exposed rock surfaces where these are not overlain by later drift. Essentially they are shallow and gravelly with well-formed surface crusts. On sloping land they are particularly prone to high runoff, causing truncation.

(3) Recent alluvial soils range from those of wadi (river) beds and terraces, to old lake deposits or playas. Wadi deposits tend to be banded and very variable in texture, though predominantly sandy. Typically there is a great deal of variation in texture, both vertically and horizontally, in flood plain soils, and small areas of clay soil may be found in old channel depressions. Frequently these may still be swampy — at least seasonally. Old lake deposits, like those fringing Lake Chad, are predominantly fine in texture, highly calcareous, and contain appreciable quantities of evaporites — gypsum and other salts. In some localities they are both heavily saline and alkaline, with high concentrations of sodium chloride and sodium sulphate. Typically they develop very flat and finely cracked playa surfaces.

(4) Clay plain soils are particularly important in the Sudan Republic though different types occur throughout the zone. Their singular characteristics and earlier origin set them apart from the more recent alluvial soils. In the Sudan Gezira and adjacent areas they are deep accumulations of dark clay, similar to some heavy playa soils but occupying higher, and sometimes rolling land surfaces. They contain horizons of calcium carbonate concentrations and are markedly alkaline, containing appreciable quantities of sodium sulphate. Towards their northern, drier, limit they become increasingly more saline and alkaline and are useless for agriculture unless reclaimed and regularly leached. In drier situations they have massive or platy upper horizons, but towards the south, notably in the Gezira proper between the White and Blue Niles, they develop a natural deep, loose granular mulch on the surface. All clay plain soils, sometimes loosely referred to as "black cotton soils", are subject to deep cracking in the dry season. In the case of the Gezira soils, some of the surface mulch is washed into the subsoil. They are therefore self-churning as well as self-mulching. This is, in part, the reason for the success of the large-scale agricultural projects established in the Gezira.

In general, management for agriculture of the soils of the sub-Sahara zone may require stabilization where they are sandy, or leaching and drainage in the case of saline-alkaline clays. They may also require mulching to prevent erosion by wind or runoff, and to reduce the high potential evaporation from the surface.

Vegetation

As the rainfall increases progressively from north to south there are corresponding changes in the vegetation, with an increase in the density as well as changes in the composition of the flora. The following ecological formations are recognized within our zone of interest: desert, semi-desert, grassland and scrub, thorn scrub and thorn savanna, and savanna woodland (Fig. 6.4). In more general terms the semi-desert and thorn formations belong to what is often termed "Sahel savanna", while the broad-leaved savanna woodland is known as "Sudan savanna".

The boundaries between the formations are rarely well-defined, one division often merging almost imperceptibly with the next. The rainfall limits indicated below for these formations are only approximate, and mainly follow those of Harrison and Jackson (1958) for the Sudan Republic.

The desert is here regarded as being a landscape in which the vegetation is virtually absent except along the drainage lines, or consists of a temporary cover of ephemeral herbs and grasses (the famous "gizzu" grazing of northern Darfur and Chad) that spring up after the rare rain showers. The mean average rainfall is less than 75 mm. What little vegetation is available is barely sufficient to support a thin scattering of hardy nomads and their camels, goats and sheep.

The soil type has a considerable bearing on the species present within a given range of rainfall since it determines the availability of soil moisture (Smith, 1949). On sandy soils, with an annual rainfall between 75 and 250 mm, the major grasses present are *Aristida* spp. and *Cenchrus* spp., with the sand-binding *Panicum turgidum* on the less stable dunes. Where shrubs and trees are virtually absent this is regarded as semi-desert grassland, while a thin scattering of such woody species as *Acacia tortilis*, *Leptadenia pyrotechnica*, *Maerua crassifolia*, *Balanites aegyptiaca* and *Ziziphus spina-christi* are the characteristic constituents of the semi-desert scrub formation.

On clay soils *Cymbopogon* spp., *Sehima ischaemoides*, *Aristida* spp

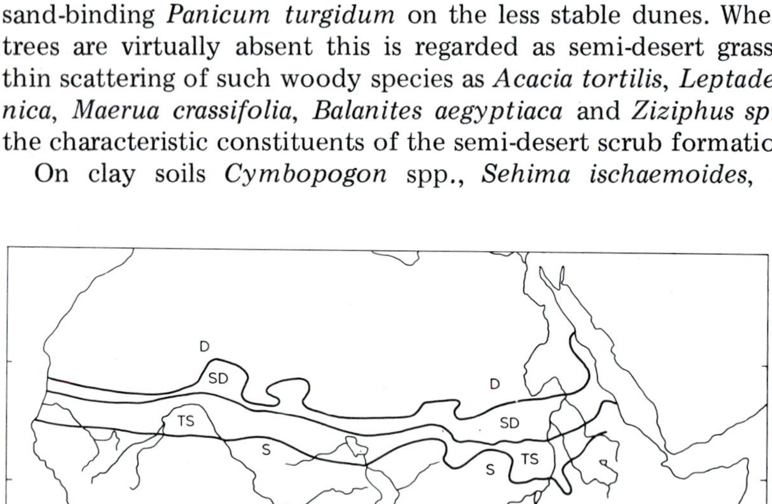

Fig. 6.4. Vegetation. D, desert; SD, semi-desert; TS, thorn scrub and thorn savanna; S savanna woodland.

and *Schoenefeldia gracilis* are amongst the more common grasses of the semi-desert grasslands within the 75—400 mm rainfall belt. Semi-desert scrub with a thin scattering of *Acacia mellifera*, *Commiphora africana* and *Boscia senegalensis* occur on both clay and skeletal soils under a rainfall ranging from 75 mm to about 250 mm. These semi-desert communities merge or interdigitate with the desert communities to the north and the thorn communities to the south, so that the boundaries between them must be regarded as somewhat arbitrary.

Acacia senegal is regarded as the characteristic tree of the sandier soils of the thorn savanna, where the rainfall ranges between 200 and 450 mm. *Aristida sieberana*, *Anthephora* spp., *Eragrostis tremula* and *Cenchrus biflorus* are amongst the major grasses present on these sandy soils.

On clay soils in the Sudan Republic, but not extending westwards into Chad, thorn scrub consisting of dense and often impenetrable thickets of *Acacia mellifera* is characteristic of the lower rainfall areas with an annual rainfall between 200 and 500 mm. *Aristida* spp., *Cymbopogon* spp. and *Schoenefeldia gracilis* are amongst the more important grass constituents. With increasing rainfall the thorn scrub gives way to thorn savanna, with *Acacia seyal* and *Balanites aegyptiaca* as the major tree constituents, and *Cymbopogon* spp. and *Hyparrhenia* spp. as the major grasses.

From about 600 mm rainfall, broad-leaved species such as *Combretum* spp., *Terminalia* spp., *Anogeissus leiocarpus* and *Sclerocarya birrea* become increasingly important. Thus the thorn savanna of predominantly *Acacia* spp. gives way to the broad-leaved trees of the savanna woodland, where *Hyparrhenia* spp., *Andropogon* spp. and *Pennisetum* spp. are amongst the more important grass constituents.

Fauna

The following is only a brief introduction to the varied fauna of the area. For more detailed information see Cloudsley-Thompson (1969).

(1) Insects. Insects are present in enormous numbers, with three orders being particularly well represented, viz. the Orthoptera (locusts and grasshoppers), Isoptera (termites) and the Hymenoptera (ants, bees and wasps). The locusts are of considerable economic importance and will therefore be treated in a little more detail than the other orders.

The desert locust (*Schistocerca gregaria*) is a potential pest over most of Africa north of the equator. A grown locust is capable of eating its own weight of food in a day. A large swarm will have eaten about ten times its own weight of vegetation while growing, and may travel up to 3200 km in a season. Large swarms may occasionally weigh as much as 50 000 000 kg (Cloudsley-Thompson, 1969).

Outbreaks are irregular owing to the existence of both solitary and gregarious forms, which differ both in behaviour and morphology. One or more generations of transitional forms link the two phases. The crowding of young hoppers will induce a trend to the gregarious and migratory phase, while a lack of crowding causes a reversion through the transitional phase to the solitary state. This behaviour pattern has enabled the migratory swarms to be controlled by either destroying large numbers of hoppers in order to reduce the population pressure, or by changing the ecological conditions in major outbreak areas in order to reduce crowding.

The migratory locust (*Locusta migratoria migratorioides*) was a major pest in most of Africa south of latitude 20°N during the years 1928–1942 (Cloudsley-Thompson, 1969). The discovery that the outbreaks errupted from a relatively small area in the flood plains of the river Niger in Mali has meant that the continuous monitoring of that and other areas, and the destruction of the swarms before they can escape, has prevented further major plagues.

Both the locust and the grasshopper are eaten by numerous animals, including snakes, birds, rats, jackals, baboons and even man.

Termite mounds are a conspicuous feature of the Sudan savanna. They are produced by relatively few species (principally *Macrotermes* spp.) and tend to be concentrated on the clay rather than the sandy soils. Some are entirely subterranean, while others inhabit excavations in dry or damp wood (Cloudsley-Thompson, 1969). Termite activity is responsible for a considerable amount of soil aeration and water penetration. They are also the cause of considerable damage to young trees, growing crops and timber constructions; the genera responsible include *Macrotermes*, *Odontotermes*, *Microcerotermes* and the harverster termite *Hodotermes*.

Ants are as abundant as the termites, and more widespread. Both are social insects, but many ants, unlike the vegetarian termite, rely on other insects for their food. Bees, wasps, hornets and dungbeetles are also fairly common.

(2) Arachnids. These include spiders, scorpions and ticks. The latter are of particular importance because of their transmission of the tick-borne diseases.

(3) Amphibia and reptiles. Amphibia are not as common in the savanna, except near permanent water, where they are able to breed. Both the desert fringes and the savanna provide an excellent habitat for reptiles, especially lizards and snakes. Snakes are mainly terrestrial and shy, they are not generally aggressive unless provoked. They are carnivorous, feeding on small mammals, birds, lizards, insects and eggs.

(4) Birds. Away from water the dominant birds are ground-birds, such as the ostrich (*Struthio camelus*), stone-curlews, coursers, pratincoles (*Glareola*

spp.) bustards, game-birds, sand grouse (*Pterocles* spp. and *Eremialector* spp.) and button quails. Predatory and scavenging birds include the vultures, eagles, falcons and the snake-eating secretary bird (*Sagittarius serpentarius*). Also present are swifts, hornbills (*Bucorvus abyssinicus*), barbets, night-jars, doves, pigeons, etc. (Cloudsley-Thompson, 1969). Large numbers of migratory Palaearctic species arrive in the autumn, either to over-winter in the area or passing through on their way further south before returning to Eurasia in the spring.

The most numerous savanna bird is the gregarious seed-eating black-faced dioch (*Quelea quelea*), which is regarded as a serious pest because of the damage it cases to ripening cereal crops — rice, millet, guinea corn and wheat. Manual bird-scaring to protect the ripening crops can amount to 30—50% of the total man hours required to produce a crop of millet (F.A.O., 1965). Feeding flocks can vary from less than a hundred to many thousands. The main feeding and nesting area of the dioch coincides with the distribution of *Acacia mellifera*, the bird's traditional nesting bush. The extent of the nesting and the intensity of the subsequent damage largely depend on the distribution and timing of the heavy rains and the presence of standing water, since the birds prefer to breed and roost in a thick cover near standing water (Dare, 1960).

There have been co-ordinated attempts to control this pest by countries in West Africa, as well as independent action in the Sudan Republic. Control may be effected by aerial spraying with fenthion (which is less toxic to other forms of wildlife than parathion), by explosives when roosting, or by the use of flame-throwers to destroy the nests (F.A.O., 1965).

Another familiar bird is the cattle egret (*Bubulcus ibis*), which is associated with all kinds of livestock, perching on their backs and flying down at intervals to either stalk just ahead or beside the grazing animal, capturing any insect disturbed in the grass. Flocks of these birds are to be found near most stock watering points.

(5) Mammals. Characteristic mammals of the Sahel savanna and the desert margins include the oryx (*Oryx tao*), addax (*Addax nasomaculatus*), ground squirrel (*Euxerus erythropus*), mongoose (*Herpestes icheunon*), gerbils (*Gerbillus* spp.) and the fennec fox (*Fennecus zerda*). A number of mammals commonly associated with the Sudan savanna migrate northwards during the rainy season into the Sahel, as and when grazing and water supplies become available. These include both grazing animals and their predators, the elephant (*Loxodonta africana*), giraffe (*Giraffa camelopardalis*), various species of antelope and gazelle, lion (*Panthera leo*), hyaena (*Crocuta crocuta*) and jackal (*Canis mesomelas*).

There is a considerable amount of evidence from the records of the early explorers, from Roman times onwards, to indicate that game are no longer as widely distributed nor in such large numbers as in the past. This is largely due to excessive hunting, together with a general deterioration of the habitat owing to an expansion of both animal and crop husbandry (Sidney, 1965; Wickens, 1975). Unless some measures are taken to control hunting it is likely that some species, e.g. the oryx, will become extinct in the not too distant future.

(6) The people. The history of the tribes occupying the Sahel region has a strong bearing on their present way of life. The former dominance of the Negro races, both in the Sahara and to the south has given way, under pressure, to the peoples of the Mediterranean brown race, who dominate North Africa and the Sahara. These are the peoples of the Berber language group, although most have now adopted Arabic as their native tongue and Islam as their culture.

The Arab conquest of Egypt and the Mediterranean Berber States, which took place during the 7th century, was rapidly followed by the domination of the then lucrative trans-Saharan trade by Arab-speaking peoples. By the 11th century the Mali Empire had accepted Islam, as did the later Kanem-Bornu Empire in West Africa. Likewise, the Kingdom of Kush and Nubia Christianity in the Nile valley gave way during the 14th century to Islam. Thus, by a pincer movement from the east and west, Islam and its culture invaded the southern fringe of the Sahara (Davidson, 1971).

The more important tribes in the zone we are considering are the Tibu, Bagarra, Gamala and the Fulani. The large and very important Hausa tribe in West Africa is largely to the south of our area.

The Tibu

Most of the Tibu have their homes along the wadis of the Tibesti Mountains, although some groups have settled as far afield as Kufrah and the southern Fezzan, the shores of Lake Chad and in Darfur. Members of a family may leave Tibesti during the winter months to wander with their camels and goats far south across Borku. If the grazing is good and the political situation favourable, they may remain in the Chad basin for several years, yet they will, during this time, continue to maintain contact with other members of the family left behind in Tibesti. Should the rains fail or the political situation deteriorate, the nomads will return to the fastness of their mountain home (Grove, 1970).

The origins of the Tibu are somewhat obscure; possibly it is with the mysterious Garamantes of the Roman era. What is established is that the mothers of many of the Bornu kings were, up to the 13th century, from the various nomadic Tibu tribes (Fisher and Fisher, 1974). The Tibu were certainly once notorious as professional bandits. Even today they cannot be regarded as the most peaceful of tribes.

The Baggara and Gamala

These are the cattle and camel-owning Arab pastoralists who live in the belt of country stretching from the White Nile westwards to Lake Chad. The cattle-owning Baggara are descended from the camel-owning Gamala, Muslim pastoralists who moved into the area some six centuries ago. They have interbred with the local negroid peoples and, as they progressively moved further south, changed from the herding of camels to cattle. Cattle are better suited to the wetter climate and heavier soils to the south of the Pleistocene sand sheet, although the predominantly zebu cattle of the Baggara are not fully adapted to conditions further south than approximately 10° latitude, and are brought north onto the better drained sandy soils during the rainy season, where the pestiferous biting flies (Tabanidae) are absent (see p. 230). This is believed to be one of the reasons why the Baggara have not penetrated further south with their cattle.

Both the Gamala, who inhabit the semi-desert regions, and the Baggara, who dwell in the savannas, practise a form of transhumance. During the early rains they leave their territory or dar in order to take advantage of the fresh grazing further north, returning to their dar at the end of the rains to harvest their crops. During the dry season they move south in search of better grazing and water, returning to their dar with the first rains in order to plant their crops. It is a system well adapted to the maximum utilization of low productivity grazing, and provided livestock numbers can be regulated according to the carrying capacity of the range the system works extremely well.

The Fulani

The people had their early home in Senegal, where the original Negro Fulani were infiltrated by northern Berber pastoralists, who inter-married and adopted the Fulani language. Because of insufficient grazing, the pastoralists were forced to move further south and west. During the past 200 years they have extended their range to northern Cameroon. Two types of Fulani are recognized today: a predominantly negroid group among the sedentary cultivators in the Senegal valley and a predominantly caucasoid group amongs the pastoralists (Ifemesia, 1965).

Their movements are comparable to the Baggara, grazing their cattle in the well-drained upland areas during the rainy season and descending to the riverine pastures during the dry season (Grove, 1970).

In recent years some tribesmen have emigrated to the Sudan Republic, where they are known as Fellata. One group has settled in southern Darfur, where they tend their crops and graze their Bororo cattle. Another group has remained wholely nomadic and penetrated even further eastwards, to the Blue Nile (Barbour, 1961).

General

In most years the rainfall within the zone is insufficient for the production of crops unless it is supplemented in some way. In the more southerly, wetter, fringes of the zone it may be possible to grow "catch crops" with low water requirements in unsupplemented sites, but more usually, cultivation of this type is restricted to sites receiving runoff from adjacent areas. Favoured sites of this kind may be wadi beds or other depressions. The fringes of rivers and lakes with seasonal fluctuations which leave expanses of moist, arable ground as the waters recede are also "favoured sites"; the "seluka" cropping of the Nile Banks which is of great agricultural importance in the central and northern Sudan is a typical example. Under these circumstances crop yields can fluctuate widely from year to year as they are directly related to both the timing and overall quantity of either the local rains or those in the distant catchment.

The crops grown are mostly grains of a subsistence economy rather than cash crops. Commercial agriculture in the Sahara fringes is heavily dependent on irrigation and a variety of methods are found, ranging from primitive "arrested runoff" techniques to modern, large-scale mechanized irrigation schemes (Fig. 6.5).

Arrested runoff techniques are mostly traditional in origin and aim at increasing the water available to the crop by concentrating local runoff onto particular sites. Simply, the idea is to manipulate the natural accumulation in depressions by intercepting and diverting drainage channels or sheet-wash.

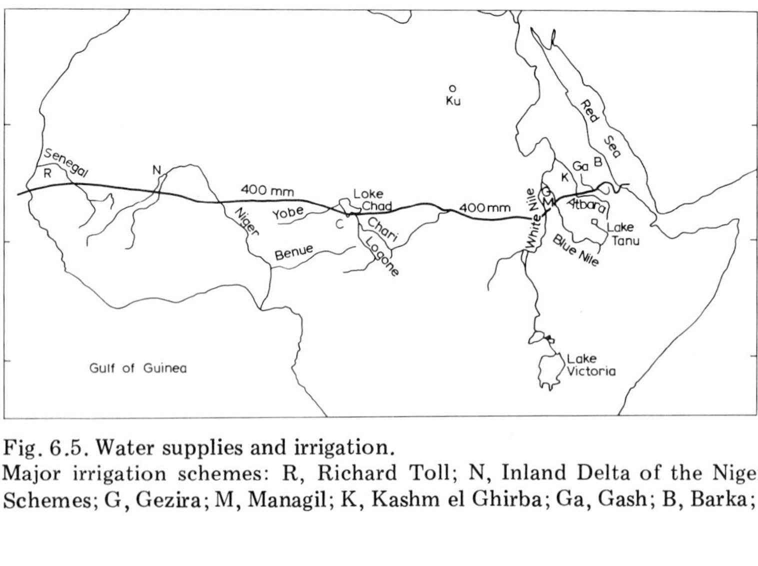

Fig. 6.5. Water supplies and irrigation.
Major irrigation schemes: R, Richard Toll; N, Inland Delta of the Niger; C, Lake Chad Schemes; G, Gezira; M, Managil; K, Kashm el Ghirba; Ga, Gash; B, Barka; Ku, Kufra.

A similar approach is the low earth banks, or bunds, constructed along the contours of some gently sloping lands to the east of the Nile. These hold back sheet-wash flow and create strips of moisture-charged ground behind them on which rapid maturing millet is sown.

An increased sophistication in water management techniques involving the diversion of wadis or the impounding of water of major river floods enables large-scale seasonal irrigation schemes to be operated. Examples of these are those of the Gash river in the Red Sea Hills and the inland delta of the Niger. Where the sometimes very large quantities of seasonal water can be adequately managed, large areas of cash crops such as cotton are grown using modern mechanized farming methods.

Irrigation from permanent rather than seasonal water supplies, enabling crops to be grown throughout the year, again varies from small-scale traditional methods to the largest of modern schemes. The most primitive method is the bucket on a rope used to obtain water from shallow wells to irrigate small vegetable gardens in oases, or along wadi beds where a fluctuating water table may still remain high enough to be within easy reach throughout the year. The bucket and counter-weighted pole — the familiar "shaduf" of Egypt — is common along the Nile and in oases. It is both a more efficient and permanent installation. Animal-powered lifting wheels such as the "saqiya" are used in some places, as are mechanical pumps.

Sources of permanent or semi-permanent irrigation water range from groundwater, which may be shallow or at considerable depths (or more rarely appearing at the surface as springs) or permanent surface water of which Lake Chad, the Nile, the Niger, the Senegal and some of their tributaries are the major sources.

Shallow groundwater is naturally the most easily exploited and is the source for most of the traditional systems of irrigation from wells. Hand-dug wells of 30 m or more are not, however, uncommon. These deep wells are constructed primarily for drinking water rather than for irrigation. In Kordofan Province of the Sudan Republic the annual cleaning of the very deep wells, which are constructed in only weakly consolidated sand, used to be carried out by men whose names were drawn by lot, and a precautionary collection was taken for their widows and dependents before they were sent down.

Shallow water tables in many areas tend to fluctuate, particularly those along seasonal water courses, and at the end of the dry season they may have dried up entirely. Under these circumstances the temporary wells and gardens watered from them are often made anew each year.

A variation of the single-shaft well is the "foggara", more commonly found on the north side of the Sahara but also found in the central desert — for instance in the Fezzan of southern Libya — and in the southern fringes. A foggara is a gently sloping tunnel, typically dug back into an alluvial fan to intercept a sloping water table at a point higher up than its level in the lower part of the fan. If the up-slope level is high enough the tunnel may emerge at the surface on the lower fan and a small irrigation canal led from it. In other circumstances the water is collected in a terminal well. Foggaras, of which there are many ancient and abandoned examples, can be traced by the lines of mounds of soil thrown out of the vertical maintenance shafts, which are usually spaced every 10 to 20 m along the line of the tunnel. The characteristic pattern formed by these lines and mounds can be clearly seen from aerial photographs, indicating that in many areas there was much more agricultural activity than at present. As with deep hand-dug wells, the effort and danger involved in their construction and maintenance has led to their decline in the face of mechanical well-digging methods and water pumps which are now available.

Modern drilled wells and pumps are replacing traditional systems, and producing much larger yields of water. They can tap much deeper sources of groundwater, some of which are very large indeed. Some of the groundwater is artesian, much is also "fossil" in that it is water trapped in the formations many eons ago and it is not being replenished (recharged) from the present-day surface sources. The danger with these groundwater reservoirs is that where the extraction exceeds the recharge they are in effect being "mined" and will eventually be exhausted. The Kufra irrigation schemes in southern Libya provide a good example of such exploitation, with 10 000 ha devoted to fodder production for livestock breeding and fattening, and a further 5000 ha of small farms. The necessity for a thorough evaluation of the aquifer characteristics before commencing large schemes is well illustrated by the fact that the water table was lowered by 15 m after the first year, as opposed to an estimated maximum decline of 35 m after 40 years (Allan, 1976). With the very large expansion of groundwater exploitation since the nineteen-fifties, problems of aquifer management are indeed becoming increasingly important and approaches to this include artificial recharge.

As we have seen, there are only a few sources of surface water capable of supplying the needs of large-scale irrigation schemes. These are, from the west to east; the Senegal and Niger rivers, Lake Chad, the White and Blue Niles, and the Nile tributary, the Atbara. The availability of water in a zone where frost is virtually unknown means that a wide range of tropical crops can be grown, as well as adapted temperate crops in the winter months. The very large quantities of water available from the individual sources also means that large-scale production of cash crops is possible.

Huge irrigation schemes such as that of the Gezira in the Sudan Republic date from the nineteen-twenties and they have been steadily expanded, with very rapid growth since the nineteen-fifties. Essentially they are highly mechanized schemes established on virgin lands and they show little similarity to the traditional native agriculture. They are usually geared to the production of export crops and are thus major earners of foreign currency.

As such they are very important in the economies of a number of Sahel countries.

Because the irrigation water originates from the southern monsoon rainlands, unlike some waters from groundwater reserves, the quality is excellent with very low concentrations of dissolved salts. The lands irrigated, however, are often saline or alkaline, requiring initial reclamation by leaching out the salts and subsequent careful management to prevent their re-appearance in the rooting zone. In many schemes the drainage systems to remove surplus salts are as important as the irrigation network.

A general problem in agriculture throughout the zone as a whole is the damage which can be caused by the desert locust (*Schistocera gregaria*). Locust outbreaks are cyclic and damage has been very severe in the past but it has now largely been brought under control through international efforts (Hemming, 1974). Loss of grain to birds can also be serious. Large flocks of the black-faced dioch (*Quelea quelea*) for instance, can be very destructive throughout the Sahel and have required major control measures.

Crops

Dry-land cropping is only possible along the southern fringe of the zone where the rainfall is in excess of about 300 mm. The main crops, sorghum (dura) and millet, grown under these conditions have short summer growing seasons. Millets of the bullrush type (*Pennisetum typhoideum* varieties) can get by with less rainfall than dura (*Sorghum bicolor*) and will grow successfully on sandy soils. Both of these are staple crops across the full width of the continent, but in West Africa "hungry rice" (*Digitaria exilis*) is also a major subsistence crop.

Under irrigation, practically the whole range of tropical field crops can be grown. The most extensively grown grains are sorghum and millet, mostly for local consumption but large quantities are exported from the Sudan as cattle feed; it is also proposed that sorghum will be the main feedstock for a new starch-producing industry. Rice (*Oriza sativa*) is also a major crop with much of it being exported. In the north of the zone, in oases and particularly along the Nile, Mediterranean wheat and barley are grown in the winter. A major effort is now being made in the Sudan Republic to convert a large part of the Gezira production from cotton to wheat to attain national self-sufficiency and an export surplus.

The premier irrigated export crop throughout the zone is cotton. It is the main crop of the Gezira and the neighbouring schemes in the Sudan, but it has been less successful in West Africa, possibly because the soils are less suitable. Tobacco, groundnuts and oil seeds are other large-scale irrigated field crops which are important as earners of foreign currency.

Though most citrus crops can be grown in the zone, apart from dates in the north, fruit is of little commercial importance. Date orchards are important to the local economy in the oases and along the Nile but they will not fruit successfully where there is significant summer rainfall (Cobley, 1956).

Agricultural practices within the zone of 150 to 400 mm rainfall are broadly similar with local variations in response to the varying land and water situation. From west to east the main features in each country are as follows.

Mauretania. Most of Mauretania lies in the true Sahara where only oasis-type agriculture is possible. In the south-west however, a strip of land — the Chemama — along the Senegal river, receives a rainfall of up to 450 mm and flood irrigation from the river allows some cultivation, predominantly of sorghum, millet and groundnuts.

Senegal. In the northern part of Senegal where the rainfall is less than 400 mm the strip along the Senegal river — the Fouta — is the equivalent of the Chemama in Mauretania. A barrage on a branch of the river near the town of Richard Toll provides water for a large irrigated rice scheme.

Mali. The main agricultural area is the inland delta of the Niger river and this is well within the sub-Sahara zone. Traditionally a natural flood irrigation area and an important fishery where schemes have been put in hand over the past 40 years to control the waters and expand the irrigated areas with dams, barrages and canalization. These were originally French colonial schemes to provide exports of cotton, but much of the land is now used for rice cultivation, which combines well with fish culture. Lower down the river, rice, tobacco and the ubiquitous groundnuts are grown on riverside lands inundated by the summer floods.

Niger. A similar situation to that in Mali exists along the Niger river. As the Republic of Niger is, however, more southerly, some grains and groundnuts are grown on rainfall alone. The bulk of the groundnut production, however, which has been the mainstay of the country's economy, occurs, as elsewhere, south of the 600 mm isohyet. Favoured sites for cultivation away from the river are found in the "Dallols" or wide valley floors of the now inactive drainage system. Here, the groundwater is in places close to the surface, allowing small-scale irrigation of garden crops. The exploitation of what is often artesian groundwater from an extensive aquifer is encouraging considerable expansion of agriculture in some areas.

Nigeria. Only the extreme north-east of the country adjacent to Lake Chad can be said to fall strictly within the semi-arid zone. The scope here is mainly for growing grain crops on lands left moist by the seasonally receding lake, and a technique has been developed of building low "bunds" to hold back part of the water to grow sorghum and other crops during the winter months. Under this system the moisture retention capabilities of the soils are crucial in the selection of lands to be cultivated. At the present time a number of large-scale irrigation schemes relying on water pumped up from the lake are being established, largely on clay plain soils. Fishing is a major industry in Lake Chad.

Chad. A similar situation to that in Nigeria pertains along the eastern lake shore, though the area of clay plain soil that can be utilized is less extensive. Away from the lake are very large tracts of land with no surface water, except in the wet season, which support only rain-fed cultivation of drought-resistant crops.

Sudan. As far east as the Nile the conditions are similar to those in Chad with scattered villages dependent on deep wells or open cisterns, "hafirs", during the dry season. Here the growing and stock-piling of watermelons as dry season fodder for livestock is practised in addition to the usual sorghum and millet cultivation. Because of their greater ease of working with primitive tools, sandy soils are usually preferred by the predominantly Arab cultivators. Migrant Negro groups however, sometimes work the heavy clays found in inter-dunal hollows, which have better moisture retention characteristics and which are generally more productive.

Along the Nile the considerable stretches of alluvial soil have for centuries been the basis of a flourishing system of mixed irrigated agriculture similar to that of Upper Egypt. Not only are extensive date orchards and small farms irrigated by animal and water-powered devices, but large areas of crop-land are watered in natural flood basins into, and out of which, the annual flood can be led by canals. Crops are also grown on the riverside lands exposed by the falling waters in winter. The introduction of motor pumps during this century has greatly increased the potential of land under cultivation, but the traditional patterns largely remain and there is now a serious problem of lack of land and over-population. A recent development has been the flooding of large areas to the north, in Nubia, by Lake Nasser which has been formed behind the new Aswan High Dam in Upper Egypt. The agricultural population from these areas has been removed to irrigation schemes further south on the Atbara.

The riverine pattern of cultivation is confined mostly to the north of Khartoum and is strongly oriented towards the production of dates, wheat and barley, though there is considerable commercial-scale cultivation of cotton, vegetables and oilseeds.

On the clay plains south of Khartoum is the best known of all irrigation schemes in Africa — the Gezira. It was started in the nineteen-twenties on virgin, practically unpopulated land between the Blue and White Niles ("Gezira" means island). Initially some 120 000 ha were irrigated from a barrage built at Sennar on the Blue Nile, holding back water from the summer floods originating in the Ethiopian Highlands. There has been steady expansion and the irrigated area is now over 800 000 ha; water is also supplied from the new Roseries dam, further upstream, completed in 1966. The scheme has involved the introduction of large numbers of settlers and at cotton-picking time there is also a huge influx of temporary labour from all over the country, but even so there is usually a labour shortage at this time in contrast to the general labour surplus for the rest of the year.

The Gezira scheme and its Managil extension are primarily devoted to the production of Egyptian long-staple cotton, which has been the mainstay of the economy of the Sudan. It is produced by tenant farmers in family-sized holdings which vary between 6 and 16 ha in size from scheme to scheme. Water and cultivation services are provided by the managing board which was originally a joint government and private venture. Only about half of each holding is put down to cotton at any one time, the rest growing sorghum and fodder for domestic and local consumption. Oilseeds, including groundnuts, are grown as additional cash crops. It is interesting to note that groundnuts, which prefer sandy soils, are here grown on heavy clays. The factor which favours this is the natural deep granular mulch, referred to earlier.

A recent switch from cotton to large-scale wheat growing has been primarily due to government reluctance to depend too far on one crop. This policy is part of the recent remarkable expansion and diversification of agriculture in the Sudan as a whole, largely funded by Arab oil money.

Apart from the Blue Nile, the Barka, Gash and Atbara rivers supply large irrigation schemes. The Barka scheme, centred on Tokar on the Red Sea coast, and the Gash scheme, centred on Kassala in the Red Sea Hills, depend on the diversion of summer flood waters. These can vary greatly in volume, which affects the total amount of land that can be irrigated from year to year. The main crops are cotton, oilseed and small grains, which are produced in quantity for export. The new scheme at Kashm el Ghirba on the Atbara river has been developed for the displaced Nubian population and involves about 20 000 ha of irrigated land. The scheme, started in the nineteen-sixties, has been initially geared towards sugarcane as its primary crop.

East of the Nile a rather unusual practice is followed on the clay plains where the land is gently sloping. Quick-maturing millet is grown behind low embankments constructed to intercept sheet-wash runoff following heavy storms. The grain is planted immediately following a storm of sufficient intensity. The crop grows and matures in 80 days, using the moisture that has been induced to infiltrate the soil behind the bunds.

THE ECOLOGICAL IMPACT OF AGRICULTURE

In the traditional land-use sector, the sparse precipitation confines what rain-fed crops there are to "favoured" sites having sufficient additional water concentrated on them and enough moisture retention capability to support a fast-maturing crop. In many cases the crop relies upon only one effective injection of water into the soil. In practice the technique of cultivation is by pre-irrigation, as the cultivators will not normally hazard the seed in the ground until a substantial wetting of the soil has occurred. The first heavy shower of the rainy season is often the occasion of frantic activity.

This form of agriculture is of necessity static, as it is tied to the "favoured" and traditional sites. Though they may not be cultivated in years of unusually low rainfall, the same sites are used more or less continually, in marked contrast to the traditional shifting cultivation pattern of most of Africa, including the savanna lands to the south.

Therefore, once cleared, the natural perennial vegetation will have no chance to re-establish itself. The natural vegetation becomes dominated by annual herbs and grasses. Shade trees providing useful fodder, like *Balanites aegyptiaca*, *Acacia tortilis* and *Adansonia digitata*, may be left as are other large trees which may be too difficult to remove. The number of residual trees however, tends to diminish, as they are rarely, if ever, deliberately replaced, and natural regeneration is prevented by cultivation and dry-season grazing. Trees are also heavily abused by lopping for firewood or fodder. Some may be deliberately or accidently destroyed by fire as it is common practice to heap weeds and crop residues against their bases for burning. Others may be deliberately ring-barked. The reasons given vary, but in some places it is claimed that it is done because the trees provide nesting sites for birds that damage the crops. Such reasons are almost certainly not sound and clearly the continual destruction and non-replacement of the trees leads to a general deterioration of the environment as well as the loss of a secondary resource.

As on the north side of the Sahara, these hydrologically favoured sites also carry some of the best seasonal grazing. Cultivation is thus, to some extent, in conflict with the interests of the independent graziers. In general, with the expansion of the population and the introduction of tractors which can be driven out to cultivate distant sites, these pressures are increasing. However, cultivation does not necessarily mean occupation throughout the year and for this reason the adjacent vegetation may not necessarily suffer the severe degradation associated with the permanent settlements.

Systematic irrigation, on the other hand, demands permanent occupation, of the oasis-type where it is based on wells, or the more dispersed homesteads or hamlets where canal irrigation is involved. As we have mentioned, oasis settlements are usually surrounded by zones of intense local desertification due to unrestricted free-grazing by the animals of the occupants and by those of nomads visiting the wells in their traditional migrations. The diameter of the zone of devastation is, however, often restricted to the distance the sheep and goats can walk out and return in a day. The heavily degraded vegetation of the immediate surroundings is in marked contrast to the heavily protected gardens of the oasis which, as well as containing many deliberately and accidentally introduced exotics, may also have well-grown examples of native species.

New, large-scale irrigation schemes mean that large populations are suddenly introduced into previously sparsely populated areas. Though degradation of the surrounding natural vegetation has occurred settlers in well-planned schemes are, by design, usually too heavily occupied with their irrigated crops to be concerned with ranging large numbers of cattle outside the schemes. Further, fodder for enough animals to meet the purely domestic requirement is often grown as part of the "official" crop rotation.

The sudden modification of sometimes very large areas to what is, in effect, riverine conditions, obviously has important ecological effects. The introduction of exotic weeds is a common occurrence, and sometimes a major problem, but local species also flourish on odd patches within the schemes such as canal banks and roadsides. Water hyacinth (*Eichhornia crassipes*) which is a major problem in many areas, is usually kept out of the schemes by vigilance. The broader, long-term effects of establishing large, moist, well vegetated areas with increasing numbers of "amenity" trees has its most obvious impacts on animal ecology and the human environment. Similar schemes, such as those in Sind in Pakistan, have produced a marked reduction in dust storms over the years. On the other hand, schemes in Africa have resulted in very large increases in the incidence of water-borne diseases such as bilharzia and guinea worm. It has been found that an emulsion of the fruits of *Balanites aegyptiaca* is lethal to both the freshwater snails and water fleas that help transmit these diseases (Suliman and Jackson, 1959). Thus, growing *Balanites* along the banks of irrigation channels where their fruits can fall into the water, could help to reduce the incidence of the disease.

Other environmental modifications, such as reduced temperatures and increased humidity could possibly affect the environment of the surrounding regions, but any such effects would probably be masked by natural climatic fluctuations. The full impact resulting from the introduction of large-scale irrigation, depending for the present upon imported skills and equipment, and the consequences of a complete break with traditional patterns of land-use, have yet to be realized. It should be pointed out however, that irrigation schemes in arid and semi-arid areas which rely on water from higher rainfall areas do not provide the most efficient use of water because of the higher water requirement and higher evapo-transpiration losses as compared with more humid areas.

The boundary of the sub-Sahara zone with the zone immediately to the south, where rain-fed agriculture can be practised on most sites where soil and landform are not limiting factors, being determined by the 400 mm isohyet, is not of course clear-cut. For this reason, the severe problems of the zone to the south resulting from over-population, over-cultivation and grazing, and in particular, the increasing inability to practise the traditional shifting cultivation or bush fallow rotations because of lack of land, overspills the southern fringe. The results of this are a widespread loss of natural fertility due to over-cropping using methods that apply little or no fertilizer. In the case of sandy soils there is a tendency for the surface to become destabilized, effectively re-activating the old dune sheets, with very severe deterioration of the environment as a whole. However, the climate itself acts as a regulator and the wiping out of marginal agriculture over large areas for several years by the drought of the early nineteen-seventies may, in fact, have given some chance of recovery to the natural vegetation, though the more vulnerable species themselves will have suffered damage.

It is not by accident that the traditional pattern of land-use is based on an inter-dependence of pasturalism and cultivation. In an area of marginal rainfall transhumance and the ability to migrate is essential for survival in the long-term drought and recovery cycle which prevails (Winstanley, 1976). The increase in the human and animal population of the last few decades, and the development of new political groupings, has severely strained the traditional patterns, leading in some areas to possible irreversible ecological damage. As a reaction to this a number of national and regional programmes to counter desertification have been implemented in the nineteen-seventies by governments and international agencies such as the F.A.O. and U.N.D.P. Though the subject has been of serious concern to those involved in the development of the Sahel since the nineteen-forties, large-scale concerted action has only now been started and in most cases practical results have yet to be seen. Major modifications to the land and its vegetation are continuing to take place and, like much of the world in the present day, the dry margin of the Sahel is in a state of flux. It remains to be seen in the next few decades if a non-destructive and continuously productive *modus vivendi* can be devised.

Livestock

The important domesticated animals in the area are camels, sheep, goats, cattle, horses and donkeys. Of these, the first four may be used for human consumption; all, except the sheep and goats, may be used as beasts of burden.

There are three distinct systems of livestock management practised in the area: sedentary, transhumance and nomadic. Changing conditions may induce individual herdsmen, for short-term management considerations, to change from one system to another. Under the sedentary system animals are kept at a permanent base throughout the year. The number of animals is limited to the carrying capacity of the adjacent dry-season grazing (including crop residues). Should this number be exceeded, the surplus stock are usually placed in the care of a migratory herdsman, either belonging to the owner's family or to a traditionally nomadic tribe.

The transhumance system also recognizes a permanent base, but the stock numbers and/or the environmental conditions are such that for part of the year grazing has to be obtained elsewhere. The older nomadic system recognizes no such permanent base, and the stock owners wander freely in search of the best available pastures for their stock.

(1) Camels. The one-humped, Arabian camel or dromedary is truly a beast of the desert or semi-desert. Very little is known about its early domestication, but its general use as a beast of burden seems not to have occurred until about 20 A.D. (Tothill, 1948). Two types of camel are recognized; firstly, the heavier pack animal capable of carrying a load of 250 kg some 25 km during a day. Local strains or "breeds" are recognized with different endurance capabilities; secondly, the lighter riding camel, which has been bred for speed at the expense of weight-carrying capacity and staying power. A fast dromedary can achieve a speed of 10 km h^{-1}, but not for very long. The normal speed is nearer 6 km h^{-1}, double that of the pack animal (Cloudsley-Thompson, 1969).

During the winter the camel-owning nomads of the Sudan Republic and Chad, the Gamala tribes, settle in their homeland or dar. At the start of the rains the camel herds move south to take advantage of the flush of new growth. Once the grazing is sufficiently available, the direction is reversed and the camels are taken as far north as possible, beyond their homeland. In some areas there may be no water, apart from the supplies carried by the camel owners, and the camels rely entirely on the moisture that they can obtain from the vegetation. Eventually they return to their homeland until the cycle is repeated the following year (Hassan, 1962).

Camels are essentially browsing animals, relying on the leaves of the trees and shrubs rather than grasses and herbs for their food. It has been noticeable in the Sudan that as grazing has deteriorated and become less capable of supporting cattle, who rely mainly on the grasses and herbs for their food, so camels have tended to move further south, into former cattle areas (Hunting Technical Services, 1974).

(2) Sheep. All the sheep belong to a long-legged "hair" type, with a number of recognizable "breeds", together with their innumerable crosses. The breeds may be recognized by various characteristics. For example, colour — white, black, brown and all ranges of parti-colours; coat-type — whether short, long, fine or coarse; variation in size; horned or hornless, etc.

The desert breeds are generally good milkers and are mostly associated with the camel-owning nomads. Others, such as the "Fulani sheep" range with the nomadic cattle further south. Flocks are also maintained by the sedentary villagers. They are primarily meat-producing animals, although some are kept for their milk, while the long-haired Maure of Mauritania and Mali are used more for their skins and fleece.

(3) Goats. The ubiquitous goat is essentially a browsing animal and is able to survive where there is insufficient pasture to support grazing animals. They have often been much maligned as destroyers of the natural vegetation. Although the increase in the goat population depends directly upon the gradual impoverishment of the vegetation, such an increase is not necessarily the cause but the consequence of such impoverishment from other causes (F.A.O., 1953). The impoverishment is, however, progressively hastened if the goat population should exceed the carrying capacity of the vegetation.

Goats are adapted to a wider range of food than either sheep or cattle. Prolific breeders and producers of considerable amounts of milk, they are hardy under arid conditions but susceptible to trypanosomiasis and do not thrive in humid areas. In Nigeria, the very short-horned "Sokoto Red" or "Maradi" and the "Kano Brown" are valued for their skins, which are used for glacé leather. Sheep and goats are sometimes herded together.

(4) Cattle. The cattle of the savannas of northern tropical Africa may be broadly classified into two groups, the northern zebu breeds of the mainly Arab-speaking peoples and the older Sanga stock of the nilotic and negroid races.

The Sanga breeds are derived from both longhorn and shorthorn cattle introduced from the Near East via Egypt by the Hamitic nomads more than 5000 years ago. The modified descendents of these animals, with their long, massive horns and relatively small hump, are generally to be found south of 10° latitude, the result of the steady pressure of the Arab and Berber Muslim invasion that persisted through the Middle Ages. The large, humpless "Kuri" or "Lake Chad" cattle, with their huge, bulbous horns, are an outlier of the Hamitic breeds. The Kuri are well adapted to the semi-aquatic conditions of their lake habitat; they are also strong swimmers. It is a breed that is unfortunately rapidly declining in numbers.

The Zebu cattle were apparently introduced into the Horn of Africa by Semitic immigrants during the last millenium B.C., and this was followed by further introductions into East Africa by Arab and Indian traders, mainly since the Arab invasion of the seventh century.

The various invaders introduced their own breeds of cattle, which interbred with the local cattle to produce a number of new breeds. Amongst the more important breeds are the lyre-horned "Red Bororo", a large beef animal capable of surviving the harsh environment of Niger and north-eastern Nigeria. They are a fast-moving breed and large herds are trekked by their Fulani (Fellata in the Sudan) owners across Chad to supply the meat markets of Khartoum (Fig. 6.6). The "White Fulani" is a similar lyre-horned breed, but less hardy. It is found between Senegal and Nigeria. The slow maturing, dual-purpose "Sokoto" breed is readily trained as a draft animal and is used for ploughing and pulling carts in northern Nigeria. The "Kenana", found in

Fig. 6.6. Red Bororo cattle in Darfur en route to Khartoum.

the Blue Nile Province of the Sudan, is famed for its milking qualities.

In the Sudan Republic and Chad the traditionally camel-owning Arab tribes, the Gamala, found their camels ill-suited to the more humid savannas. As a consequence, offshoots from the Gamala changed over to cattle and became known as the Baggara (cattle owners). Their cattle are predominantly of Zebu origin, although some admixture of Sanga blood has also taken place. The failure of the Arabs to penetrate further south than 10° latitude and settle there may be partly attributed to their Zebu cattle, which unlike the Sanga, are unable to tolerate the muddy conditions and innumerable biting flies (Tabanidae) during the rainy season (Lewis, 1952, 1953). The local native cattle are also more resistant to trypanosomiasis and other insect-borne diseases.

The most common system of cattle management is that of transhumance. This enables the maximum use to be made of the semi-arid grazing during the rainy season, and is supplemented by the dry-season grazing in the better watered areas further south. Thus, in the Sudan Republic the cattle-owning Baggara have their dar or homeland in the broad-leaved savanna to the south of the semi-arid thorn savanna and scrub. It is here that the cattle graze during the early rains and the crops are planted. Later, during the height of the rains, the mud and the biting flies (Tabanidae) make life extremely unpleasant. The Baggara then move northwards to the healthier sandy soils of the thorn savanna. Here they will remain with their cattle until the rains slacken, and then the return journey begins. By the end of September or early October all the animals are back in the tribal area, where they will remain while the crops are harvested. By November or December the grazing and supplies of water from the seasonal pools are finished. The animals are then moved even further south, to the Bahr el Arab, where grazing and water are still available. Later in the dry season both surface water and grazing become scarce, so that shallow wells have to be dug and the cattle travel further afield for their grazing. With the beginning of the rains in May the Baggara move in a somewhat random fashion to take full advantage of the first flush of grazing and of the surface pools. As soon as grazing conditions permit they return to their dar and the annual cycle is completed.

Such was the generalized pattern when described by Harrison (1955). Conditions have changed somewhat in recent years. Numerous watering points have been established, which enable the Baggara to remain longer in certain areas and/or exploit areas that were previously uninhabitable during the dry season. Widespread over-grazing has now somewhat reduced the advantages to be gained by the seasonal migrations to particular areas for pasture (Hunting Technical Services, 1974).

The primary factor governing the distribution of cattle (and other stock) during the dry season has been the availability of water. Cattle can survive on a low plane of nutrition for long periods, but are unable to survive more than a few days without water. The low level of nutrition during the dry season is largely due to the low feeding value of the available fodder. Conditions are even further aggravated by the shortage of fodder due to over-grazing, requiring additional energy being wasted acquiring food. The effects of this depleted fodder intake can be seen in the recent fall in calving rate together with a rise in mortality (Hunting Technical Services, 1974). The human population, however, has continued to increase, resulting in a decrease in the ratio of cattle to the human population. The increased marginality of cattle production has led to a new generation of Baggara turning to cultivation rather than cattle in order to survive (Hunting Technical Services, 1974).

In West Africa the nomadic pattern of the pastoral Fulani is similar to that of the Baggara Arabs in the Sudan Republic. It is, in general terms, a southward movement during the dry season and a northward movement with the rains, the limits of the migration being the tsetse-infested savanna country to the south and the deserts to the north. The Fulani of northern Sierra Leone differ somewhat in being shifting cultivators as well as cattle herdsmen, following a system of shifting pasturage. An area is heavily stocked for two or three years, after which they move to another area, following a cycle that brings them back to the first area after a period of 15 to 20 years.

In a situation, common to our zone of interest, where the animals are privately owned and the grazing, subject to local tribal rights, is free, there can be incentive for the grazier to balance livestock numbers to the carrying capacity of the grazing, since there is the certain possibility that others will take advantage of any remaining grazing.

Traditional agriculture offers very little capital growth potential, since profits cannot be satisfactorily re-invested in land or labour. Agricultural land is generally non-negotiable and hired labour in rural areas is generally not readily available since most villagers are already pre-occupied with the growing of their own food crops.

Thus, the only outlet is to invest in livestock and graze them on the open range. Even sedentary agriculturalists will, when stock numbers exceed the carrying capacity of the village grazing area, employ nomadic graziers in order to obtain a satisfactory return on their investments. Unproductive stock are generally sold when prices are favourable, while the sale of productive females is generally avoided, except in times of financial necessity, since such females would represent a reduction in the reproductive capacity of the herd and therefore a loss of potential financial gain (Adams, 1974).

It has been widely believed that the nomadic pastoralists have been reluctant to sell their stock on the grounds that stock numbers symbolize their owner's status and wealth. If this were true then it would be expected that herds would contain about equal numbers of males and females. However, it is clear from analyses of herd structures in northern Nigeria and the central Sudan Republic (Table 6.3) that pastoralists do maintain a suitable breeding herd structure, which can only have been achieved by the selective culling of surplus males and barren females. The higher proportion of males in the Sudan could be attributed to their use as baggage animals by the nomadic herdsmen.

Cattle management has been treated at some length because it is believed that the same, or similar, principles apply to the camel herds, etc., for which such data are not available.

TABLE 6.3

Composition of cattle herds (data from Hunting Technical Services, 1973, 1974)

Locality	Sample	Females (%)			Males (%)		
		Breeding	Immature	Total	Mature	Immature	Total
Nigeria, Lake Chad	21 480	66	19	85	7	8	15
Sudan Republic, Darfur	5414	43	26	69	10	21	31

(5) Horses. The Dongolawi horse occurs along the banks of the Nile to the north of Khartoum. It is believed to have reached its present habitat from Egypt in dynastic times. In the opinion of Bennett et al. (1948) it is "large, flashy, ugly and useless. . . . Its originally low, and now rapidly diminishing numbers are no cause for regret". The only other native breed in the Sudan Republic, the Kordofani, is believed to have been introduced during the Middle Ages, from the north-west. It is a rather common breed and is quite suitable as a pack animal.

In West Africa most breeds appear to have been introduced from the seventh century onwards, from both North Africa and the Sudan. Although horses are used to a much greater extent in West Africa than in the Sudan, they are used mainly for ceremonies and sport.

In general, horses are regarded as prestige animals; the soft, yielding sands and rough, rocky lands as well as the deep, seasonal mud are quite unsuitable terrain for horses. The shortage of succulent grazing is an additional factor limiting their widespread usage.

(6) Donkeys. The donkey is the most common transport animal throughout the area. In the south its widespread use is confined to soils that do not become too muddy during the rainy season. The northern limit for its use is determined by available drinking water since, unlike the camel, the donkey is unable to go for days without water.

Grazing

Providing water supplies are adequate, the two most important factors regarding grazing are the amount available and its nutritional value. Availability, in terms of abundance, varies from year to year depending on the annual rainfall and, to a lesser extent, upon the constituent species. Thus, many annual grasses such as *Aristida* spp., *Eragrostis* spp. and *Schoenefeldia gracilis* will have died and disintegrated before the end of the dry season. Some species will have germinated with the first rains and disappeared before the last rains, while others, such as *Eragrostis tremula* will germinate with the late rains. Perennial species, such as *Aristida sieberana* and *Panicum turgidum* remain at least partially green and semi-succulent throughout the year, and during the dry season animals select only the more succulent portions for food.

Availability may also be expressed in terms of palatability and accessibility to the grazing animal. Thus, *Cymbopogon* spp. are generally regarded as unpalatable because of their aromatic oils, yet they may be lightly grazed when young or following a fire.

Most of the grasses have rather wiry stems and the grazing animal tends to be selective, especially with the perennial grasses, eating only the green and succulent shoots. The herb *Blepharis* spp. is another example, forming widely-spaced stands, generally on bare soil. The nomads regard this as excellent grazing, despite its obvious low yield and difficulties in grazing such widely spaced herbs. Cattle are able to eat this plant when green, but during the dry season it becomes tough and prickly, palatable only to camels, sheep and goats.

The nutritional value of the grazing falls rapidly during the dry season (Table 6.4). Expressed in terms of dry matter the figures in Table 6.4 are even more startling, 6.32% crude protein during the rainy season dropping to 0.86% in the dry season. It has even been demonstrated in Khartoum that grass alone during the dry season will not keep stock alive (Fisher, 1944). Herbs and browse from trees and shrubs are necessary to supplement the diet. Such trees as *Acacia* spp. (especially *Acacia albida* — see Wickens, 1969), *Piliostigma* spp., *Balanites aegyptiaca*, etc., are lopped during the dry season in order to provide browse.

Because of the inherent poor quality of the diet and the low carrying capacity of the range, nomadism (or transhumance) is a necessary and an efficient way of managing livestock. In the low rainfall areas with inherently poor soils, crop husbandry cannot be expected to sustain the entire population living in the area.

An apparent increase in productivity is achieved by stimulating an early flush of grass by means of grass fires. Unfortunately the continued use of fire leads to an increase in fire-resistant species, which are generally less palatable than the original constituent species, leading to a general deterioration of the grazing. The use of firebreaks and a system of fines for indiscriminate burning has been attempted in some areas but has generally proved unsuccessful, mainly because of the difficulty of adequate supervision.

In recent years, partly because of the drought and partly because of the effects of over-grazing, grass fires are becoming less frequent and are also more restricted in the size of the area burnt. Grass-burning is not restricted to the southern Sudan zone but is widespread throughout Africa and is

TABLE 6.4

Percentage changes in seasonal composition of grasses (from Brémaud and Pagot, 1962)

Chemical analysis (%)	Middle of the rainy season		Start of the dry season		End of the dry season	
	Aug.	Sept.	Nov.	Dec.	April	May
Moisture	75.00	74.00	51.00	48.00	8.20	9.50
Crude protein	1.58	1.47	1.70	1.80	0.70	1.27
Ether extract	0.42	0.45	0.71	0.69	0.55	0.68
Carbohydrates	11.83	13.24	24.00	25.60	50.26	48.60
Fibre	8.23	9.64	16.50	18.50	31.00	31.90
Minerals	2.29	2.54	4.60	5.40	9.20	7.30

believed to be a very ancient practice — if not a natural one. The Phoenician and Carthaginian voyagers have placed on record the sighting of numerous widespread fires as they sailed around the coast of West Africa (Warmington, 1964; West, 1965). The problem of how to prevent indiscriminate burning is likely to remain with us for some time yet.

Forest products

A number of trees and shrub species are of local economic importance because of their edible fruits, medicinal properties, bark suitable for cordage, browse for stock, etc. The baobab, (*Adansonia digitata*) has all these properties; the artificially-hollowed trunk also has a water storage capacity of 250—400 gallons (Blunt, 1923; Fisher and Fisher, 1971) and has thus been used in areas where potable water is difficult to obtain.

The only indigenous tree species producing a commercially exportable product is *Acacia senegal*, the source of gum arabic. A number of *Acacia* species produce gums that are sold commercially but the best grade is that obtained from *Acacia senegal*; commercial quantities of lower grade gums are obtained from *A. seyal* and *A. nilotica.*

Acacia senegal is widespread through the savannas of tropical Africa, extending from West Africa eastwards to the Somali Republic and south through East Africa to Tanzania. It occurs mainly on sandy soils within the 250—450 mm rainfall belt, but may also occur on clay soils in the higher rainfall areas.

The Republic of Sudan is still the major producer of gum arabic for the world's markets; it is the Sudan's third most important agricultural export commodity worth nearly £S 8—£S 9 million per annum. Senegal, Mauritania, Mali, Niger and Nigeria produce negligible quantities of gum by comparison but are keen to increase their production. The world demand, however, has decreased in recent years.

In the Sudan gum was originally obtained from natural exudations; tapping was only introduced at the beginning of this century. Normally a small axe is used to break the outer bark, which is then peeled back, one strip up and one strip down the branch, to expose a length of about 75 cm. The gum forms in 3—8 weeks, depending on the weather, exuding from the exposed abscission layer of former leaves or other damaged tissues (Fig. 6.7). The physiological factors affecting gum exudation are still not known; cold weather will check the exudation.

The disadvantages of using an axe are that it is slow, causes excessive damage to the tree and is inefficient in manpower, as well as being uncomfortable for the tapper. For instance, it is extremely difficult to tap a tree without several false cuts and misses, all of which are accompanied by showers of dust and dead foliage upon the tapper, who is also subjected to the numerous prickles on the branches. The Sudan Forests Department has now devel-

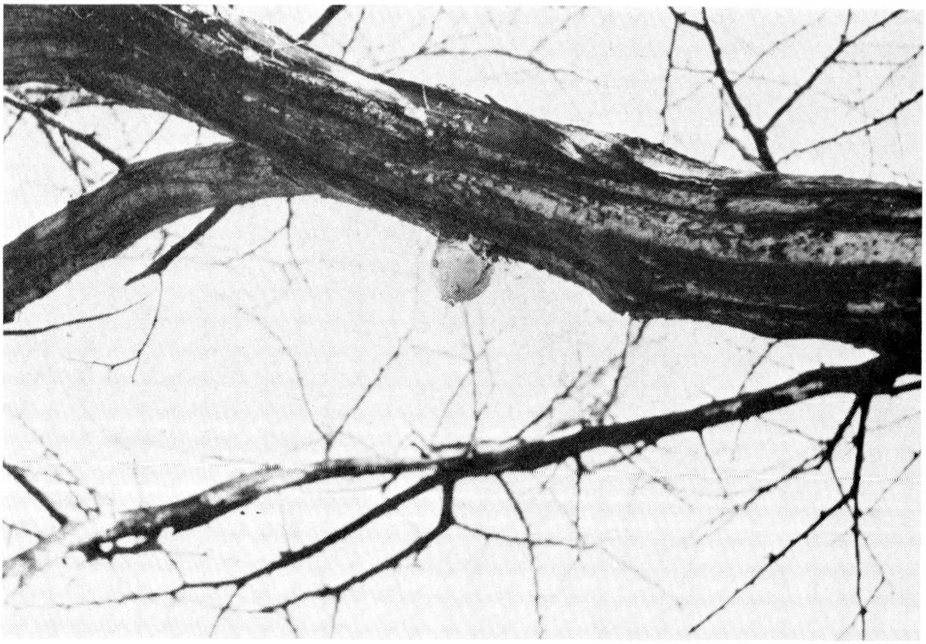

Fig. 6.7. A branch of *Acacia senegal* exuding gum.

oped a tapping implement somewhat resembling a halberd, which is a considerable improvement upon the axe; its use has spread to Nigeria.

A tree is tapped when it is 4 to 5 years old; premature tapping reduces the life of the tree. The yield of gum generally increases with the age of the tree; although some trees may produce little or no gum, a single large tree can produce up to 2000 g. The average yield from a small tree may vary between 100 and 300 g, and that from a large tree around 450 g (Hunting Technical Services, 1964).

A tapper may tap up to 100 trees in a day, and if tapping is his sole occupation, he can deal with approximately 30 ha of gum garden (plantation) during a season and pick from 200 to 300 kg of gum. Tapping takes place from October to March, except for the cooler and mid-winter months when the low temperatures check the exudation of gum. A tree is ready for tapping when the leaves fall on shaking the branches. Collection is by removal of the gum droplets about every 10 days during the season.

Within the cultivated areas around a village community the gum *Acacia* plays, or used to play, a very important role. The cultivation cycle of 4 years cropping followed by 14 years under tree fallow, as was once practised in Kordofan Province of the Sudan Republic (Hunting Technical Services, 1964) was both useful and beneficial. The fallow period rested the soil and helped to reduce erosion, yet provided a small income from the gum. The long period without cropping also helped to reduce *Striga* infestation, a parasitic weed which is regarded by some agriculturalists as the most important factor limiting yields of millet and sorghum. However, there is no evidence to suggest that the nitrogen status of the soil was significantly increased during this period (Jewitt and Manton, 1954). Though a member of the Leguminosae, *Acacia senegal* apparently does not fix nitrogen, although failure to do so may possibly be associated with the low phosphate status of the soil, or perhaps some other factor.

The tree fallow is almost exclusively *Acacia senegal*, which has either regenerated from stumps deliberately left during the cultivation cycle, or from seedlings planted towards the end of the cultivation cycle. Such gum "gardens", when established, have the great advantage of producing a cash crop during the dry season, when labour is readily available, from fallow land whose other productive use is for grazing, or, as an alternative to gum, firewood.

Although gum yields are increased following a season of heavy rainfall, less gum tends to be marketed. This is because there is usually sufficient food for the villagers and consequently little incentive for them to collect gum for sale in order to purchase additional food, especially since tapping is such a laborious and dirty occupation.

The increasing demand for arable land in the immediate vicinity of ever-expanding townships has resulted in a tendency for a reduction in the period under tree fallow and an extension of the period under cropping, often to the detriment of soil fertility and structure. The widespread thickets of the Sodom Apple, *Calotropis procera* bears silent witness to this malpractice.

The extended period under cropping also means that the regeneration of the gum gardens from the coppice regrowth is lessened, necessitating the re-establishment from seed, with a consequential loss of revenue due to the delay before the trees are old enough to tap. In some areas the tree fallow period has been abandoned altogether and the soils are cropped continuously.

Stocking rates range from 1 to 500 trees ha^{-1}, with an average of about 110 trees. With an average yield of 225 g per tree, this would produce about 22.5 kg ha^{-1}. The Sudan Forests Department's recommended spacing at 4×4 m, i.e., 500 trees ha^{-1}, would produce a more economic return per hectare. The relative inefficiency of the present system of gum production can be gauged from the Forest Department's estimate that the entire annual production for the Sudan Republic could be obtained from a properly managed plantation of 55×55 km (Hunting Technical Services, 1964).

USE AND ABUSE OF THE ENVIRONMENT

The general principle of a seasonal utilization of the natural vegetation for grazing and browse, whether by nomadic or the more rationalized transhumance systems, is without doubt well adapted for obtaining maximum utilization with minimum damage, provided the population densities are within the carrying capacities of the range. Governments tend to disapprove of these systems because of difficulties of administration, tax collecting, education, health, etc. Nevertheless, as a system of livestock and range management for an arid or semi-arid environment, it is unlikely to be bettered.

During the last 50 years, advancement in both the medical and veterinary sciences has led to an increased life expectancy for both man and his associated livestock. The increasing provision of supplies of potable water by means of boreholes, etc., in areas where water was previously unobtainable except during the rains, has enabled the herdsman to stay longer and concentrate his stock in the area around the watering point, almost always to the detriment of the grazing. Enforced rotation of watering points by shutting-off supplies is, unfortunately, rarely practised.

The provision of watering points has also enabled permanent settlements to be established in areas where either it was not previously possible or on a much reduced scale. Such settlements have been accompanied by a natural desire of the people to take advantage of the increasing medical, educational and marketing facilities, etc., to be gained by living in the steadily expanding townships.

Urbanization has naturally increased the pressure on the arable lands in the vicinity, leaving the more distant lands abandoned to the grazier. Such increased pressure has unfortunately necessitated an intensification of the cropping, at first by shortening the fallow period and eventually by abandoning any attempt to rest the land. In some countries there has been little or no attempt to maintain fertility and soil structure by means of organic manures even when they are readily available.

The shortening or loss of the fallow period during which the land would be left to revert to bush has also meant a decrease in the availability of trees and shrubs for timber, fuel and shade. The absence of shade means higher soil temperatures, less ground cover and higher wind erosion and soil wash. Forestry plantations in arid and semi-arid environments are at present uneconomical in terms of monetary gain, but their amenity value is considerable. Despite the slow growth rates far more attention should be paid to establishing plantations for fuel and local timber.

Soil conservation techniques, as practised in southern Africa are virtually unknown in this area. In the past, brushwood fences have been placed along the field boundaries to keep stock out of the growing crops. These have also provided some protection from wind and sheet erosion. The recent increase in the use of the plough, whether by camel, ox or tractor, has introduced an additional erosion hazard by requiring larger fields with the consequential reduction in protective boundaries. Today man is clearly out of balance with the environment. The semi-arid areas are over-populated and over-stocked and over-grazing is widespread. Even if it were practical, it would be totally uneconomic to even attempt to restore the balance by use of artificial fertilizers, reseeding, fencing, etc. Indeed there is little incentive for improvement when the state is the landowner.

The 1965—1974 drought

The recent widespread drought has served to highlight the climatic limitations of semi-intensive land-use in regions immediately to the south of the Sahara. The six Sahelian countries (a term wrongly used by the international news media) of Mauritania, Senegal, Mali, Upper Volta, Niger and Chad, are the worst affected. The Sudan Republic, although similarly affected, has been able to cope with the situation because the migration of the people from the drought-stricken areas has been unhindered by international frontiers (Adams, 1975). Ethiopia and Somalia, which were also badly affected, are outside the scope of the present discussion. Meteorologists have as yet been unable to decide amongst themselves whether the drought represents a short-term weather fluctuation or the portent of a longer climatic change.

Previously, the short life expectancy of man, however morally undesirable such an expectancy might have been, plus the ravages of pandemic diseases, such as rinderpest, etc., on the livestock population, had the desirable effect of enabling man and his livestock to be kept sufficiently low in numbers as to be more or less in balance with the environment. In times of drought, and there have been equally severe droughts in the past, it has been a case of survival of the fittest.

According to Winstanley (1976), prior to 1965 the total food production for Mauritania, Senegal, Mali, Upper Volta, Niger and Chad was increasing at the same rate as for the rest of the world (Fig. 6.8A). Furthermore, the total food production was increasing at a faster rate than the population (Fig. 6.8B). After 1965 food production decreased rapidly as the drought began to take effect, although the population continued to rise steadily. The average decrease in the monthly rainfall was between 20 and 40%, with the largest deficits at the beginning and the end of the wet season, when rainfall is of particular importance for the sowing and maturation of the crops.

The drought has had a devastating effect upon the internal economies of the countries dependent upon rainfed subsistence agriculture and nomadic herding. During 1972—1973 the grain shortage in Mauritania, Senegal, Mali, Upper Volta, Niger and Chad was estimated to be 1.2 million tons (Sheets and Morris, 1974); the total loss of cattle during the same period was estimated at 3.5 million head.

Although international agencies were able to aid the human population, no such rescue operation was possible for their livestock, since the neighbouring countries were already over-stocked and incapable of absorbing any additional animals as well as being affected by a general decrease in rainfall.

What of the future? In many areas we have apparently already passed the

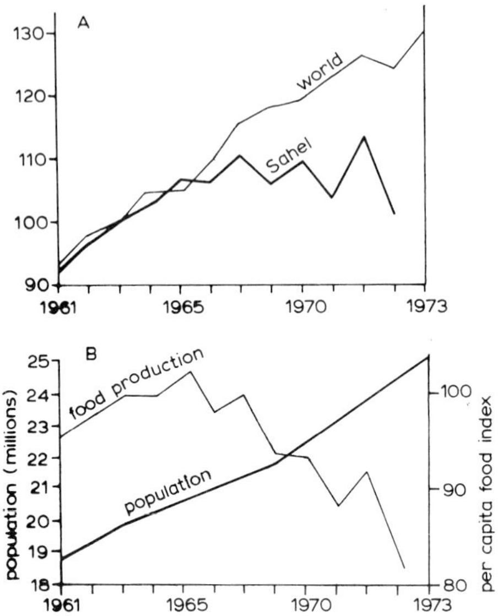

Fig. 6.8. Some recent trends in the six Sahelian countries of Mauritania, Senegal, Mali, Upper Volta, Niger and Chad (from Winstanley, 1976). A, index of total food production (1963 = 100). B, Population (millions) and index of per capita food production (1963 = 100) in the six Sahelian countries.

point where an increasing population has exceeded the carrying capacity of an already deteriorating natural environment. In a world where food is in short supply, imported foods cannot be the solution. The only alternative must be a very strict control of the human and livestock populations, together with increased agricultural production by the use of improved and probably uneconomic techniques. Since the benefits of medical and veterinary sciences are so much more rapidly effective than those which agriculturalists are able to effect (namely increases in crop and animal production) the emphasis must be on population control.

Range recovery following the return to "normal" rains is likely to be extremely slow. Indeed, it will only be possible at all if the re-stocking is kept well below the carrying capacity of the improving range until such time as near-climax conditions are reached. Unfortunately, it is extremely doubtful if any country will be strong enough to withstand the internal political pressures for a premature expansion of their livestock industry.

REFERENCES

Adams, M.E., 1974. Development Planning in the Savannah Regions of Western Sudan. Second International Seminar on Changes in Agriculture, Reading, 9—19 September 1974. Mimeographed 10 pp.

Adams, M.E., 1975. A development plan for semi-arid areas in Western Sudan. Explor. Agric., 11: 277—287.
Allan, J.A., 1976. The Kufrah agricultural scheme. Geogr. J., 142: 48—56.
Barbour, K.M., 1961. The Republic of Sudan. University of London Press, London, 292 pp.
Bennett, S.C.J., John, E.R. and Hewison, J.W., 1948. Animal Husbandry. In: J.D. Tothill (Editor), Agriculture in the Sudan. Oxford University Press, London, pp. 633—667.
Blunt, H.S., 1923. Tebeldis — a note. Sudan Notes Rec., 6: 114.
Boudet, G., 1972. Désertification de l'Afrique tropicale sèche. Adansonia II, 12: 505—524.
Brémaud, O. and Pagot, J., 1962. Pâturages, Nomadisme et Transhumance en Zone Sahélienne. In: Recherches sur la zone aride. Les problèmes de la Zone Aride. U.N.E.S.C.O., Actes du Colloque de Paris, pp. 339—352.
Buchanan, K.M. and Pugh, J.C., 1969. Land and People in Nigeria. University of London Press, London, 252 pp.
Cloudsley-Thompson, J.L., 1969. The Zoology of Tropical Africa. Weidenfeld and Nicolson, London, 355 pp.
Cobley, L.S., 1956. An Introduction to the Botany of Tropical Crops. Longmans, London, 357 pp.
Dare, P.J., 1960. Ornithology: *Quelea quelea aethiopica* Sundeval. Annu. Rep. Entomol. Sect. 1959/60. Gezira Research Station, Sudan Government.
Davidson, B., 1971. Africa in Historical Perspective. In: Africa South of the Sahara, 1971. Europa Publications Ltd., London, pp. 3—17.
De Fabregues, B.P., 1965. Étude des pâturages naturels sahéliens de la région de Nord Gouré (République du Niger). Inst. Élev. Med. Pays Trop., Maison-Alfort, 163 pp.
F.A.O., 1953. Grazing and Forest Economy F.A.O. Forestry and Forestry Products Studies No. 4, Rome, 161 pp.
F.A.O., 1965. Report of the F.A.O. Conference on Quelea Bird and Water Hyacinth Control in Africa. Meeting Report PL/1965/14, Rome, 29 pp.
Fisher, A.G.B. and Fisher, H.J. (Translators), 1971. Gustav Nachtigal, Sahara and Sudan. IV. Wadai and Darfur. C. Hurst and Co., London, 439 pp.
Fisher, A.G.B. and Fisher, H.J. (Translators), 1974. Gustav Nachtigal, Sahara and Sudan. I. Fezzan and Tibesti. C. Hurst and Co., London, 460 pp.
Fisher, C.P., 1944. Notes on the livestock of the Sudan for the Soil Conservation Committee. Appendix XV, Report of the Sudan Conservation Committee. Sudan Government.
Grove, A.T., 1958. The Ancient Erg of Hausaland and Similar Formations on the South Side of the Sahara. Geogr. J., 124: 528—533.
Grove, A.T., 1970. Africa South of the Sahara, 2nd edn. Oxford University Press, London 280 pp.
Grove, A.T. and Warren, A., 1968. Quaternary land forms and climate on the south side of the Sahara. Geogr. J., 134: 194—208.
Harrison, M.N., 1955. Report on a Grazing Survey of the Sudan. Sudan Government, Khartoum, mimeographed, 250 pp.
Harrison, M.N. and Jackson, J.K., 1958. Ecological classification of the vegetation of the Sudan. Sudan For. Dep. Bull. No. 2, New Series.
Hassan, H.I., 1962. The environments of the nomads. In: The Effect of Nomadism on the Economic and Social Development of the People of the Sudan. Proc. 10th Ann. Conf. 11th to 12th January 1962. Khartoum, pp. 22—28.
Hemming, C.F., 1974. The Locust Menace. Centre for Overseas Pest Research, London, 28 pp.
Hunting Technical Services Ltd, 1964. Report of the Survey of Geology, Geomorphology and Soils, Vegetation and Present Land Use. Land and Water Use Survey in Kordofan Province of the Republic of the Sudan. DOX-SUD-A26, Doxiadis Assoc., Athens, 349 pp.
Hunting Technical Services Ltd, 1973. Annex IV, Agriculture. In: Sir M. MacDonald and Partners and Huntings Technical Services Ltd, Investigations and Feasibility Study of an Irrigation Project South of Lake Chad, Nigeria. Feasibility Report AGL/:SF/NIR 43, London, 141 pp.
Hunting Technical Servies Ltd, 1974. Annex 3, Animal Resources and Range Ecology. In: Hunting Technical Services Ltd and Sir M. MacDonald and Partners, Southern Darfur Land-Use Planning Survey. London, 177 pp.
Ifemesia, C.C., 1965. The peoples of West Africa around A.D. 1000. In: J.F. Ade Ajayi and I. Espie (Editors), A Thousand Years of West African History. Ibadan University Press and Nelson, London and Edinburgh, pp. 39—54.
Jarrett, H.R., 1962. Africa. MacDonald and Evans, London, 350 pp.
Jewitt, T.N. and Manton, J.S., 1954. Soil exhaustion of the Goz sands of the Suda Trans. Vth Int. Congr. Soil Sci., Leopoldville.
Kendrew, W.G., 1961. The Climates of the Continents. Clarendon Press, Oxford, 608 pp.
Lebon, J.H.G., 1965. Land Use in Sudan. Bude Geographic Publications, 191 pp.
Lewis, D.J., 1952. Early Travellers' account of Surret flies (*Tabanidae*) in the Anglo-Egyptian Sudan. Sudan Notes Rec., 33: 276—298.
Lewis, D.J., 1953. *Tabanidae* of the Anglo-Egyptian Sudan. Bull. Entomol. Res., 44: 175—216.

Rapp, A., 1974. A review of Desertification in Africa — Water, Vegetation and Man. Secretariat for International Ecology, Stockholm, Rep. No. 1, 77 pp.
Sheets, H. and Morris, R., 1974. Disaster in the desert. Issue, 4: 24—43.
Sidney, J., 1965. The past and present distribution of some African ungulates. Trans. Zool. Soc. Lond., 30: 1—396.
Smith, J., 1949. Distribution of tree species in the Sudan in relation to rainfall and soil texture. Sudan Forests Dep. Bull. No. 4.
Suliman, A. El. G.M. and Jackson, J.K., 1959. The Heglig tree (*Balanites aegyptiaca* (L.) Del.). Sudan Silva, 9.
Tothill, J.D. (Editor), 1948. Agriculture in the Sudan. Oxford University Press, London, 974 pp.
Warmington, B.H., 1964. Carthage. Penguin Books, Harmsworth, 286 pp.
Warren, A., 1970. Dune trends and their implications in the central Sudan. Ann. Geogr. J., 10: 154—180.
West, O., 1965. Fire in vegetation and its use in pasture management. Commonw. Bur. Pasture Field Crops 1/1965, mimeographed, 53 pp.
Wickens, G.E., 1969. A study of *Acacia albida* Del. (*Mimosoidae*). Kew Bull., 23: 181—202.
Wickens, G.E., 1975. Changes in the climate and vegetation of the Sudan since 20 000 B.C. Boissiera, 24: 43—65.
Winstanley, D., 1976. Climatic changes and the future of the Sahel. In: M. Glantz (Editor), The Politics of a Natural Disaster: The Case of the Sahel. Praeger, New York, N.Y., pp. 189—213.

Chapter 7

ECOLOGY OF THE SEMI-ARID REGIONS OF INDIA WITH EMPHASIS ON LAND-USE

J.S. SINGH * and M.C. JOSHI **

* *Department of Botany, Kumaun University, Nainital-263002 (India)*
** *Biological Sciences Discipline, Birla Institute of Technology and Science, Pilani - 333031 (India)*

INTRODUCTION

Semi-arid zones cover around 956 750 km^2 of India, occurring chiefly in the states of Rajasthan, Punjab, Haryana, Uttar Pradesh, Gujarat, Madhya Pradesh, Maharashtra, Karnataka, Andhra Pradesh and Tamil Nadu. In this chapter we review and collate the available information, which is a pre-requisite for developing a rational management plan for the semi-arid area of India.

GEOGRAPHICAL LIMITS

Unfortunately there has not been a consensus of opinion regarding the criteria to be used for classification of arid and semi-arid regions. According to the values of aridity based on precipitation and temperature (Köppen, 1918), primary thorn forest is the expression of the semi-arid clime, desert vegetation of the arid clime, deciduous forest of the sub-humid clime and semi-evergreen and evergreen forests, respectively, of humid and perhumid climes. According to Meher-Homji (1967), Köppen's formula correctly classes Agra, Delhi, Hissar, Jaipur, Ajmer, Jodhpur, Bhuj, Dwarka, Veraval, Jamanagar, Rajkot, Bhavnagar, Deesa, Khandwa, Malegaon, Ahmadnagar, Miraj, Bijapur, Sholapur, Raichur, Kurnool, Bellary, Gadag and Coimbatore as semi-arid with thorn forests as their vegetation type. However, semi-arid stations of the thorn forest tract such as Kotah, Mainpuri, Ahmedabad, Dohab, Neemuch, Akola, Jhansi, Indore, Amraoti, Aurangabad, Poona, Hyderabad (Deccan), Gulburga and Pamban fail to be classified as semi-arid.

The application of the concept of De Martonne (1926) omits thorn forest stations such as Mainpuri, Neemuch, Khandwa, Akola, Jhansi, Indore, Amraoti and Pamban from the category of semi-arid. Meher-Homji (1967) has also examined the utility of Thornthwaite's (1948; P/T and P/E ratios) indices. The former (P/T ratio) excludes a number of stations from the class of semi-arid; many stations come into sub-humid rather than semi-arid, while the latter (P/E index) method groups most of the likely stations as semi-arid with the exception of Neemuch, Hyderabad (Deccan) and Coimbatore, but also includes Chitaldrug, Madurai, Trichinopoly, Salem and Nellore in the same class.

Meher-Homji (1965) has used five criteria from a phyto-climatic point of view to define aridity and semi-aridity, namely climatic, morpho-ecological, floristic, vegetational and agronomic. He evolved an "index of aridity-humidity" and delineated the geographic limits of the arid and semi-arid regions

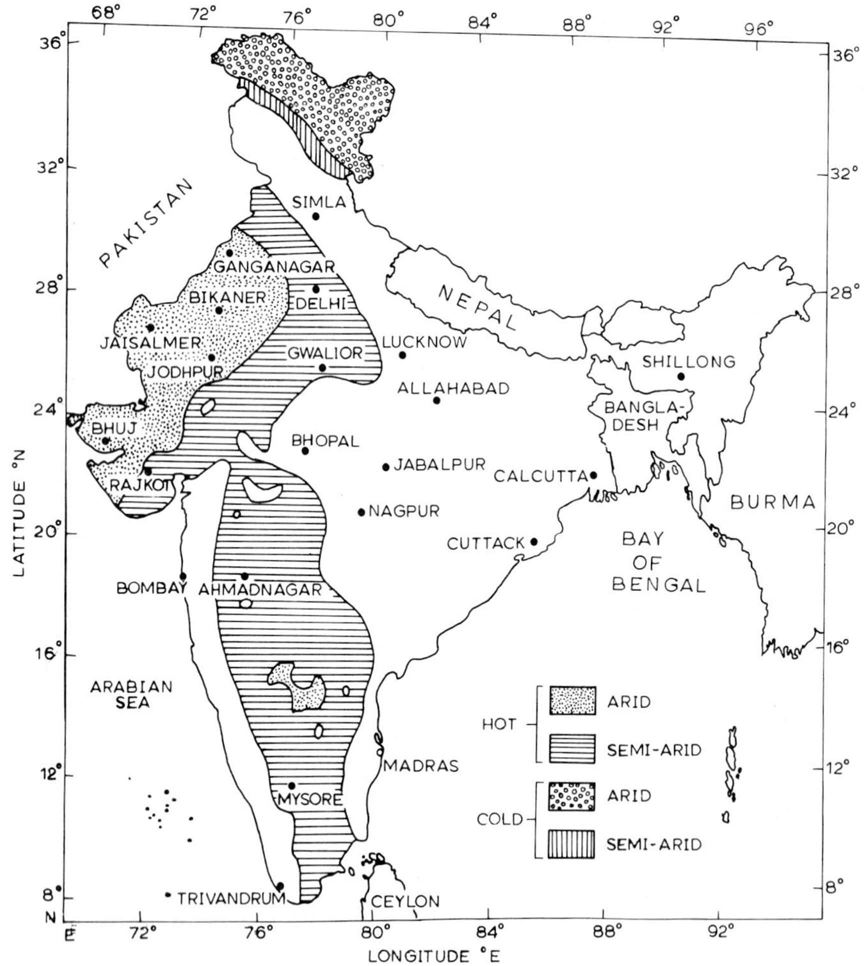

Fig. 7.1. Geographical limits of arid and semi-arid regions of India (after Krishnan, A., 1969).

of India. According to this classification, there are two main belts of semi-arid climes in India. One in the north, contiguous with the desert of Thar, and the other in the south. The northern zone lies between latitude 22° and 32°N and between longitude 70° and 79°E. The southern semi-arid zone in the Deccan, situated between latitude 15° and 21°N and longitude 73.5° and 79.5°E includes the Deccan plateau together with the northern portion of Karnataka State and the western part of Andhra Pradesh. There are two additional small patches of semi-arid climate in the south; one in the region of the Coimbatore situated in the shadow of the Nilgiri and Palni hills; the other in the extreme south-east corner of India comprising the Ramanathapuram and Tirunelveli districts of Tamil Nadu. The semi-arid regions of the north and south are separated by a narrow strip which comprises the mountainous region of the Satpura Range and the plain of the Tapti river.

Recently, Meher-Homji (1975) has suggested that species—area statistics used in combination with the climatic characteristics provides a better indication of the bio-climatic degree of aridity—humidity of a region than the climatic data alone. According to him, the problem posed by some areas of Rajasthan, Haryana, Punjab and Gujarat as to their classification in the arid, semi-arid and sub-humid category, could be solved by resorting to species—area statistics worked out by him.

Krishnan, A. (1969) has analyzed the various features of climate and delineated the geographic limits of the arid and semi-arid regions according to Thornthwaites' (1948) index, as shown in Fig. 7.1. According to this classification, Rajasthan, Maharashtra, Andhra Pradesh and Karnataka represent a larger proportion of area under hot, semi-arid clime (Table 7.1). Thus

TABLE 7.1

The distribution of arid and semi-arid regions in India, with percentage values of the various states given in parentheses

State	Area (km^2)	
	Arid	Semi-arid
Andhra Pradesh	21 550 (7)	138 670 (15)
Gujarat	62 180 (20)	90 520 (9)
Haryana	12 840 (4)	26 880 (3)
Karnataka	8570 (3)	1 339 360 (15)
Madhya Pradesh		59 470 (6)
Maharashtra	1290 (0.04)	189 580 (19)
Punjab	14 510 (5)	31 770 (3)
Rajasthan	196 150 (61)	121 020 (13)
Tamil Nadu		95 250 (15)
Uttar Pradesh		64 230 (7)
Total	317 090	956 750

about 59% of the total semi-arid area is represented in peninsular India. In further discussions in this chapter we have used the geographic limits of the semi-arid regions as defined by Krishnan, A. (1969).

CLIMATE

The climatic conditions have been analyzed in detail by Krishnan (1969), Meher-Homji (1971), Subrahmanyam (1971), Subrahmanyam and Sastry (1971) and Mamoria (1975). The year can be divided into four distinct seasons on the basis of temperature and rainfall:

(1) the hot dry weather or summer season (from mid-March to end of May);

(2) the warm wet or rainy season (from June to September);

(3) the season of retreating south-west monsoons (from October to November);

(4) the cold dry weather or winter season (from December to mid-March).

The range of mean temperatures for the hottest and the coldest months of the year is summarized in Table 7.2. In general, the mean maximum temperature exceeds 35°C in the south of the Vindhyan range and over 37°C in parts of Deccan in March. In April temperatures of 38 to 43°C are common in Gujarat, southern Madhya Pradesh and south-west Punjab. In May the temperature commonly reaches 48°C and may even exceed 54°C, as recorded for Sriganganagar in Rajasthan. These high temperatures are associated

TABLE 7.2

Ranges of mean temperature in the semi-arid regions for the hottest and the coldest months of the year

State	Temperature (°C)	
	Max.	Min.
Andhra Pradesh	39—42 (May)	14—19 (December)
Gujarat	32—42 (November, coast; May, inland)	11—15 (January, inland to coast)
Jammu and Kashmir	30—32 (July)	—12 to —13 (February)
Karnataka	35—41 (April, May)	15—18 (December)
Madhya Pradesh	40—42 (May)	9—11 (December, January)
Maharashtra	37—42 (May)	11—15 (December)
Punjab (including Haryana)	40—41 (May)	6—7 (January)
Rajasthan	39—42 (May)	8—11 (January)
Tamil Nadu	33—39 (April, May)	18—24 (January)
Uttar Pradesh	41—42 (May)	6—8 (January)

with a low relative humidity ranging between 30 and 50% in west Deccan, Gujarat and south-west Rajasthan.

Mean annual rainfall within the semi-arid zone varies from 31.1 to 72.8 cm. It may be recalled, however, that Bharucha and Shanbagh (1957), while correlating the vegetation types of India with climatic factors, retained as the limit of the arid zone the 250 mm isohyet and the limit of the semi-arid zone the 750 mm isohyet. A more conservative inclusion of areas as semi-arid would show a range of annual rainfall from 330 to 640 mm (Mamoria, 1975).

In more northern latitudes from 70 to 96% of the annual rainfall is contributed by the south-west monsoon (June to September). In the southern

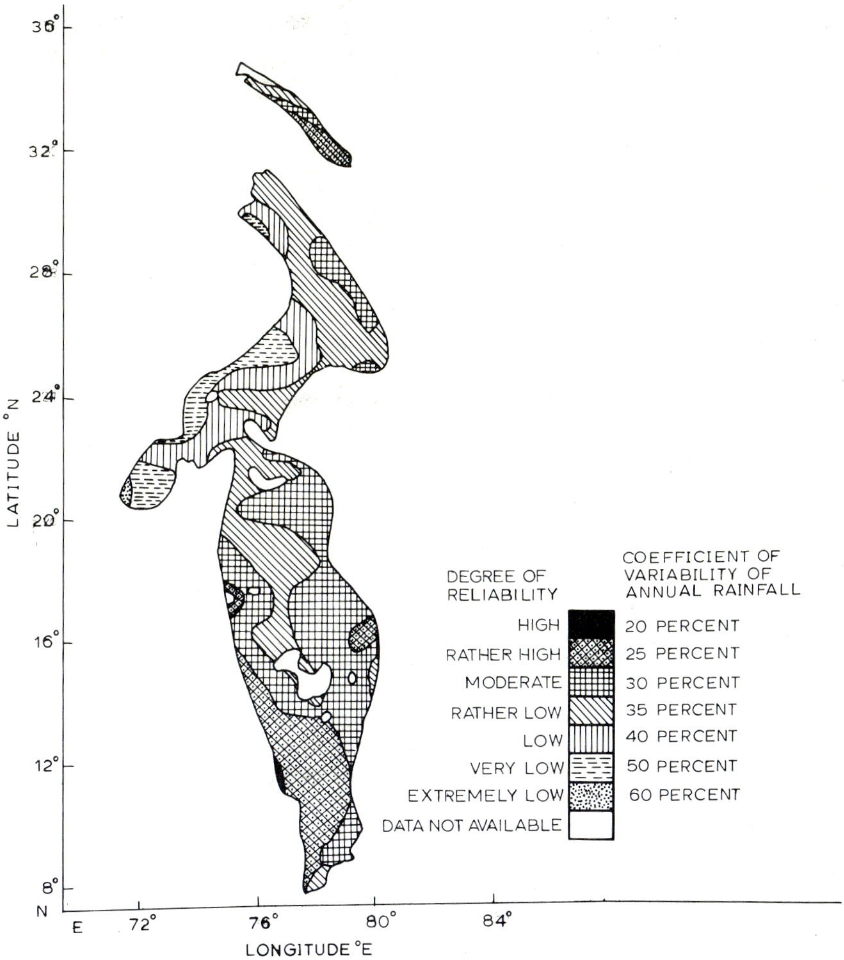

Fig. 7.2. Coefficient of variability in annual rainfall of the semi-arid regions for the period 1901—1950 (based on Government of India Census Report, 1961).

localities 7 to 63% of the annual precipitation results from the north-east monsoon during October to December.

There is a marked variability in the amount of rainfall from year to year. The coefficient of variability on the basis of data from 1901 to 1950 indicates that a large proportion of the area under semi-arid clime shows 25 to 35% variability in the annual rainfall, although the actual range is from 20 to 60% (Fig. 7.2). In general, the lower the rainfall the greater is the variability. Variation in rainfall by as little as 10% is serious in areas of moderate precipitation, particularly when it is on the negative side. Any management plan of such areas, therefore, has to take cognizance of the fact that the reliability of rainfall usually ranges from low to moderate within the semi-arid regions of India. For example, floods hit parts of the Rajasthan in the early monsoon period of 1968, while a drought occurred later in the same season. Again while heavy floods occurred in Rajasthan in 1973, severe drought was registered in major parts of Maharashtra, Gujarat and Madhya Pradesh.

The ombrothermic diagrams of certain representative stations are reported by Meher-Homji (1963, 1967). These indicate a dry period of 9—11 months each year. The high evapotranspiration leads to high net water deficits in most of the semi-arid localities (Table 7.3).

Water balances worked out following Thornthwaite and Mather (1955) show that none of the stations in the semi-arid zone have any water surplus, even on a monthly basis. Contributions to the ground-water reservoir from this region are thus normally absent. In certain years, however, when precipitation during the rainy season is much greater than normal, a local water surplus may occur for brief periods only and these surpluses not only produce enormous surface runoff, resulting in inundation, but also significantly contribute to the ground-water resources.

TABLE 7.3

Mean annual potential evapotranspiration and water deficiency for semi-arid locations in a few states of India (based on Krishnan, A., 1969)

State	Normal annual rainfall (mm)	Mean annual potential evapo-transpiration (mm)	Mean annual water deficiency (mm)
Rayalaseema region (Andhra Pradesh)	674	1723	1049
Jammu and Kashmir	241	678	437
Bellary-Bijapur (Karnataka)	599	1457	858
Maharashtra	712	1515	802
Rajasthan	776	1397	621
Tamil Nadu	917	1756	859

GEOLOGY

The three well-defined geological regions in India with marked geomorphological components are:

(1) the peninsular region which occupies nearly 10% of the total land area. The peninsular shield has old, stable rocks of igneous origin;

(2) the extra peninsular region which occupies nearly 15% of the total land area. It is a folded mountainous region;

(3) the Indo-Gangetic plain separating the peninsular and extrapeninsular regions. This alluvial plain is of Pleistocene and recent formation.

The semi-arid zone is spread across all three geomorphological regions (Krishnan, M.S., 1960, 1974; Wadia, 1961).

Peninsular India is a remnant of the old Gondwanaland complex and is probably as old as life itself. The whole terrain is made up of several large or small plateaus, built of granite-gneiss capped by bauxite and laterites. To the west and north-west of the Central Highlands (Malva Plateau) are the Aravalli ranges running for about 800 km from SSW to NNE in Rajasthan, terminating in the Delhi ridge. Drier parts of the plains of Punjab and Haryana extend southwards and merge with Rajasthan. This tract extends for about 640 km with an average width of 300 km. Most of this plain was under the sea from the Permocarboniferous period until it was uplifted during the Pleistocene, with indications that the region was once fertile.

The Deccan trap originated from basaltic lava flows. It has flat topped hills and deep cut valleys. The Deccan plateau is composed of Pre-Cambrian or indeterminate Palaeozoic crystalline and metamorphic rocks belonging to the Dharwar, Aravalli, Cuddapah and Vindhyan systems. A great variety of mineral resources associated with metamorphosed Archaean terrain are found in these areas.

The Aravalli system dates back 3500 million years, though later sediments seem to have been incorporated around 1000 million years ago. The Cuddapah system, which developed between 600 and 1400 million years ago, occurs in the Cuddapah and Kurnool districts of Andhra Pradesh, Maharashtra, Karnataka and Rajasthan. The Vindhyan system is sub-divided into (a) the lower Vindhyan which developed about 1100 to 1300 million years ago, and (b) the upper Vindhyan, dating back 1000 to 1600 million years, and is in the form of intrusive rocks (dykes). These are found in Tamil Nadu (North Trichinopoly, Salem, North and South Arcot areas), Karnataka (Bellary district) and Andhra Pradesh.

SOIL

The semi-arid regions support a variety of soils, the spatial distribution of which is given in Fig. 7.3.

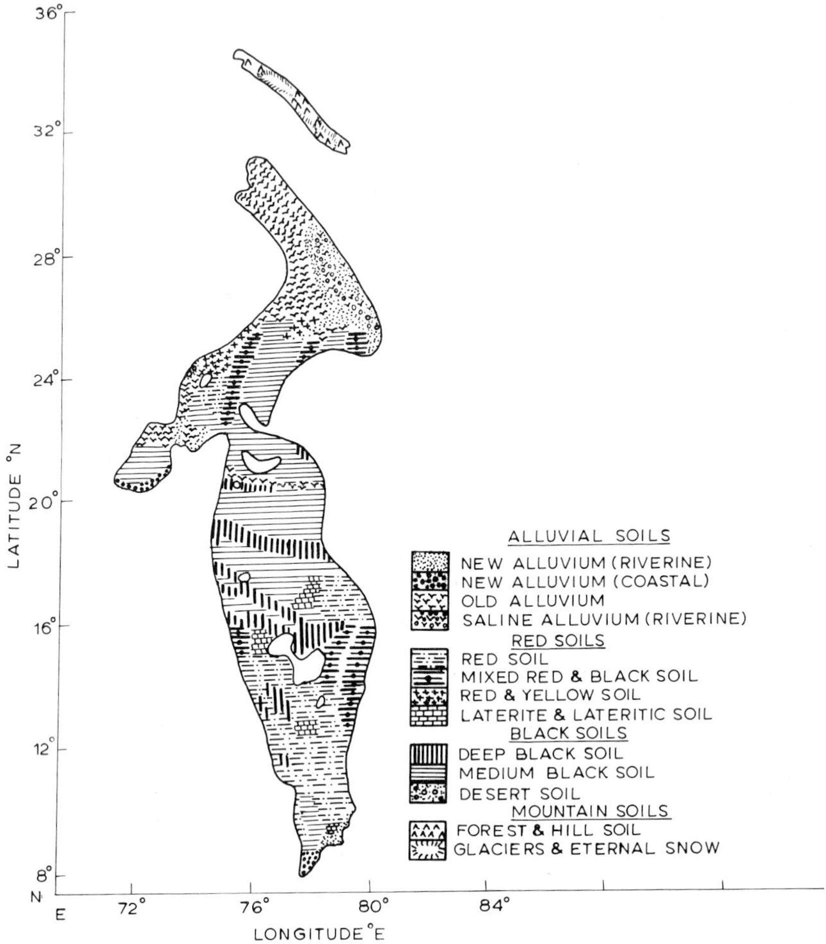

Fig. 7.3. Spatial distribution of major soil types in the semi-arid regions of India.

Alluvial soils occupy extensive tracts. The new alluvium (riverine) occurs close to the banks of rivers while the new alluvium (coastal) is formed along sea coasts. It is sandy to loamy sand in texture. The old alluvial areas are no longer subjected to alluviation because they occur in the valleys above the level of inundation. Their texture varies from clay to sandy loam. They are low in lime, phosphorus and humus. The productivity of these soils is dependent upon their texture, the amount of plant nutrient contained in them and their conditions of drainage. An occasional feature of these soils is the presence of hard pans at a certain level in the soil profile caused by infiltration of silica or calcareous matter, and the presence of calcareous nodules, known as "Kankar". These pans impede penetration of plant roots and affect the percolation of water. Nevertheless, under irrigation these soils support luxuriant crops.

The desert soils occur in parts of Rajasthan, southern Punjab and adjacent arid lands. The sand with which the area is covered is partly derived from the disintegration of rocks. They contain a high percentage of soluble salts, varying amounts of calcium carbonate and are usually low in organic matter. The soils in depression areas contain a high percentage of salts, are notably impervious and have as a result, impeded drainage. The yellow colour in soils is probably due to a higher degree of dehydration, as found in Rajasthan and adjacent areas. Chestnut or grey-brown soils cover an area of 36 400 km^2 in Rajasthan.

Saline and alkaline soils also occur in the semi-arid zone. Saline alluvium (riverine) occurs in drier tracts of the north where saline and alkaline efflorescence is very common. These tracts, once fertile, have now become impregnated with these salts, as a result of poor drainage and a high water table.

The semi-arid regions of peninsular India contain black, red and lateritic soils, saline and alkaline soils, alluvial soil, red and yellow soils and mixed red and black soils. The largest soil group is that of red soils and covers the Archaean basement of peninsular India. Red soils are light in texture and devoid of lime and free CO_3. Black soils originated from basaltic rocks of the lava plateau and are rich, residual soils. They contain less clay and are highly retentive of moisture. Laterite and lateritic soils are poor in nitrogen, phosphorus and potassium.

Mountain and hill soils are formed in depressions within the mountains in the valleys and basins and on the less steeply inclined slopes. In general, north-facing slopes have a thicker soil cover compared to the more precipitous southern faces.

Most of the soils in the semi-arid region are comparatively dry. They are deficient in nitrogen and organic matter, less so in phosphate, while potash deficiency is rare. Considerable use of proper commercial fertilizers is therefore essential as most of the soils are not suitable for sensitive crops.

BIOTA

Man

According to the 1961 census, the human population density varies from 50 to 800 persons km^{-2} within the semi-arid zone. A larger proportion of the hot, semi-arid region registered population densities ranging from 51—200 persons km^{-2} compared to the national average of 138 persons km^{-2}. Pockets of higher density (201—500 persons km^{-2}) exist both in the southern extremity as well as the north-eastern extremity of the region. The cold

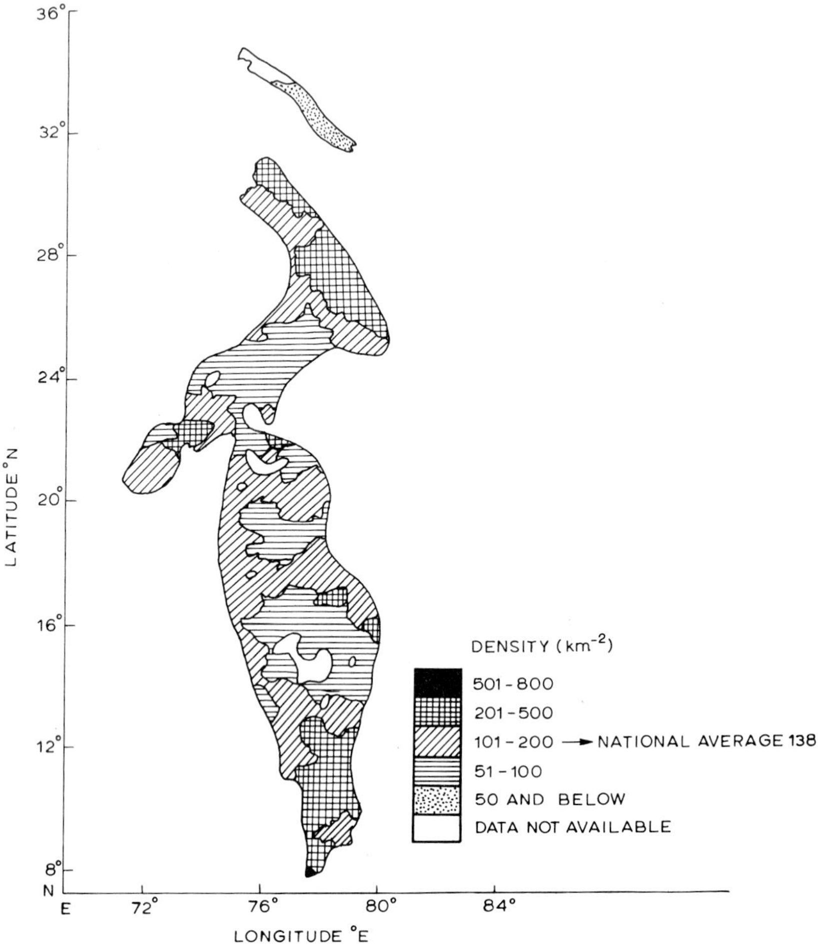

Fig. 7.4. Density of human population in the semi-arid regions of India (based on Government of India Census Report, 1961).

semi-arid region has a low density of 50 and below (Fig. 7.4).

More remarkable has been the growth of the population within this area during the 60-year period that elapsed between the 1901 and 1961 censuses. The total increase over the 1901 population was 201 million during the 60 years, showing a national average growth of 84.05%. Within the semi-arid region a larger proportion of the area exhibited a growth of 40 to 100%, while a significant part of the area registered an increase of 100 to >130% within the 60-year period (Fig. 7.5). The growth within this period was partly due to migration from the surrounding areas, emergence of new settlements and an increase in canal irrigation. This high density and the high rate of population accretion hardly needs emphasis in the context of the stress

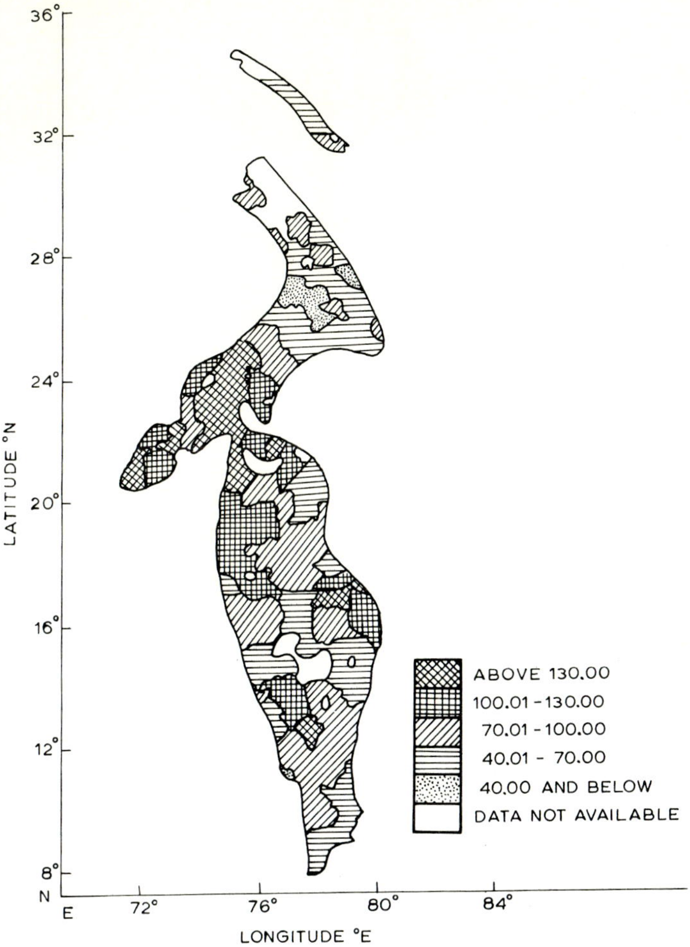

Fig. 7.5. Growth of population in semi-arid regions between 1901 and 1961 (based on Government of India Census Report, 1961).

on the delicately balanced, water-limited, semi-arid ecosystems with their inherently low carrying capacity.

Livestock and poultry

According to the 1966 livestock census, India had 175 million head of cattle, 52 million buffalo, 64 million goats and 42 million sheep. This amounts to nearly 11% of the total livestock population of the world (Mamoria, 1975). The distribution of livestock and poultry in some of the states which are at least partly semi-arid is given in Table 7.4.

In the semi-arid zone the livestock density varies from 50 to 250 livestock

TABLE 7.4

Distribution of livestock and poultry in the semi-arid regions, as per 1966 Census (figures in thousands)

State	Cattle over 3 years		Buffaloes over 3 years		Sheep	Goats	Horses and ponies	Total live-stock	Poul-try
	Total males	Breeding cows	Total males	Breeding cows					
Andhra Pradesh	5407	3830	1460	3114	8004	3758	49	31 594	14 715
Gujarat	3101	1762	35	1801	1652	2771	70	14 338	2324
Jammu and Kashmir	559	647	42	343	1152	606	66	4079	1535
Karnataka	3666	3118	289	1671	4748	2784	65	23 485	8277
Madhya Pradesh	9203	7136	1237	2303	1015	6606	150	31 478	5739
Maharashtra	6525	4468	339	1704	2200	5105	100	25 366	9887
Punjab (including Haryana)	2462	1627	297	2603	1196	1391	68	14 319	2209
Rajasthan	4029	4779	153	2136	8806	10 323	63	37 476	865
Tamil Nadu	4947	2852	502	1231	6622	3771	17	24 569	11 226
Uttar Pradesh	13 231	6710	1429	5955	2623	8136	230	49 972	3771

units per 100 ha cropped area (Singh, 1974). The livestock units have been derived on the basis of recommended animal feed for various livestock types of different age groups in terms of digestible crude protein and total digestible nutrients. Highest and medium densities occur in the southern part of the Rajasthan uplands and the Aravalli range, the western parts of Karnataka and the north-eastern part of Tamil Nadu. In the drier parts of Rajasthan and Gujarat the density is lower (less than 50—100 per 100 ha of cropped area).

Approximately 60% of the sheep are concentrated in Rajasthan, Andhra Pradesh, Tamil Nadu and Maharashtra, while 50% of the goats are found in Rajasthan, Maharashtra and Uttar Pradesh. Goat-breeding has been specially concentrated in semi-arid regions on account of the low cost of maintenance and their hardy nature, notwithstanding the deterioration of land. The greatest concentration of camels (11 to >20%) has been recorded in Marushthali, Rajasthan and south-west Haryana with a medium concentration (3—10%) in the Kutch peninsula, the northern parts of the Gujarat plain, north-eastern Haryana and southern Punjab (Singh, 1974).

Livestock in India suffers not only from climatic vagaries and poor grazing but is handicapped by poor genetic potential, and low supplies of feed and fodder. The scarcity of feed is the greatest obstruction in the way of improving the productivity of livestock in the semi-arid regions of India. The decreasing coverage of grazing lands and cultivable wasteland obviously puts the pressure of livestock population on cropped land. The lowest densities, under 100, of the livestock units per 100 ha of cropped area are observed in south-western dry parts of Punjab through Rajasthan and Gujarat to the rainshadow region of the Deccan plateau with its farthest limit in the Rayalassemma region. Primarily, the low densities are on account of the low carrying capacities of the land and the low agricultural efficiency. In Rajasthan and Gujarat the large number of sheep and goats are also responsible for pulling down the weighting of livestock units. In major parts of the semi-arid zones the low livestock densities are due to erratic and meagre rainfall, sterile and sandy soils and low fertility of agricultural land.

Wild fauna

Not much literature is available on the wild fauna of the semi-arid regions, although those of the adjacent arid zone have been studied in some detail (Prakash, 1974). Among the fairly abundant large mammals are the Indian gazelle (*Gazella gazela bennetti*) [1], blue bull (*Boselaphus tragocamelus*) and the Indian black buck (*Antelope cervicapra rajputanae*). Those which were abundant once but are now either restricted to certain favourable pockets or are fast disappearing include the caracal (*Felis caracal*), wild ass (*Equus*

[1] Animal nomenclature follows Prakash (1974).

hemionus khur), tiger (*Panthera tigris*), leopard (*Panthera pardus*), sloth bear (*Melursa ursinus*), sambhar (*Cervus unicolor*), chital (*Axis axis*) and chausingha (*Tetracerus quadricornis*). The hare (*Lepus nigrocollis dayanus*) is common in the rangelands where on the one hand it does considerable damage to the natural range and on the other is well sought after for game (Prakash and Taneja, 1969). The most common rodents in the grass—scrub vegetation are *Funambulus pennanti*, *Gerbillus nanus indus*, *Gerbillus gleadowi* and *Tatera indica indica*, out of which the Indian gerbil (*Tatera i. indica*) is the most abundant (Prakash and Jain, 1971).

Krishnamurthy (1976) has studied the avifauna biomass of the semi-arid grasslands at Khirasara. The total number of species recorded during the study period was 16. The maximum biomass was contributed by *Columba livia* (47.01%) during January 1973. *Streptopelia decaocta* and *Pterocles indicus* contributed 30.52 and 30.30%, respectively, during January and July 1973.

The termite, *Trinervitermes biformis* is reported to be a serious pest of a number of crops such as cotton, wheat, egg plant and groundnut and attacks fruit trees such as *Mangifera indica*, *Citrus*, *Psidium guyava*, *Punica* and *Achrus* in the state of Maharashtra (Patel and Patel, 1954). Ayyar (1940) has reported on the extensive damage to newly planted sets of grapevine by termites in south India. The true harvester termite, *Anacanthotermes macrocephalus* and the nasute harvester termite, *Trinervitermes biformis*, cause considerable damage to pastures in both arid and semi-arid regions (Sen-Sarma and Mathur, 1961). Hoon (1962) has reported that the topsoil (about 30 cm thick) in a semi-arid locality, lying on a harder moorum-stratum is abundantly penetrated by the termites. In areas where the soil consists of deep alluvial deposits the termite tracks penetrate as deep as 3 m.

Soil microbes

In the semi-arid grasslands of Khirasara, Krishnamurthy (1976) recovered relatively more fungal biomass in the upper soil layer (0—10 cm) throughout the year, whereas on the same site, the bacterial biomass was relatively higher in the 10—15 cm soil layer.

VEGETATION AND LAND-USE

Plant climate

Meher-Homji (1962a, b) has reported on the biological spectra of certain semi-arid localities of India (Table 7.5). The values indicate a range of plant climates from phanerophytic to therochamaephytic. Therophytes are able to

TABLE 7.5

Biological spectra of certain semi-arid regions

Region	Ph	N	Ch	H	G	HH	Th	L	P	E	Plant climate
Normal spectrum	28	15	9	26	4	2	13				
Semi-arid parts of north-west India	11	12	18	10	5	3	33	6			Thero-chamaephytic
Semi-arid parts of the Deccan	15	15	12	11	6	28	7	1			Thero-chamaephytic
Poona region	22	16	20	4	1	2	26	7			Nanophanerophytic with predominance of chamaephytes
Andhra region	33	18	12	4	1	3	16	10			Phanerophytic

Ph, phanerophytes; N, nanophanerophytes; Ch, chamaephytes; H, hemicryptophytes; G, geophytes; HH, hydrophytes, helophytes; Th, therophytes; L, lianas; P, parasites; E, epiphytes.

tide over the unfavourable dry conditions through seeds, while chamaephytes are able to resist the browsing and grazing pressure to a considerable extent due to the location of their perennating structures.

Woodland

The major woodland types of the semi-arid region are shown in Fig. 7.6. The tropical thorn forest and the tropical dry deciduous forest constitute the chief vegetation expression in the hot semi-arid region, while the sub-alpine forest is the major forest cover in the cold semi-arid zone. Localized pockets representing the tropical moist deciduous forest and the tropical dry evergreen forest occur adjacent to areas possessing more favourable water regimes. The southern tropical dry deciduous forest is represented by dry deciduous forest, savanna forest, teak pole forest and *Hardwickia* forest in Maharashtra; by dry mixed forest with teak and dry mixed forest without teak in Madhya Pradesh and Andhra Pradesh; and by dry mixed deciduous forest, Red Sanders forest and Sandal-bearing scrub forest in Karnataka and Tamil Nadu. The northern tropical dry deciduous forest is represented by riverain Shisham forest and scrub forest in Punjab and by dry Sal forest, dry mixed deciduous forest and dry deciduous forest in Uttar Pradesh (Champion and Seth, 1968).

The tropical thorn forests are found in Kutch and Gujarat, south-western Punjab, western Haryana, western and northern Rajasthan and the Deccan plateau in peninsular India. These forests are open, low and predominantly

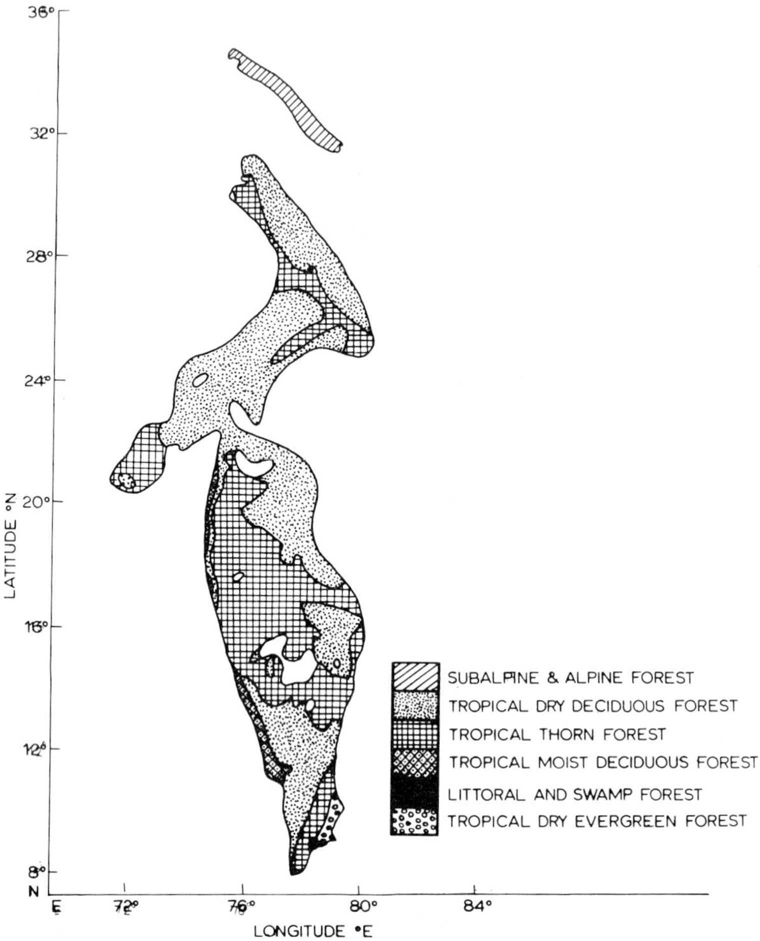

Fig. 7.6. Major woodland types in the semi-arid regions of India (based on Champion and Seth, 1968).

xerophytic, and are dominated by species of *Prosopis*, [2] *Tamarix*, *Acacia*, *Albizzia*, *Capparis*, *Maytenus*, *Salvadora*, *Zizyphus* and *Balanites*.

Meher-Homji (1975), while analyzing the floristic composition and main vegetation elements in the semi-arid and arid regions, has recognized the following types in north Gujarat and Rajasthan:

I. *Calligonum polygonoides:* mostly on sand dunes and sometimes on loose sand in western Rajasthan.

[2] Plant nomenclature follows Champion and Seth (1968) and Bor (1960).

II. Thorny types:

(a) *Prosopis—Acacia—Capparis—Zizyphus—Salvadora:* sandy plains of Rajasthan and northern Gujarat.

(b) *Acacia leucophloea—Balanites* "series": Amthala-Abu Road. (Mathur and Verma, 1964).

(c) Mixed xeromorphic thorn forest: alluvial plain, Mt Abu (Gupta and Saxena, 1968).

(d) Thorny scrub: lower piedmont slopes, Mt Abu (Gupta and Saxena, 1968).

(e) Foot of the hill vegetation: Harsh Nath (Nair and Nathawat, 1957).

III. Low forests of *Anogeissus pendula*:

(a) *Anogeissus pendula* low forests: Aravallis.

(b) *Anogeissus pendula—Acacia* "series": dry deciduous forest, Deldar Abu Road (Mathur and Verma, 1964).

(c) *Anogeissus pendula* shrubby steppe: hills around Sirohi (Gupta and Saxena, 1968).

(d) *Anogeissus pendula—Acacia* "series": scrub—woodland, Chibagaon Sirohi (Mathur and Verma, 1964).

IV. Dry deciduous forest types:

(a) *Anogeissus pendula*, *Anogeissus latifoia*, *Terminalia*, *Diospyros* type: southern Rajasthan.

(b) *Anogeissus latifolia—Wrightia* "series": dry deciduous forest, Chipaveri, Mt Abu (Mathur and Verma, 1964).

(c) *Anogeissus latifolia—Wrightia* "series": scrub—woodland, Ranora-Abu Road (Mathur and Verma, 1964).

(d) Dry mixed deciduous forest: Rajasthan (Mathur, 1960).

(e) Dry mixed deciduous forest: hills of north-east Rajasthan (Vyas, 1967).

(f) Dry mixed deciduous forest: Mt Abu (Gupta and Saxena, 1968).

(g) Dry deciduous teak forest: southern Rajasthan (Mathur, 1960).

(h) Dry deciduous teak forest: hills to the east of Baroda.

According to Rao (1970), the vegetation of Kutch is chiefly dominated by thorny scrub forests. In over-grazed areas the scrubby vegetation is highly degraded, while in protected and moister areas vegetation tends to develop into deciduous types like that of the Deccan trap with species of *Sterculia*, *Lannea*, *Bauhinia*, *Ehretia*, *Bombax*, *Cordia* and *Moringa*.

The flora and vegetation of the Bellary district have been studied in detail by Shankarnarayan (1962—1963) and Satyanarayan and Shankarnarayan (1964). The forests in this area represent two major formations, viz., southern thorn forest and southern dry deciduous forests with the following communities:

(1) *Prosopis spicigera—Acacia latronum*

(2) *Albizzia amara—Acacia latronum*

(3) *Euphorbia* scrub

(4) *Gyrocarpus jacquinii*
(5) *Tectona grandis—Anogeissus latifolia*
(6) *Anogeissus latifolia—Hardwickia binata*
(7) *Hardwickia binata*
(8) Bamboo brake
(9) *Boswellia serrata*
(10) *Butea monosperma—Tectona grandis*

The pattern of vegetation succession in response to biotic operations such as cutting and grazing is summarized in Table 7.6, as reported by Shankarnarayan (1962—1963).

From the foregoing references to woodland vegetation one might gain the impression that the semi-arid regions in India are thickly wooded. But far from it, most of the vegetation is highly degraded owing to the heavy pressure of human and livestock populations. Fig. 7.6 should in fact be regarded as representing potential forest types only.

TABLE 7.6

Succession of vegetation at Bellary

Southern dry deciduous forest of *Tectona grandis* and *Anogeissus latifolia*		Southern thorn forest of *Prosopis spicigera—Acacia latronum*	
	Hills and slopes Red soils		Plains Red soils—black soils
Cutting	↓	Cutting (→ *Chrysopogon montanus*)	↓
	Butea monosperma *Tectona grandis* (Stunted)		*Acacia latronum* *Jatropha glandulifera*
Cutting	↓	Cutting	↓
	Sehima nervosum *Cymbopogon coloratus*		*Dichanthium annulatum* *Chloris bournei*
Grazing	↓	Grazing	↓
	Iseilema laxum *Heteropogon contortus*		*Iseilema anthephoroides* *Heteropogon contortus*
Grazing	↓	Overgrazing	
	Chrysopogon montanus ← *Andropogon pumulis* →		*Brachiaria eruciformis* *Aristida adscencionis*
Overgrazing	↓	→ Closure to grazing and cutting	↓
	Aristida setacea *Melanocenchris jacquemontii*		*Prosopis spicigera* *Acacia latronum*
Closure to grazing, burning and cutting	↓		
	Tectona grandis *Anogeissus latifolia*		

Grassland

The grasslands of the semi-arid areas have mainly originated as a result of de-forestation and intense browsing and grazing. However, they constitute a very important aspect of the landscape because of their significance in animal husbandry. Information on the grasslands of India, including the semi-arid regions, has been reviewed in detail by Dabadghao and Shankarnarayan (1973), Singh (1976), Yadava and Singh (1977) and Singh and Joshi (in press).

The major grass cover types of the semi-arid regions are shown in Fig. 7.7. By far the largest portion of the hot semi-arid area is covered by *Sehima—Dichanthium*, followed by *Dichanthium—Cenchrus—Lasiurus* and *Phrag-*

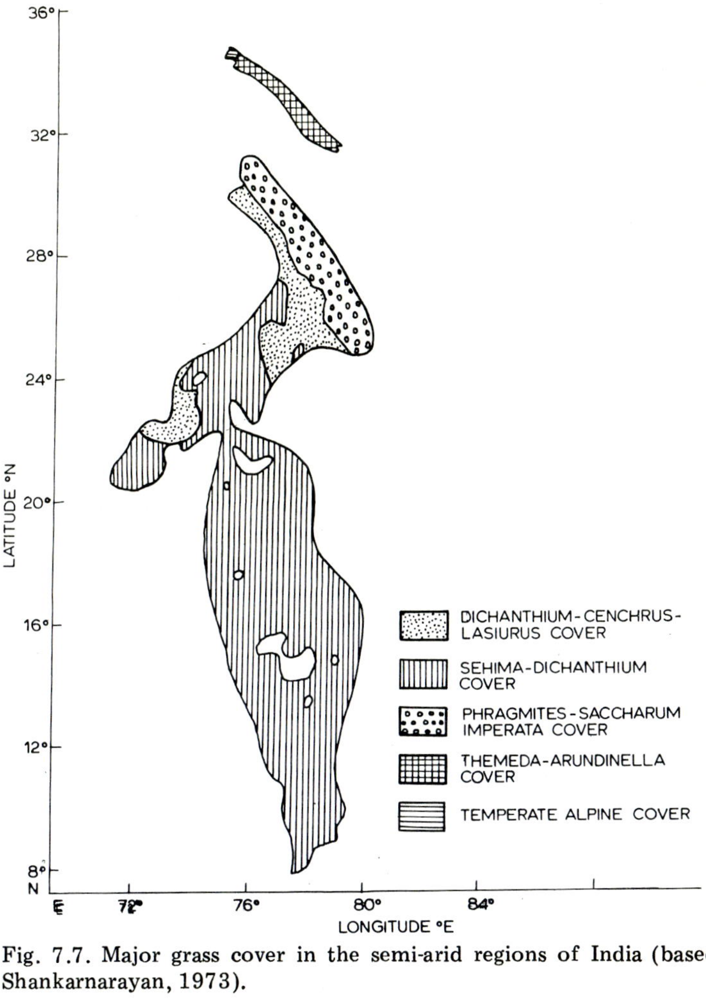

Fig. 7.7. Major grass cover in the semi-arid regions of India (based on Dabadghao and Shankarnarayan, 1973).

mites—Saccharum—Imperata. The cold semi-arid region is covered by *Themeda—Arundinella* and temperate-Alpine cover types. Floristically, the areas shown under each cover are not homogeneous but represent a myriad of communities (for a full review see Yadava and Singh, 1977).

Table 7.7 includes values for above-ground plant biomass on selected sites. The two typical semi-arid sites, viz., Pilani and Khirasara, indicate a range of 0 to 201 g m^{-2} of live shoot biomass over a year and 0 to 178 g m^{-2} of dead shoot biomass. Localities in higher rainfall areas (also classed as sub-humid grasslands elsewhere) indicate a range of peak values of live shoots from 457 to 1974 g m^{-2}. The peak of the below-ground biomass in typical semi-arid grasslands ranges between 86 and 205 g m^{-2} while in areas with greater rainfall the peak may range from 333 to 1167 g m^{-2} (Table 7.7). This table includes values from protected grasslands only; heavily grazed sites will have much lower values, particularly in low rainfall areas within the semi-arid region.

The net primary production values and the annual net flow rates into, and losses from, vegetation compartments are given for three sites in Table 7.8. The net production increases, as would be expected, with an increase in the rainfall. In low rainfall areas (e.g. Pilani and Khirasara) the below-ground production is greater than the above-ground production, while the reverse is true for Kurukshetra grassland. Because of the low shoot production the carrying capacity of these grasslands is very low.

Energy flow and nutrient cycling (nitrogen and phosphorus) have been determined for three grasslands occurring in the Pilani area, on the basis of data collected by Gill (1975). A brief description of his results follows.

The annual rate of energy capture in the form of net primary production varies from 1.44 kcal m^{-2} day^{-1} to 3.10 kcal m^{-2} day^{-1} among the three sites. On average, the semi-arid grasslands fix energy in terms of total net production at rates of 4.63, 1.20 and 1.84 kcal m^{-2} day^{-1} during the rainy, winter and summer seasons, respectively, out of which 2.35, 0.17 and 0.29 kcal m^{-2} day^{-1} are directed to above-ground production while the balance accumulates in the below-ground parts. Net energy captured by the vegetation over the year represents 0.05% of the total incident solar radiation (Table 7.9). Maximum transfer to the standing dead compartment occurs during the winter season (1.34 kcal m^{-2} day^{-1}), followed by the rainy and summer seasons (0.71 and 0.29 kcal m^{-2} day^{-1}, respectively). The pattern of litter production closely follows that of the transfer to the standing dead compartment.

The rates of root disappearance are higher than the rates of litter disappearance, particularly during the summer season. This is not surprising since the standing crop of roots is markedly higher than that of the live shoots or litter biomass. While all the litter produced during the year disappears, the roots conserve 23% of the energy channelled below-ground which ensures fresh initial growth at the advent of the next growing season.

TABLE 7.7

Variation in shoot biomass in selected grasslands of the semi-arid region (Values are in g m^{-2})

	Pilani [a]			Khirasara [b]			Kurukshetra [c] *			Ujjain [d] *			Jhansi [e] *		Delhi [f] * (Live + Dead)
	Live	Dead	Roots	Live	Dead	Roots	Live	Dead	Roots	Live	Dead	Roots	(Live + Dead)	Roots	
June	42	4	42	48	25	137	773	203	955	36	37	675	—	—	106
July	56	—	—	201	65	99	1103	196	1000	126	40	650	568	268	128
August	72	0	86	178	45	142	1526	318	671	252	53	600	850	204	436
September	76	10	53	100	50	145	1974	142	611	435	85	550	898	140	597
October	39	21	16	24	44	108	1707	400	675	457	127	610	1408	333	771
November	39	20	18	22	69	77	486	1268	1167	313	220	715	952	298	407
December	35	17	27	22	69	205	105	845	643	51	422	925	795	191	465
January	38	27	44	21	178	197	187	782	871	66	290	775	1132	204	400
February	53	20	48	7	113	120	261	631	768	70	227	750	1106	206	200
March	71	1	55	0	109	109	445	553	802	53	185	765	547	170	—
April	71	0	65	0	90	121	560	466	634	28	167	800	1187	251	9
May	64	2	43	0	70	99	800	471	809	24	112	790	496	233	—

[a] Kumar and Joshi (1972): dominant plant taxon *Cenchrus biflorus.*
[b] Krishnamurthy (1976): dominant plant taxon *Heteropogon—Dichanthium.*
[c] Singh and Yadava (1974): dominant plant taxon *Panicum miliare.*
[d] Misra (1973): dominant plant taxon *Dichanthium annulatum.*
[e] Shankar et al. (1973): dominant plant taxon *Heteropogon—Sehima.*
[f] Varshney (1972): dominant plant taxon *Heteropogon contortus.*
* These sites are also classed under a sub-humid clime (see Singh and Joshi, in press).

TABLE 7.8

Organic matter changes in the compartments of semi-arid grassland ecosystems

Parameter	Pilani [a] (391 mm rainfall)	Khirasara [b] (518 mm rainfall)	Kurukshetra [c] (790 mm rainfall)
Net primary production	239	605	3538
Shoot production	84	200	2407
Dead shoots	84	181	2044
Litter	94	88	1728
Below-ground parts	155	405	1131
Disappearance from			
above-ground litter	116	88	1666
below-ground parts	155	421	1185
Total	271	421	2851

[a] Gill (1975).
[b] Krishnamurthy (1976).
[c] Singh and Yadava (1974).

In the rainy and summer seasons energy-fixation exceeds loss but in winter the reverse is true as a considerable amount of energy carried over from the preceeding season is rapidly lost. On an annual basis the energy turnover amounts to 89% (Table 7.9).

In these grasslands there is an average uptake of nitrogen during the rainy season of 2.07 g m^{-2}. Out of this 1.21 g m^{-2} is reflected in the above-ground parts and only 0.86 g m^{-2} is retained by the roots. The low transfer of nitrogen from live shoots to the dead shoot compartment in this season indicates a net accumulation in live shoots. Possibly a considerable amount of nitrogen is transferred to live parts from senescing tissues. About 43% of the nitrogen retracted by plants in the rainy season is re-cycled to the soil during the same period. A portion of the nitrogen released may, however, come from the decomposition of carry-over herbage. The annual total uptake of nitrogen (2.93 g m^{-2}) is only slightly higher than the rainy season uptake, pointing to the relative inactivity of vegetation during the post-monsoon period. A comparison of rainy season and annual values indicates that during the post-monsoon period more nitrogen is retained by the roots. This may also be indicative of some transfer of nitrogen from live shoots to roots before the shoots senesce. The annual release of nitrogen through root and litter decomposition amounts to 2.30 g m^{-2} which more or less balances the total annual uptake.

The total uptake of phosphorus during the rainy season amounts to 0.30 g m^{-2}, of which 53% is retained by the roots. The rates of transfer of phosphorus from above-ground live to standing dead and from standing dead to the litter compartments are comparable (0.04 and 0.06 g m^{-2}) during the

TABLE 7.9

Seasonal and annual energy flow ratios (input to output) for semi-arid grasslands (based on Gill, 1975)

Transfer	Rainy season	Winter season	Summer season	Annual
ISR to TNP	0.001	0.0003	0.0002	0.0005
TNP to ANP	0.51	0.14	0.14	0.36
TNP to BNP	0.49	0.86	0.84	0.64
ANP to SD	0.30	1.27	1.00	0.92
SD to L	1.29	0.88	1.31	1.03
ANP to L	0.39	0.69	1.31	0.96
L to LD	1.21	1.07	1.00	1.12
BNP to RD	0.68	1.29	0.41	0.77
TNP to TD	0.59	2.21	0.52	0.89

ISR, total incident solar radiation.
TNP, energy captured in total net production.
ANP, energy used in above-ground net production.
BNP, energy used in below-ground net production.
SD, energy transferred from live shoots to standing dead shoots.
L, energy transferred from standing dead shoot to litter.
LD, energy dissipated via litter disappearance.
RD, energy dissipated via root disappearance.
TD, total energy dissipated from the system.

rainy season. On an annual basis, about 69% of the total uptake remains in the below-ground parts. The annual release of phosphorus through root and litter decomposition amounts to 0.48 g m^{-2}, which is slightly higher than the total uptake value indicating the influence of carry-over plant material.

The animal carrying capacity of a majority of the grasslands is low. On average it is around 2.5 sheep per ha or one ox or cow per 2.4 to 4 ha (Chakravarty, 1971). However, the carrying capacity of sown pastures may be as high as 7.6 sheep per ha (Chakravarty, 1968).

By and large most of the grasslands of the semi-arid regions suffer from over-grazing. The period of active vegetation growth is limited to 3—4 months during which most of the rainfall occurs. For the remainder of the year there is very little herbage due to a scarcity of soil water and too high temperatures. Grazing is often practised year round with no control. This results in the degeneration of grasslands and consequent soil erosion and total loss of fertility.

Burns et al. (1933) and Kumar and Godbole (1939) have shown that rotational grazing increases the carrying capacity of grasslands in the *Sehima—Dichanthium* cover type. Prajapati (1970), while studying the effect of different systems of grazing by cattle on *Lasiurus—Eleusine—Aristida* grassland, has reported that 2 months alternate grazing without any specific provision for seed production and 2 or 4 months early grazing are detrimental to the range vegetation as compared to early deferment of grazing during the growing season and during spring. Early deferment of 4 months (August—November) was more beneficial because all the plant species could complete their life-cycle and produce seeds for future regeneration. Upadhyay et al. (1971) compared the continuous versus deferred-rotational grazing systems and found that the latter was more advantageous as far as total animal grazing days are concerned (2925 in deferred-rotational as against 2097 in the continuous system).

Bhimaya et al. (1967) have observed that a 2-year protection and controlled grazing in dry grassland zones resulted in an increase in the above-ground herbage by 148.3, 91.9 and 116.3% on "poor", "fair" and "good" condition classes of rangelands. For the improvement of *Lasiurus sindicus* grasslands, Gupta and Saxena (1970) recommended fencing, weeding, re-seeding and deferred-cum-rotational grazing. During the first year of establishment of the seeded pasture, no grazing should be allowed. The range should be divided into compartments for efficient utilization and animals should be grazed through a system of rotational grazing.

In a study to elucidate the relations between animal production and the type of grazing system in rangelands of western Rajasthan, Ahuja (1964) found that gain in body weight both in heifers and ramlambs under continuous-controlled grazing (4.1 ha per heifer or 1.3 ha per ramlamb on a "fair" range) was highest (1656 kg per 100 ha in heifers, and 752 kg per 100 ha in ramlambs) while that under a system of deferred grazing for 16 weeks from 1 July followed by grazing at a controlled rate, was lowest (683 kg per 100 ha for heifer and 476 kg per 100 ha for ramlamb). When grazing was deferred for only 8 weeks from 26 August preceded and followed by grazing at a controlled rate, the results in body weight gain followed that of continuous controlled grazing. The above studies indicate that either the grasslands should be allowed to be grazed continuously at a controlled rate or deferment during a pre-determined suitable period be practised. Both the "controlled rate" as well as the duration and timings of deferment need to be investigated for different types of grazing lands within the semi-arid region.

Cropland

The semi-arid regions exhibit a complex seasonal pattern of land-use including arable cropping by Man and grazing by different classes of livestock. Bhardwaj (1965) and Whyte (1965) have traced the history of land-use in the Indo-Pakistan region. Throughout the Palaeolithic period the chief means of subsistence were hunting and food gathering. In the second glacial phase, people lived in small isolated groups along river terraces. In the fourth and fifth millenium B.C. agriculture was introduced, according to Piggot (1950) from the Middle East through Baluchistan into Sindh and further throughout the Indus valley. It appears that the climate of the region at that time was more humid than at present.

During the Vedic period (C. 2000—600 B.C.) the peasants were very active utilizing the ground water through wells and constructing canals to supply water to crops (Mazumdar et al., 1958).

In the subsequent period up to 700 A.D., which corresponds to the Hindu period of Indian history, irrigation systems were expanded and an equitable distribution of irrigation was ensured (Dunbar, 1949). During the British rule, the linking canal systems were set up.

The extent of cultivated and other land as recorded for 1968—1969 is shown in Table 7.10, for those states which include at least some semi-arid areas. While the forested areas in these states range from 2 to 33%, the area under cultivation ranges from 39 to 78%, the rest belonging to the category of uncultivated land, fallow land and land not available for cultivation. Land not available for cultivation comprises such habitats as barren outcrops of rocky hills, sandy deserts, alkaline and saline beds. From 16 to over 20% of such land occurs in arid and semi-arid areas. Significantly very high (over 20%) and high (16—20%) proportions of other uncultivated land (including cultivable wasteland) embraces the Marusthali, Aravalli ranges, the eastern Rajasthan uplands, the Malwa plateau, the western parts of the Vindhyan scarplands, the uplands of Uttar Pradesh, the Karnataka high plateau and a few districts of Andhra Pradesh and Tamil Nadu which have a semi-arid clime (Singh, 1974).

TABLE 7.10

Land-utilization in certain Indian States, 1968—1969 (in millions of ha)

State	Total [a] reporting area	Forest	Land not available for cultivation	Total other cultivable land except fallow land	Total fallow land	Net area sown
Andhra Pradesh	27.48	6.12	4.24	2.80	3.40	10.92
Gujarat	18.53	1.63	4.88	1.54	0.69	9.80
Haryana	4.40	0.01	0.48	0.12	0.43	3.27
Jammu and Kashmir	4.53	2.77	0.55	0.40	0.13	0.68
Karnataka	18.96	2.82	1.70	2.68	1.71	10.05
Madhya Pradesh	44.36	14.61	4.32	5.64	1.63	18.16
Maharashtra	30.77	5.42	2.47	2.29	2.22	18.37
Punjab	5.03	0.11	0.64	0.13	0.21	3.94
Rajasthan	34.04	1.24	5.93	8.14	5.43	13.31
Tamil Nadu	13.01	1.93	2.24	1.22	1.72	5.90
Uttar Pradesh	29.59	3.98	4.49	2.28	1.42	17.42

[a] Total reporting area means the area for which data were available.

Singh (1974) has analyzed the intensity of cropping and has divided the country into four groups:

(1) low intensity index (100—110%)

(2) medium intensity index (111—120%)

(3) high intensity index (121—130%)

(4) very high intensity index (over 130%)

The intensity of cropping here refers to the number of crops raised on a field in one agricultural year; for example, if one crop is raised on a field in a year, the index is 100% and if two crops are raised it is 200%.

The category of low intensity of cropping embraces Marushthali, Rajasthan, the Kutch peninsula, the Gujarat plains, the Kathiawar peninsula, the Malwa plateau, the Satpura range, the Chhota Nagpur plateau and Maharashtra plateau, southern Andhra Pradesh and Karnataka. Medium intensity of cropping is exhibited by the Tamil Nadu uplands, the extreme south of the Karnataka high plateau and the Vindhyan scarplands. Thus the low (100—110%) and medium (111—120%) intensities predominate in arid and semi-arid regions while the high intensities are confined to flood plains and deltas with optimal irrigation facilities and/or sufficient rainfall (Singh, 1974).

The following are the major crop regions within the semi-arid clime.

Cotton (*Gossypium* spp.): Gujarat, central Maharashtra, southern Punjab, south-western Madhya Pradesh, north-central Karnataka and southern Tamil Nadu.

Groundnut (*Arachis hypogea*): western Gujarat, Tamil Nadu, Inland Andhra Pradesh, north-western Maharashtra, northern Karnataka, south-western Maharashtra, central Punjab, central Uttar Pradesh, eastern Gujarat and Coastal Andhra Pradesh.

Sugarcane (*Saccharum officinarum*): western and north-central Uttar Pradesh, northern Haryana-Punjab, south-western Maharashtra, coastal Andhra Pradesh, northern Tamil Nadu and central Karnataka.

Maize (*Zea mays*): southern Rajasthan-Madhya Pradesh, northern Punjab, northern Andhra Pradesh, western and central Uttar Pradesh.

Jowar (*Sorghum vulgare*): northern Karnataka, central Maharashtra, south-western Maharashtra, south-western Uttar Pradesh, central Madhya Pradesh, Andhra Pradesh, central-eastern Rajasthan, central Inland Tamil Nadu and Gujarat.

Wheat (*Triticum sativum*): Punjab, Haryana, western, west-central, north-eastern, south-western and southern Uttar Pradesh, central Madhya Pradesh and Madhya Pradesh-Rajasthan.

Rice (*Oryza sativa*): eastern Tamil Nadu, western Karnataka, western Maharashtra, central Andhra Pradesh, eastern Madhya Pradesh and eastern Uttar Pradesh.

Millets (such as *Pennisetum*, *Eleusine*, etc.): Tamil Nadu, west-central and western Maharashtra, Gujarat, southern and northern Karnataka, south-

Areas (in millions of ha) and production (in millions of tonnes) of some major crops in the semi-arid regions during 1972—1973 (based on Mamoria, 1975)

State	Cotton		Sugarcane		Rice		Millets	
	Area	Production (million bales of 180 kg each)	Area	Production	Area	Production	Area	Production
Andhra Pradesh	0.303	0.115	0.121	9.209	2.578	3.649	0.561	0.126
Gujarat	1.790	1.383	0.038	1.801	0.405	0.148	0.138	0.067
Haryana	0.265	0.442	0.136	6.000	0.292	0.466	—	—
Jammu and Kashmir	0.001	0.014	0.001	0.015	0.204	0.348	—	—
Karnataka	0.882	0.418	0.104	8.458	1.010	1.749	0.368	0.107
Madhya Pradesh	0.699	0.334	0.546	1.430	4.482	3.345	0.156	0.328
Maharashtra	2.531	1.053	0.181	11.198	1.285	0.755	0.177	0.054
Punjab	0.504	1.016	0.105	4.840	0.477	0.955	—	—
Rajasthan	0.343	0.287	0.034	1.350	0.136	0.080	0.058	0.007
Tamil Nadu	0.297	0.352	0.113	9.770	2.641	5.384	0.446	0.339
Uttar Pradesh	0.050	0.041	1.342	58.743	4.413	3.280	0.425	0.247

State	Jawar		Bajra		Ragi		Gram		Wheat	
	Area	Production	Area	Production	Area	Production	Area	Production	Area	Production
Andhra Pradesh	2.527	1.057	0.433	0.134	0.200	0.177	0.072	0.019	0.017	0.006
Gujarat	1.015	0.213	1.716	0.884	0.050	0.025	0.034	0.028	0.371	0.548
Haryana	0.188	0.055	0.900	0.436	—	—	0.929	0.522	1.231	2.160
Jammu and Kashmir	—	—	0.174	0.092	—	—	0.003	0.008	0.180	0.168
Karnataka	1.794	1.082	0.266	0.080	0.937	0.811	0.242	0.041	0.299	0.110
Madhya Pradesh	1.956	1.514	0.202	0.128	0.018	0.004	0.132	0.020	3.529	2.447
Maharashtra	4.994	1.246	1.279	0.204	0.195	0.085	1.690	1.124	0.691	0.249
Punjab	—	—	0.129	0.108	—	—	0.321	0.208	2.401	5.361
Rajasthan	0.997	0.304	5.234	0.864	—	—	1.204	0.803	1.396	1.749
Tamil Nadu	0.659	0.505	0.458	0.276	0.290	0.269	0.004	0.002	0.002	0.008
Uttar Pradesh	0.659	0.446	1.053	0.693	0.265	0.310	1.883	1.415	6.362	7.633

ern Haryana-Punjab, western and north-eastern Rajasthan, southern Andhra Pradesh, north-central Madhya Pradesh, south-western Uttar Pradesh.

Pulses (leguminous crops): Haryana, Rajasthan-Madhya Pradesh, western Rajasthan, eastern Madhya Pradesh, central Madhya Pradesh, central Maharashtra, eastern, western and south-central and central Uttar Pradesh, north-eastern Andhra Pradesh.

Oilseeds: Punjab-Haryana, Haryana-Rajasthan, eastern and south-central Rajasthan, Rajasthan-Madhya Pradesh, south central, north-central and eastern Madhya Pradesh, Uttar Pradesh-Madhya Pradesh, northern Uttar Pradesh, eastern Maharashtra, north-eastern Andhra Pradesh.

The area and production of major crops in states which are at least in part

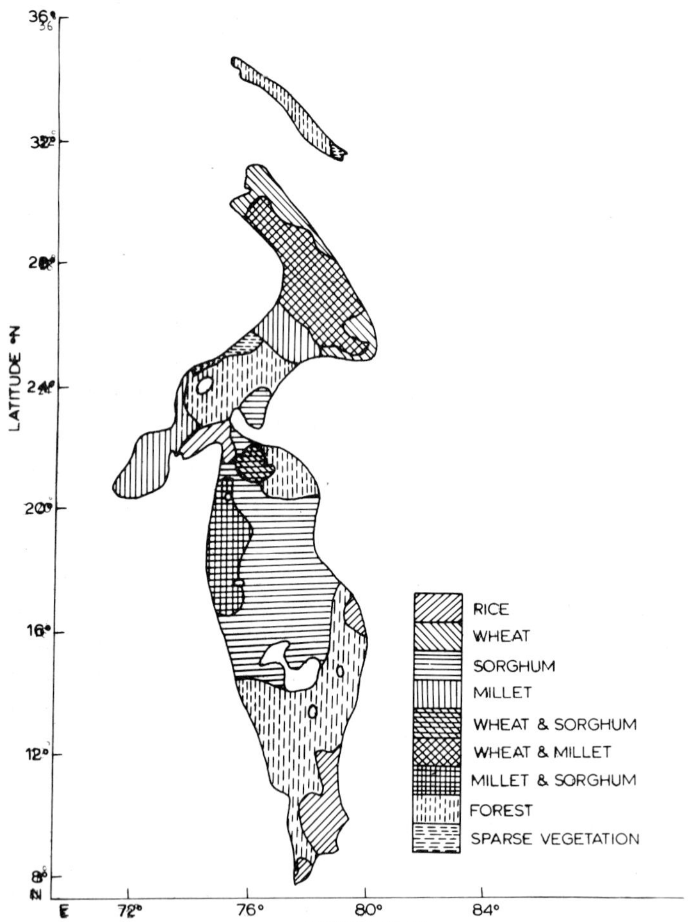

Fig. 7.8. Areas under major food grain crops in the semi-arid regions (based on Mellor, 1976).

TABLE 7.12

Possibility of droughts in a period of ten years in some of the drought-affected areas (based on Vyas, 1968)

Meteorological sub-divisions	Early-drought	Mid-drought	Late-drought	Total
Chhota Nagpur	—	—	1.5	1.5
U.P. (east)	—	—	2.6	2.6
U.P. (west)	—	1.0	—	1.0
Punjab (north-east)	1.1	1.0	—	2.1
Rajasthan (west)	1.3	2.2	1.1	4.6
Rajasthan (east)	—	1.1	—	1.1
Gujarat	—	1.4	—	1.4
Berar	2.2	—	3.9	6.1
Central Provinces (west)	1.1	—	—	1.1
Bombay (Deccan)	—	1.5	—	1.5
Hyderabad (north)	—	—	1.3	1.3
Mysore	—	1.0	1.5	2.5
Madras (south-east)	1.1	—	1.5	2.6
Madras (Deccan)	—	1.6	1.3	2.9
Madras coast	1.1	1.0	—	2.1

semi-arid are summarized in Table 7.11 for the year 1972—1973. Jowar (*Sorghum vulgare*) and Bajra (*Pennisetum typhoides*) are the major food crops where irrigation facilities are lacking. The harsh climate and lack of irrigation facilities result in often marginal cultivation and production in the semi-arid area. The area in the semi-arid regions under the four major food grains is shown in Fig. 7.8.

In semi-arid regions, 75 to 95% of the cultivation is rainfed; and since the rainfall is highly erratic besides being low, the frequency of crop failures owing to recurrent drought is high. According to information in Vadilal (1968) Madhya Pradesh, Konkan and Maharashtra may experience drought once in 5 years, Mysore and Vidarbha once in 4 years and Gujarat, east Rajasthan, west Uttar Pradesh and Madras once in 3 years. The drought spells leading to crop failures may occur either in the first part, middle or in the later part of the rainy season. Possibility of droughts in a period of 10 years in some of the drought-affected areas is shown in Table 7.12.

CONCLUSIONS AND RECOMMENDATIONS

The ever increasing human population and associated increases in livestock have put tremendous pressure on the semi-arid ecosystems which are already water-stressed and fragile. Large-scale deforestation coupled with grazing results in grasslands. Heavy grazing pressure decreases the plant cover and carrying capacity of these grasslands and encourages invasion by unpalatable desertic plants with low conservation value. This, together with increasing soil loss due to erosion, results in a loss of perennial plant cover moving the land towards a desertic condition. This calls for immediate attention for evolving a rational management plan.

One of the most important areas of concern is water management. The rainfall is unpredictable and erratic. The amounts and intensities of rainfall per storm are frequently too great to allow complete infiltration; the rainfall intensities often exceed the combined capacity of surface storage and infiltration, the latter being further reduced due to destruction of vegetal cover. Further, the rainfall is concentrated in 3—4 months each year according to the monsoonal pattern.

In such a water-limited, pulsed-system the grasslands are further stressed by uncontrolled grazing resulting in desertification. This emphasizes the need for detailed studies on the carrying capacity of grasslands and planning for optimal grazing strategies through regulation of grazing space, time and stocking density. Many areas may have to to re-seeded and protected for long periods in order to restore them to the status of healthy pastures. Provision of top-feed (tree/shrub fodder) through farm-forestry may play an important role in rejuvenating the grazing lands by reducing the present pressure on them. Ganguli et al. (1964) have pointed out that leaves of *Acacia nilotica*, *Prosopis spicigera*, *Salvadora oleoides*, *S. persica* and *Zizyphus nummularia* are quite palatable while those of *Acacia senegal*, *Albizzia lebbeck*, *Anogeissus pendula* and *Calligonum polygonoides* are moderately so.

Techniques for dry farming and effective conservation need to be studied and popularized. On account of a constant hunger for forest soils there has always been a continued struggle between agriculture and forestry. There has been constant endeavour to advance one or the other reason for the destruction of a good forest and converting it into agricultural land and ultimately into cultural waste. This can be checked by intensifying agriculture on the existing agricultural land through provision of irrigation, improved seeds and judicious application of fertilizers and by discouraging marginal cultivation. There can be no two opinions as to the necessity of increasing agricultural production to meet the demands of the ever increasing population. This increase could result either from extension of crop acreage or from increasing the intensity of cropping with a view to increasing yield per unit of land area. According to Ralph (1970), in the absence of improved water management, the possibility of increasing the area under cultivation in India is rather limited, hence more reliance must be placed on increasing productivity per unit area through improved cropping techniques.

Some pressure on forest can be reduced by encouraging farming for fuel and fodder trees which would also act as shelter belts for wind erosion. Besides the species listed above, others that hold promise in this respect are

Acacia leucophloea, *Aegle marmelos*, *Anogeissis latifolia*, *Bauhinia purpurea*, *Dalbergia sissoo*, *Pongamia pinnata* and *Lannea coromandelica*. Besides farm-forestry, the concept of mixed farming is worth considering. Mixed farming involves the complementary use of livestock and crops (including pasture) to reap the maximum benefits from both. Its ultimate aim is a self-sufficient unit where all the primary needs of the rural family will be met, by growing food, pasture and cash crops, rearing dairy or beef animals, sheep, pigs or poultry, or combinations of these, and selling off any surplus produce or animals. The advantages of the system lie in the fact that the land is used intensively, fertility improves following the growth of leguminous crops or pasture leys in a rotation, in the latter case grazing animals will return dung and urine to the soil. Crop residues are fully utilized for soil building or fodder and diversification provides a most stable source of income. This calls for micro-level planning starting from a village unit and subsequently covering the entire region.

This review indicates a general paucity of information on ecological conditions operating within the semi-arid zone. There has been a complete lack of any co-ordinated total system study of natural or of man-modified ecosystems. Investigations on the impact of socio-economic factors on the semi-arid ecosystems in India are also wanting. Such studies integrated with ecosystem-level studies of these areas will be of tremendous help in future planning.

ACKNOWLEDGEMENTS

We are grateful to the various scientists for allowing us the free use of their data and publications. Mr. A.K. Saxena assisted us in making rough sketches of some of the maps.

REFERENCES

Ahuja, L.D., 1964. Livestock management on rangelands in western Rajasthan. In: Proc. Symp. Problems of Indian Arid Zone. UNESCO and Ministry of Education, Jodhpur, India, pp. 231—240.

Ayyar, T.V.R., 1940. Handbook of Entomology for South India. Supdt. Government Press, Madras, pp. xviii + 528.

Bhardwaj, O.P., 1965. The arid zone of India and Pakistan. In: L. Dudley Stamp (Editor), Arid Zone Research XVII. A History of Land Use in Arid Regions. UNESCO, Paris, pp. 143—174.

Bharucha, F.R. and Shanbagh, G.Y., 1957. Precipitation-effectiveness in relation to the vegetation of India, Pakistan and Burma. Bot. Mem. No. 3, University of Bombay, 157 pp.

Bhimaya, C.P., Kaul, R.N. and Ahuja, L.D., 1967. Forage production and utilization in arid and semi-arid rangelands of Rajasthan. In: Proc. Joint Symp. of India's Natl Inst. Sci. India and Indian Counc. Agric. Res. on India's Food Problems, New Delhi, pp. 215—223.

Bor, N.L., 1960. Grasses of Burma, Ceylon, India and Pakistan. Pergamon Press, London, 767 pp.

Burns, W., Kulkarni, L.B. and Godbole, S.R., 1933. Grassland problems in western India. Bull. Agric. Dep. Bombay, 171: 1—151.

Chakravarty, A.K., 1968. Role of some perennial grasses of western Rajasthan in development of grasslands of the arid zone of India. In: R. Misra and B. Gopal (Editors), Proc. Symp. Recent Advances

in Tropical Ecology Part II. International Society for Tropical Ecology, Varanasi, India, pp. 620—630.
Chakravarty, A.K., 1971. Pasture production and its use in the arid and semi-arid areas of Rajasthan (India) and Kazakhstan (USSR). Ann. Arid Zone, 8: 251—256.
Champion, H.G. and Seth, S.K., 1968. A Revised Survey of Forest Types of India. Government of India Press, Nasik, 404 pp.
Dabadghao, P.M. and Shankarnarayan, K.A., 1973. The Grass Cover of India. Indian Council of Agricultural Research, New Delhi, 713 pp.
De Martonne, E., 1926. Areisme et indice d'aridite. C.R. Acad. Sci., 182: 1395—1398.
Dunbar, G., 1949. A History of India (2 Vols). Nicholson and Watson, London.
Ganguli, B.N., Kaul, R.N. and Nambiar, T.N., 1964. Preliminary studies on a few top-feed species. Ann. Arid Zone, 3: 33—37.
Gill, J.S., 1975. Herbage Dynamics and Seasonality of Primary Productivity at Pilani, Rajasthan. Ph.D. Thesis, Birla Institute of Technology and Science, Pilani, India.
Gupta, R.K. and Saxena, S.K., 1968. Some ecological observations on the vegetation of Mt. Abu in western Rajasthan. Indian For., 94: 315—323.
Gupta, R.K. and Saxena, S.K., 1970. Some ecological aspects of improvement and management of Sewan (*Lasiurus sindicus*) rangelands. Ann. Arid Zone, 9: 193—208.
Hoon, R.C., 1962. The incidence of white ants (termites) in the region of the Hirakund Dam Project. Proc. Int. Symp. on "Termites in the Humid Tropics". UNESCO, New Delhi, pp. 141—149.
Köppen, W., 1918. Klassifikation der Klimate nach Temperatur, Niederschlag und Jahresverlauf. Peterm. Mit., 64: 193—203; 243—248.
Krishnamurthy, L., 1976. Systems Analysis and Optimization of Land Use systems with reference to a Grazing Land. Ph. D. Thesis, Saurashtra University, Rajkot, India.
Krishnan, A., 1969. Some aspects of water management for crop production in arid and semi-arid zones of India. Ann. Arid Zone, 8: 1—17.
Krishnan, M.S., 1960. Geology of India and Burma. Higginbotham Ltd, Madras-2, 604 pp.
Krishnan, M.S., 1974. III. Geology. In: M.S. Mani (Editor), Ecology and Biogeography in India. Dr. W. Junk b.v. Publishers, The Hague, pp. 60—98.
Kumar, A. and Joshi, M.C., 1972. The effect of grazing on the structure and productivity of vegetation near Pilani, Rajasthan, India. J. Ecol., 60: 665—675.
Kumar, L.S.S. and Godbole, S.R., 1939. Rotational (Intermittent—Intensive) system of grazing in Bombay Province. Agric. Livestock, India, 9: 667—675.
Mamoria, C.B., 1975. Geography of India (Agricultural Geography) S.L. Agarwala and Co. Educational Publishers, Agra-3, 700 and I—III pp.
Mathur, C.M., 1960. Forest types of Rajasthan. Indian For., 86: 734—739.
Mathur, C.M. and Verma, S.K., 1964. Vegetation types of Sirohi Forest Division. Indian For., 90: 413—430.
Mazumdar, R.C., Raychaudhuri, H.C. and Kalikinkar, D., 1958. An Advanced History of India. Part I. Ancient India. McMillan, London, 1122 pp.
Meher-Homji, V.M., 1962a. Phytogeographical studies of the semi-arid regions of India. Ph.D. Thesis, Bombay University.
Meher-Homji, V.M., 1962b. The bioclimates of India in relation to the vegetational criteria. Bull. Bot. Surv. India, 4: 105—112.
Meher-Homji, V.M., 1963. A contribution to the study of plant succession and climax vegetation with special reference to the dry parts of North-West and South-East India. Trop. Ecol., 4: 40—54.
Meher-Homji, V.M., 1965. Aridity and semi-aridity. A phytoclimatic consideration with reference to India. Ann. Arid Zone, 4: 152—158.
Meher-Homji, V.M., 1967. On delimiting arid and semi-arid climates in India. The Indian Geogr., XLII: 1—6.
Meher-Homji, V.M., 1971. Bio-climatic variability with special reference to India. Trop. Ecol., 12: 155—176.
Meher-Homji, V.M., 1975. Phytogeography of the Thar Region. In: R.K. Gupta and Ishwar Prakesh (Editors), Environmental Analysis of the Thar Desert. English Book Depot, Dehradun, pp. 237—273.
Mellor, J.W., 1976. The agriculture of India. Sci. Am., 235: 154—163.
Misra, C.M., 1973. Primary Productivity of a Grassland Ecosystem at Ujjain. Ph.D. Thesis, Vikram University, Ujjain, India.
Nair, N.C. and Nathawat, G.S., 1957. Vegetation of Harsh Nath, Aravalli Hills. J. Bombay Nat. Hist. Soc., 54: 281—301.
Patel, G.A. and Patel, H.K., 1954. Seasonal incidence of termite injury in the Northern Parts of the Bombay State. Indian J. Entomol., 15: 376—378.
Piggot, S., 1950. Prehistoric India. Harmondsworth. Penguin Book Limited, London, 293 pp.
Prajapati, M.C., 1970. Effect of different systems of grazing by cattle on *Lasiurus—Eleusine—Aristida* grassland in arid region of Rajasthan vis-a-vis animal production. Ann. Arid Zone, 9: 114—124.

Prakash, I., 1974. The ecology of vertebrates of the Indian desert. In: M.S. Mani (Editor), Ecology and Biogeography in India. Dr. W. Junk b.v. Publishers, The Hague, pp. 369—420.

Prakash, I. and Jain, A.P., 1971. Some observations on Wagner's gerbil, *Gerbillus nanus indus* (Thomas) in the Indian desert. Extrait Mammal., 35: 614—628.

Prakash, I. and Taneja, G.C., 1969. Reproduction biology of the Indian desert hare, *Lepus nigrocollis dayanus* Blandford. Extrait Mammal., 33: 102—117.

Ralph, W.C., 1970. Food self-sufficiency in India. In: S.C. Mathur (Editor), Agricultural Policy and Food Self-Sufficiency in India. Assoc. Publ. House, New Delhi, pp. 32—46.

Rao, R.S., 1970. Studies on the flora of Kutch, Gujarat State (India) and their utility in the economic development of the semi-arid region. Ann. Arid Zone, 9: 125—142.

Satyanarayan, Y. and Shankarnarayan, K.A., 1964. Vegetation of Bellary district, Mysore State, II. Vegetation types. Ann. Arid Zone, 3: 54—62.

Sen-Sarma, P.K. and Mathur, R.N., 1961. *Trinervitermes biformis* (Wasmann) a mound building termite in South India. Indian For., 87: 252.

Shankar, V., Shankarnarayan, K.A. and Rai, P., 1973. Primary productivity energetics and nutrient cycling in *Sehima—Heteropogon* grassland. I. Seasonal variation in composition, standing crop and net production. Trop. Ecol., 14: 238—251.

Shankarnarayan, K.A., 1962—1963. Parallelism between grassland and forest types of Bellary district. Ann. Arid Zone, 1: 132—141.

Singh, J., 1974. An Agricultural Atlas of India. A geographical Analysis. Vishal Pub. Univ. Campus, Kurukshetra (Haryana, India), 356 pp.

Singh, J.S., 1976. Structure and function of tropical grassland vegetation of India. Pol. Ecol. Stud., 2, 2: 17—34.

Singh, J.S. and Joshi, M.C., in press. Primary production. In: R.T. Coupland (Editor), Grassland Ecosystems of the World: Analysis of Grasslands and Their Uses. Cambridge University Press.

Singh, J.S. and Yadava, P.S., 1974. Seasonal variation in composition, plant biomass and net primary productivity of a tropical grassland at Kurukshetra, India. Ecol. Monogr., 44: 351—375.

Subrahmanyam, V.P., 1971. Climatic water balances of the Indian arid zones. Proc. Symp. Problems of Indian Arid Zone. Jodhpur, pp. 405—411.

Subrahmanyam, V.P. and Sastry, C.V.S., 1971. A climatic appraisal of the ground water potential of the semi-arid region of South India. Geogr. Rev. (India), 33: 81—87.

Thornthwaite, C.W., 1948. An approach towards a rational classification of climates. Geogr. Rev., 38: 55—94.

Thornthwaite, C.W. and Mather, J.R., 1955. The water balance. Publications in climatology Drexel Institute of Technology (Lab. of Climat. Genterton, N.J., U.S.A.) 8: 1, 1—104.

Upadhyay, V.S., Dabadghao, P.M. and Shankarnarayan, K.A., 1971. Continuous versus deferred-rotational grazing system. In: Annual Report, Indian Grassland and Fodder Research Institute, Jhansi, p. 48.

Vadilal, D. (Editor), 1968. Foundations of Indian Agriculture. Vora and Co., Pub. Pvt. Ltd, Bombay, 378 pp.

Varshney, C.K., 1972. Productivity of Delhi grasslands. In: P.M. Golley and F.B. Golley (Editors), Papers from a Symposium on Tropical Ecology with an Emphasis on Organic Production. Institute of Ecology, University of Georgia, Athens, U.S.A., pp. 27—42.

Vyas, L.N., 1967. Contribution to the flora of North-East Rajasthan. J. Bombay Nat. Hist. Soc., 64: 191—231.

Vyas, V.S., 1968. Some aspects of irrigation and power development. In: D. Vadilal (Editor) Foundations of Indian Agriculture. Vora and Co. Pub. Pvt. Ltd, Bombay, pp. 135—151.

Wadia, D.N., 1961. Geology of India, 3 edn. McMillan, London (Revised), 536 pp.

Whyte, R.O., 1965. Evolution of land use in south-western Asia. In: L. Dudley Stamp (Editor), Arid Zone Res. XVII. A History of Land Use in Arid Regions. UNESCO, Paris, pp. 57—118.

Yadava, P.S. and Singh, J.S., 1977. Grassland Vegetation: Its Structure, Function, Utilization and Management. Today and Tomorrow's Printers and Publishers, New Delhi, India, 132 pp.

Chapter 8

ECOLOGY AND MANAGEMENT OF SEMI-ARID ECOSYSTEMS IN BRAZIL

GEORGE EITEN * and ROBERT GOODLAND **

* *Universidade de Brasília, Brasília, DF (Brazil)*
** *World Bank, Washington, D.C. (U.S.A.)*

INTRODUCTION

This chapter treats two large-scale vegetation types in Brazil, the northeastern caatinga, and the central cerrado. The other large-scale vegetation types in the country are the Amazon forest, the Atlantic forest, and the southern Brazil campo limpo (Fig. 8.1).

The caatinga experiences a true semi-arid climate, but although the cerrado would not normally be classified as a semi-arid type it is included here for the following reasons. It has a long dry season in which, even with

Fig. 8.1. Distribution of vegetation types in Brazil. S = Amazon savanna, F = upland forest, C = cerrado, HAC = high altitude campo, CL = campo limpo (napeadic campo).

the use of fertilizers, a crop cannot be raised reliably every year without irrigation. In addition, in perhaps one year in five, the wet-season crop is materially reduced or fails over much of the cerrado as a result of insufficient rain in the wet season. Furthermore, dry spells of up to 20 days or more in even a generally good wet season reduce yields in shallow-rooted crops so that the crop ecosystem is in fact water-limited although the natural vegetation is adapted to this. Because of these problems the cerrado is here regarded as a semi-arid ecosystem.

CAATINGA

Climate

The only true semi-arid climate in Brazil is that of the caatinga vegetation province in the north-east. This covers almost all of Ceará, the eastern edge and southern half of Piauí, more than half of Bahia (excluding its eastern and western sides), north-central Minas Gerais, and the western parts of Rio Grande do Norte, Paraíba, Pernambuco, Alagoas and Sergipe. In this region the rainfall varies from an annual average of about 300 mm or slightly lower in the driest interior parts, to 750 or 800 mm where it meets cerrado along its western and south-western edges (and also along a small portion of its north-eastern corner in Rio Grande do Norte), and to 1000 mm where it meets coastal Atlantic forest along its eastern edge. The northern edge of the caatinga province reaches the ocean, separated only by a narrow littoral vegetation. The dry season lasts for 7—10 (—11) months. There is great variation in rainfall from year to year and sometimes a year or more may pass with no rain at all in the interior drier portion, the so-called "Drought Polygon". These droughts have always caused great social problems because of the inevitable emigrations from the region; yet when it does rain, the soil is generally fertile because leaching is not excessive, and crops are good.

Vegetation and soils

Caatinga vegetation occurs at the lower elevations. The natural undisturbed vegetation consists of a series of physiognomic forms of which the most widespread (where too shallow a soil depth is not a limiting factor) is a closed scrub 3—5 m tall, often but not always with scattered low trees up to 6—8 m tall whose crowns overtop the scrub layer. Both scrub and trees are of many families but the most common in individuals or cover belong to the Leguminosae. Most woody species are spiny. Cacti are common, particularly spherical and cylindrical forms, as well as *Opuntia* among the lower cacti, and cylindrical species of the *Cereus* tribe among the tree cacti. In the drier areas, the scrub opens out to a lower, 1.5—2 m tall open scrub or a savanna with a short-grass layer 20—30 cm tall in flower and scattered spiny shrubs 1.5—3 m tall plus low cacti clumps. The open savanna form is called "seridó". Where the soil is absent, such as on rock outcrops and pavements, and on substrates made up of small stones and pebbles with very little fine soil, the vegetation is open or scattered, made up of low cacti and spiny, spherical-bushy dwarf shrubs (*Mimosa*, etc.) (Figs. 8.2 and 8.3). On deeper soils, where rainfall is sufficient, an open arboreal woodland or closed forest of caatinga low-tree species occurs, or used to occur before they were cleared. Almost all caatinga trees and shrubs are completely deciduous, but in some areas, such as southern Piauí, on deep soil over sandstone, the scrub caatinga is semi-deciduous, that is, at any one time at the height of the dry season some individuals have leaves, new or old, and some are completely bare.

It is important to emphasize that most of the original area of natural caatinga was, and still is, closed scrub through which it is difficult to walk because the branching is dense and spiny. Most photographs that have been published of the caatinga show an open vegetation of scattered shrubs and cacti; they may given an incorrect idea of what caatinga is really like. These photographs, however, were taken either in cleared areas or on rock outcrops. Rock outcrops make up a very small proportion, less than 1%, of the total caatinga area.

Fig. 8.2. Low cactus and leguminous shrub caatinga vegetation on pure pebbly substrate, south-western Pernambuco.

Fig. 8.3. Dense scrub caatinga with overtopping low trees, western Pernambuco.

The caatinga occurs on soils derived from many types of rock, igneous, metamorphic or sedimentary (including sandstone in southern Piauí and other areas), but limestone is one of the extremely common surface rocks, covering a larger percentage of the caatinga area than it does in any other vegetation type in Brazil. Caatinga also covers areas of alluvial soil. On limestone especially, as well as on other rocks, the soil is usually quite shallow, not more than about 1 m deep. This is in contrast to the other vegetation provinces in Brazil, the Amazon and Atlantic forest regions and the cerrado, for instance, where most of the soil, not only alluvial soil and loess but also autochthonous soil, is several tens of metres deep. The shallow soil depth is not limiting to caatinga physiognomy under the rainfall over most of the region, although it might be limiting to the forest where this form is permitted in the higher rainfall areas. The thin caatinga soil aggravates water supply to the plants in the dry season, for not much water can be stored, but of course the native flora is adapted to this. In a few areas salts accumulate in the soil but this is not a general problem in the caatinga.

Past and present land-use

The eastern edge of the caatinga region receives more rainfall and has the largest portion cleared for agriculture, including sown pastures. This moister caatinga is called "agreste". The driest areas in the interior have only recently been cleared for cultivation with sisal or spineless opuntia for cattle feed. The dry interior caatinga is often called "sertão". This word in other parts of Brazil does not refer to a vegetation type but means the sparsely-inhabited backlands ("the bush", "the outback") in general.

There has never been any study on management of the caatinga vegetation for economic purpose (although individual caatinga species have been studied). Large parts of the caatinga have been cleared or partially cleared for unseeded pasture. The uncleared caatinga is also used for extensive grazing, especially by cattle. Goats and some sheep are usually also present near cities, villages and scattered houses. A continuous grass layer is not common in the natural dense caatinga so that here most of the feed is browse, including legume pods and other fruits that fall from the plants. In the natural "seridó" savanna form of caatinga, and in cleared areas, grass is more common. Even so, fire is not often used in the caatinga, unlike its universal use in the cerrado.

At present, agronomists are recommending that caatinga pasture be seeded to buffel grass (*Cenchrus ciliaris*), which is highly resistent to drought.

Mountain tops in the caatinga region catch more rain (orographic precipitation) and this is more constant over the year. The caatinga is replaced there by semi-deciduous mesophytic forest or, where the rainfall is greater, by evergreen mesophytic forest ("tropical rain forest"). There is a surprising number of tree species in these evergreen forests which are identical to Amazon forest species. These areas of forest on mountain tops are called "brejos" in the north-east. In the rest of Brazil this word means "marshes". Almost all the north-east "brejos" have been cleared for agriculture.

There are an appreciable number of land vertebrates restricted to caatinga and cerrado vegetation, or to one of these. Several other species live in the gallery forests and their streams and in the occasional disjunct upland forests that occur throughout the caatinga and cerrado regions. These animals often wander into caatinga or cerrado vegetation itself (Vanzolini, 1963; Sick, 1965; Von Ihering, 1968; Cabrera and Willink, 1973; Melo Carvalho et al., 1978). Among those hunted for food and skins in caatinga vegetation are: mammals; the armadillo, *Priodontis giganteus* (tatu canastra) and the rodent, *Kerodon rupestris* (moco). The armadillos, *Tolypeutes matacos* and *T. tricinctus* (tatu bola) occur in both caatinga and cerrado vegetation. Birds; *Zenaida auriculata* (avoante). Living in forest but wandering into caatinga are the rodents, *Cavea aperea* (preá), and *Dasyprocta agouti* (cutia).

CERRADO

Climate and vegetation

The cerrado province of central Brazil occurs in an intermediate rainfall region, from about 750—800 mm mean annual precipitation where it meets the caatinga province along its north-eastern edge, to about 2000 mm where it meets forest provinces, namely, continuous mesophytic semi-deciduous or evergreen Amazonian forest along its western edge and Atlantic forest along its south-eastern edge. The climatic boundary based on annual rainfall is only approximate, for there is really a range of precipitation values along the dry and wet sides of the cerrado region in which, respectively, both cerrado and caatinga or both cerrado and mesophytic forest can occur. Which does occur depends on (1) the length of the dry season (how many dry months there are, and the exact amount of rainfall in the dry months) for the same annual precipitation; (2) water retention capacity of drained upland soils, and closeness of the water table to the surface in the dry season on slopes and in valleys, and (3) fertility of the soil and amount of available aluminum present. Thus, although the forest—cerrado boundary in northern Mato Grosso and Maranhão more or less coincides with the 2000 mm isohyet, the western edge of the cerrado region along the Araguáia River (separating Pará from Goiás), gives way to Amazon forest where the rainfall is approximately 1700 mm. This drier climate also enters the Amazon forest region as a wide north-west-pointing wedge, covering south-eastern, central, and north-western Pará, and eastern Roraima Territory. The rainfall over this wedge varies from 1700 mm to almost 2000 mm per year and reduces to almost 1500 mm in north-eastern Roraima. Thus, the drier wedge separates Amazonia into a larger western part and a smaller eastern portion consisting of Maranhão, the north-eastern corner of Pará, and Amapá. While the driest part of the wedge in north-eastern Roraima is a large savanna region (Campos do Rio Branco, continguous with the Rupununi Savanna of southern Guyana), the rest of the wedge is covered with continuous forest but in which there are scattered large areas of savanna. These savannas, as in Amazonia in general, are not cerrado since the vascular species richness per hectare is far too low, and half or more of the species are not cerrado species. However, some of the cerrado-like savannas of south-eastern Pará (which is near to the cerrado region of central Brazil) may possibly yet prove to fall into that vegetation type.

The cerrado—Amazon forest transition region is further complicated by the presence of grassy seasonal marsh, usually with scattered raised soil hummocks whose tops are covered with cerrado plants. On flat land covered with a thin layer of water during the wet season or not covered but with the regular water table or a perched water table near to the surface (a few centimetres to about a metre deep), this seasonal marsh, when it contains hummocks, is known as "pantanal". Gentle slopes with the water table coming near to the surface or flushing out and running downslope in a thin film, when containing the hummocks, are known as "campos de murunduns". Pantanal may occur in areas of only a few hectares up to continuous areas 100 km wide as on the Ilha do Bananal on the Araguáia River. Campos de murunduns occur on gentle valley sides and are usually from a fraction of a hectare to a few hectares in area in any one place. A very small percentage of the total area has transitions between cerrado and pantanal, or between cerrado and campo de murunduns.

The main part of the cerrado region in central Brazil has mostly an annual average rainfall of between 1300 and 1700 mm. Cerrado covers about 85% of the region, the rest being gallery forest on drained or swampy soil and "vereda" (grassy permanent marshes with buriti palms, *Mauritia vinifera*) along streams, and, on the uplands, small areas of mesophytic forest on more fertile soils, and non-cerrado campos on thin soils.

Undisturbed primary cerrado itself occurs in a number of different physiognomies, from closed-canopy forests and open-canopy woodlands (Fig. 8.4) (both called "cerradão"), to tree-and-scrub woodlands (trees scattered, visual dominance shared with shrubs) (Fig. 8.5) and open or closed scrub (these three all being called "cerrado" in a restricted sense of the word), savannas with scattered trees, or trees-and-scrub, or just scrub (Figs. 8.6 and 8.7) (called "campo cerrado" and when very open, "campo sujo"), and finally, areas covered with only a grass—herb layer ("campo limpo") (Eiten, 1968). Taller denser stands give way to lower and more open stands when cut and/or burned annually. It must be emphasized that density and height of cerrado does not depend on variations in rainfall but on depth and fertility of the soil.

The native cerrado ground layer is a mixture of grasses and forbs. Sedges

Fig. 8.4. Well developed cerrado woodland (= cerradão), west-central Maranhão.

Fig. 8.5. Tree/shrub cerrado woodland (= one of the forms of cerrado in the strict sense), São José dos Campos, São Paulo.

Fig. 8.6. Cerrado savanna with scattered shrubs (= one of the forms of campo cerrado), Brasília, DF.

Fig. 8.7. Cerrado shrub savanna on lithosol slope (= campo sujo), dry season, left half burned, Brasília, DF.

occur with relatively small biomass. Leguminous herbs and semi-shrubs are rich in species, usually making up the single largest family, and provide an appreciable part of the biomass both in the ground layer (i.e., field layer) and in the tree and taller shrub layer. Although the floristic composition of cerrado changes from place to place, the common plants are widespread. (Southern Maranhão, however, has a much smaller species similarity with the rest of the cerrado region even for common plants.) As an example of the native pasture plants of the cerrado, we can mention the more important grasses and small legumes for the Triângulo Mineiro cerrado of western Minas Gerais (Goodland, 1969, 1970). They are as follows.

Grasses: *Andropogon condensatus*, *A. lateralis*, *Aristida implexa*, *A. megapatomica*, *A. pallens*, *Axonopus pressus*, *Echinolaena inflexa*, *Eragrostis maypurensis*, *Gymnopogon foliosus*, *G. spicatus*, *Melinis minutiflora*, *Panicum campestre* and other spp., *Paspalum carinatum* and other spp., *Trachypogon mollis*, *Tristachya chrysothrix*, *T. leiostachya*.

Small legumes: *Aeschynomene* spp., *Andira humilis*, *A. nana*, *Bauhinia rufa*, *Cassia cathartica*, *C. langsdorfii*, *C. rugosa* and other spp., *Centrosema* sp., *Clitoria guianensis*, species of *Crotalaria*, *Desmodium*, *Eriosema*, *Erythrina*, *Galactia* and *Indigofera*, *Mimosa platyphylla* and other spp., *Phaseolus* sp., *Rhynchosia* sp., *Stylosanthes montevidensis*, *S. viscosa*, *Zornia* sp.

A grass cover is universal in the cerrado except in the densest cerradões. *Paspalum* has the largest number of species of all grass genera in the cerrado, as it has in tropical America in general. Other large cerrado grass genera,

Panicum, *Andropogon*, *Aristida* and *Eragrostis* are common in the world flora. *Gymnopogon*, *Echinolaena*, and *Ichnanthus* are small cerrado grass genera restricted to the neo-tropics. *Echinolaena inflexa*, a grass 3—4 dm tall, with one to a few culms per clump, often has 100% cover in the denser tree and scrub woodland cerrados in Brasília. *Thrasya* and *Leptocoryphium*, also small neo-tropical genera, are occasionally found in the cerrado but are even more typical of open moist ground and are common in the Amazon—Orinoco—Guiana savannas. *Tristachya* and *Trachypogon* are common to the African and American tropics. Both are widespread in cerrado. *Tristachya leiostachya* is the tallest cerrado grass in central Brazil, almost always exceeding 2 m and often exceeding 3 m as erect culms in flower. *Trachypogon plumosus* is the most common cerrado grass in southern Maranhão, with the greatest constancy and cover (followed by *Aristida longifolia*), and dominates large areas of the llanos of Colombia and Venezuela, but does not dominate the central Brazil cerrados. In southern Maranhão, it grows to 1.5—2 m tall in flower where the cerrado has been burnt in the previous dry season, and to 3 m tall or more where the cerrado has not been burnt for several years.

Chloris and *Setaria*, large genera of tropical and sub-tropical grasslands, are common in the cerrado. *Melinis minutiflora*, molassas grass, has invaded the southern half of the cerrado region. In cerrados whose soil has not been disturbed, it often chokes out everything else, especially if not regularly burned. It also invades pasture derived from forest in São Paulo, Minas Gerais and southern Goiás, as well as being a common waste ground and roadside plant in this region. It has not yet been ascertained whether molassas grass is native to the neo-tropics or not, but its spread and dominance in Brazil is only within the last few decades.

Hyparrhenia rufa and *Rhynchelytrum repens* have been introduced from Africa. Both are roadside plants common in the southern part of the cerrado region, while the former has also invaded many areas of cerrado. In Brasília to date, *Melinis*, *Hyparrhenia* and *Rhynchelytrum* are confined to scraped and piled soil and are mostly absent or only sparse and scattered in cerrados whose soil has not been disturbed. *Panicum maximum*, *Eriopogon densiflorus* and *Arundo donax* are introduced weedy grasses in bulldozed cerrado soil in Brasília but do not spread aggressively.

No cerrado grass is carpet-forming, almost all grow as clumps from a few cm to rarely more than 30 cm basal diameter. Some species grow as single-stemmed plants, such as *Gymnopogon foliosus* or as plants with one to a few culms from a narrow base, such as *Echinolaena inflexa* and *Tristachya* spp. The mature grass layer in general (not counting scattered taller culms of minute total cover) in the cerrado varies from 3—12 dm in flower, never less and very rarely more.

After burning, the ground layer of herbs, recurrent semi-shrubs and recurrent shrubs, grows back to its full cover by the end of the next wet season, or at the latest, by the end of the following season. When first recuperating in the early part of the wet season after a burn, either grasses or broad-leaf plants may predominate in the ground layer. At its full cover, the ground layer is almost always dominated by grasses, with the broad-leaf plants of the ground layer making up to 10—20 (—50%) of the total ground cover. Except in the densest cerrados, where the grass proportion is reduced both absolutely and relatively, there is no correlation between the openness of the woody layer and the proportion of the ground layer that is made up of grasses as compared to broad-leaf plants.

Table 8.1 summarizes the differences between forest, cerrado and caatinga regions in Brazil. A general account of the cerrado environment and natural vegetation can be found in Eiten (1972).

Useful animals

The following vertebrates are hunted in the cerrado region for food and skins. Specific to cerrado vegetation are the mammals: the armadillo, *Dasypus novemcinctus* (tatu-galinha), and the deer, *Ozotocerus bezoarcticus* (veado-campeiro). In addition, as stated before, the armadillos *Tolypeutes matacos* and *T. tricinctus* (tatu-bola) occur in both cerrado and caatinga vegetation. Birds: the ostrich-like *Rhea americana* (emu), the quail-like *Nothura* spp. (codorna), *Rhinchotus rufescens* (perdiz), *Cryptarellus* spp. (inhambu), and the pigeon-like *Columbina passerina* (rolinha) and *Lettotila verreauxi* ssp. *decitiens* (juriti). Reptiles: the lizard, *Tupinarubis nigropunctatus* (teuí), and the tortoise, *Testudo* spp. (jaboti). Living in forests in the cerrado region but often wandering into cerrado vegetation are the mammals, the tapir, *Tapirus terrestris* (anta), the rodents, *Hydrochoeris capybara* (capivara), *Coelogenis paca* or *Agouti paca* (paca), *Cavea aperea* (preá), *Dasyprocta agouti* (cutia), the peccaries, *Tayassus tayassu* or *T. tacaju* (caititu) and *T. albirostris* or *T. pecari* (queixada), and the monkeys, *Callitrix* spp. (mico, food and as pets) and *Cebus apela* (macaco-prego, food and as pets), etc. In addition, non-food forest mammals such as the jaguar (*Panthera onca*, called "onça pintada" when spotted and "onça preta" when black) and the small jaguaritirica (*Felis pardalis*), prized for their skins, are often found in cerrado vegetation hunting their prey.

Past and present land-use

Although, climatically, cerrado does not occur in a strictly semi-arid climate, it does have a period of 4 to 5 months during which agriculture of annual crops, with a few exceptions, is not possible without irrigation. Cassava, pineapple and sugarcane grow through the dry seasons (although the yield may be greater if irrigated) but maize, wheat, beans, soybeans, and vegetables are not grown in the dry season except where irrigated. Trees,

TABLE 8.1

Comparison of the main ecological features of upland forest, cerrado, and caatinga (tropical Brazil)

	Forest ("mata") (mesophytic; on drained soil)
1. Geographic region where this form is the main vegetation	Atlantic coast, Amazon valley
2. Surface relief of main part of region	Atlantic coast: mountainous with rounded contours Amazon: alluvial plain; low plateau; rolling terrain in south; hilly in north (tabular sandstone topography in north-west (Maranhão)
3. Main rock types	Rich sandstones (micaceous or calcareous), basalt and gabbro; some shales, granite, gneiss, limestone, etc., in drier climate; almost any rock type if soil is deep enough in wetter climate
4. Soils (in undisturbed vegetation)	Mineral portion poor to rich, acidic, abundant humus
5. Effective soil depth	Shallow to very deep
6. Average rainfall per year (mm)	1000—4500 (forest only, without cerrado, generally in regions with over 2000, or in São Paulo, over 1500)
7. Average rainfall of driest month (mm)	(10—) 40—200
8. Gaussen's xerothermic index (and number of months of effective dry season)	0 or between 0 and 40 (0—2 months)
9. Average annual evaporation (mm)	From less than 600 up to 1200 (Amazon); 600 to 1000 (Atlantic coast)
10. Rainfall/evaporation quotient	More than 2.5 at boundary of Amazon forest with cerrado; more than 1.0 at boundary of Atlantic forest with caatinga
11. Martonne's index of aridity	40 and over
12. Knoche's index of aridity	From more than 20 up to 60 along boundary of Atlantic forest with caatinga; up to more than 100 within Amazon forest
13. Main Köppen climate	Am, Af, As

TABLE 8.1 (continued)

Cerrado (xeromorphic; including arboreal "cerradão" and lower more open forms)	Caatinga (semi-desert; deciduous broad-leaf and succulents)
Central Brazil Plateau	North-east Brazil (behind coastal forest)
Tabular topography (plateaus with steep sides separated by flat-bottomed valleys); large tablelands with long gently-sloping sides to lower levels	Inter-montane and inter-plateau slightly dissected pediplain with inselbergs (tops and sides of sandstone plateaus in Piauí and north-west Pernambuco)
Under young soils: poor sandstones (siliceous, ferrugineous, clayey); under old soils: almost any rock type but very rare on basic rocks like basalt, diabase and gabbro	Limestone, calcareous sandstone; granite, schist, gabbro and other crystallines (non-calcareous sandstone in Piauí)
Mineral portion very poor (low cation exchange capacity and low base exchange), acidic; small amount of humus in cerradão, less in open forms	Mineral portion poor to relatively rich, circum-neutral to slightly alkaline; some humus to almost none
Almost always very deep	Shallow to deep
750—2000 (in São Paulo, 1000—1300 (—1500))	300 in small areas inside region to 750—800 at western limit where joins cerrado, or to 1000 at eastern and southern limits where joins forest
(5—) 10—30 (—40)	0—10
40—100 (3—5 months)	100—250 (6—8 (—11) months)
600—1400	1000 at boundary with Atlantic forest and 1400 at boundary with cerrado, going on to higher values within caatinga region
0.5 to 2.5 (the latter value at cerrado—Amazon forest boundary)	Less than 0.5
10 at boundary with caatinga, to 40 or over at boundary with forest	From less than 10 inside caatinga region to 10 at boundary with cerrado, or to 40 at boundary with Atlantic forest
20 at boundary with caatinga, to 100 at boundary with forest	Less than 20 inside caatinga region and at cerrado boundary; 20 to 60 at boundary with Atlantic forest
Aw (Cw at southern edge)	B (several varieties)

TABLE 8.1 (continued)

	Forest ("mata") (mesophytic; on drained soil)
14. Depth of permanent water table on uplands	Shallow to deep
15. Soil water storage over dry season in typical conditions	In evergreen forests, soils moist to surface; in semi-deciduous forests, upper layer more or less dries out but lower layers within reach of some tree roots stay moist; in deciduous forests on shallow soils, usually whole profile to bedrock dries out completely for a short time
16. Structure of non-disturbed vegetation	Closed-canopy arboreal; canopy tall, (15—) 20—35 m (emergents may exceed 40 m)
17. Deciduousness of stands	Mostly evergreen, sometimes semi-deciduous (rarely completely deciduous and then usually on very shallow soils)
18. Characteristic growth forms other than broad-leaved trees and shrubs	Canopy and undergrowth tree-palms; lianas; vascular epiphytes (orchids, bromeliads, aroids, ferns) especially dense in high rainfall areas; undergrowth bamboo in some areas
19. Leaf consistency of upper-layer broad-leaf plants	Mesomorphic to slightly coriaceous
20. Xeromorphic anatomical characters of leaves of plants exposed to the regional climate	Rare (when present, occurs in canopy trees, especially in secondary stands, and in epiphytes)
21. Stomate activity in dry periods during the day for most canopy species	Close relatively rapidly (observed in high rainfall area on crest of Serra do Mar, São Paulo)

TABLE 8.1 (continued)

Cerrado (xeromorphic; including arboreal "cerradão" and lower more open forms)	Caatinga (semi-desert; deciduous broad-leaf and succulents)
(3—) 6—30 (—more than 50) m	Usually none (perhaps present and deep in some sand-stone plateaus in Piauí)
Upper layer (to about 2 m deep) dries out; lower layers remain moist, with several years rainfall stored in soil above permanent water table	On shallow soils whole profile to bedrock dries out completely and stays dry during long dry season; on deep soils dries out completely in upper layers within reach of roots; (probably deeper layers remain moist on some sandstone plateaus in Piauí where there is semi-deciduous caatinga)
Closed- to open-canopy arboreal, medium-tall (7—20 m); tree-and-scrub woodland, the trees medium-tall (7—15 m) to low (3—7 m); open scrub or closed scrub (all heights); medium—tall (1.5—3 m) to low (0.5—1.5 m) scrub savanna; pure grassfield or grass-herb field	Closed- to open-canopy low- to medium-tall arboreal; medium-tall to tall (usually 2—5 m) closed scrub, often with emergent low trees (—7 or 8 m); medium-tall open scrub; scrub savanna
Semi-deciduous (cerradão rarely evergreen; scrub forms very rarely completely deciduous)	Completely deciduous (semi-deciduous on some sandstone plateau tops in Piauí
Acaulescent and low-tree palms	Cacti in all forms, including arboreal; terrestrial bromeliads; many shrubs thorny; acaulescent and low-tree palms in some areas; barrel-trunked trees in some areas
Xeromorphic, stiff and siliceous or coriaceous; (sometimes soft and densely hairy)	Mesomorphic, thin, soft
Very frequent (in canopy trees in carradão; in all trees and shrubs and in many herbs in open forms)	Rare
Very slow; stomates are open over long periods even in dry season; restriction of transpiration rare (observed in several areas in center and south of cerrado region)	Rapid; stomates open only for short periods even in rainy season; restriction of transpiration frequent even in rainy season (observed in North Bahia)

especially eucalyptus and pine are extensively planted, and if the correct species and variety are used, they grow well. Therefore, while up to 20 years ago cerrado was used in its natural form only for extensive cattle rearing, large areas of former cerrado vegetation are now planted with trees, crops and sown pasture. Areas still in cerrado vegetation are burned every year, although on average any one spot gets burned only about once in 2 years. This burning rate is for areas settled by Brazilians. The rate in the few areas still occupied only by non-civilized Indians is (and was) somewhat lower and depended on how close the area was to the Indians' hunting routes, the average rate varying perhaps from one burning every 3 to 10 years. Today, burned areas are easily seen as black patches on satellite images of certain reflectance channels.

Virtually the only management practice in natural cerrado vegetation is burning, to produce tender shoots not mixed with tough dry grass stems and leaves, and to reduce snakes and ticks. The cerrado is adapted to a low but regular degree of periodic burning. For instance, if burned only one in 3 years or less, the basic physiognomy of the cerrado is barely disturbed. The herbaceous layer returns to full or almost full density in a year, and the thin-stemmed shrubs return to their full density in a few years. The "basic physiognomy" is determined by the same substrate characteristics as determine the presence or absence of cerrado in general as against mesophytic forest. Thus, lower soil fertility and correlated high Al availability make for lower, more open cerrado; in some regions, drained upland soil with lower water-retaining capacity has the same effect. In smaller areas, surface or sub-surface waterlogging opens out and lowers the woody layer in cerrado, as does very shallow soil where few shrubs and fewer trees can find a root-hold in the bedrock. Different amounts of rainfall within the cerrado region have no effect on cerrado height and density. Only the substrate has this effect.

More frequent burning kills thick-stemmed shrubs and trees, and since these are slower-growing than herbs and thin-stemmed shrubs, the cerrado as a whole becomes lower and more open. An exception is that in southern Maranhão, burning usually does not affect the density of the larger trees but considerably opens out the shrub layer, producing a park-like aspect. Another effect of burning, of course, is that it also diminishes even more the already low soil fertility.

In areas where cerrado vegetation is more intensively pastured, another management practice is the complete or partial clearing of trees and shrubs of the taller denser cerrados but with the ground layer left as it is. Frequently certain trees are left for shade or because they supply fruits which the cattle eat when they fall from the tree. The leaves of the tree crowns and taller shrub crowns within reach of cattle and horses are very rarely browsed; even goats avoid them. The leaves of most species of trees and shrubs are hard and siliceous, even more so than the ground plants. The pulpy fruits and the legume pods and seeds, however, are often eaten.

Domestic animals. Traditionally, land-use in the cerrado has consisted of uncontrolled grazing by domestic livestock. Large areas have been cleared or partially cleared for unseeded pastures. Brazil has the world's fourth largest cattle herd, approximately 100 million head, and most (38 million) of these are raised in the three "cerrado" states of Minas Gerais, Goiás and Mato Grosso. Productivity, however, is extremely low. In order to maintain a viable herd, only 10% are slaughtered annually, whereas the United States can cull over one-third annually with no loss of performance.

The reasons for this situation are well known. Nearly all the pasture is open range with no "improvement" whatever. The nutritional content of the native grasses is low in the wet season and much lower in the dry season, during which time many herds actually lose weight. The distance between sources of water and the distance to the slaughter house are usually lengthy, which intensifies the loss of weight. Many herds are driven for several weeks before reaching their destination.

Where pasture is improved and enclosed and where stock-breeding can be controlled, productivity is vastly improved and quality is so high that Brazil can export breeding stock. On open range only about one-third of the cows produce a calf in any given year. Of these calves, probably fewer than two-thirds survive. In Brazil, cattle are slaughtered at 4 or 5 years of age compared with 1.5 to 2 years for other countries. Initial calving is delayed to the third and fourth year in Brazil, whereas the norm elsewhere is 1.5 to 2 years. Successive births delay over 20 months in Brazil instead of the customary 14—16 months.

All these problems are being actively researched in Brazil since the government and ranchers consider meat export to be economically promising and nationally appropriate and important. Nelore, the large grey to white cattle, and the smaller Gir dairy variety, with more brown colouration, are two Indian breeds reaching top quality in Brazil which is the world's only exporter, selling over 20 000 annually for breeding purposes. Zebu (*Bos indicus*), the heat and tick-resistant Indian species, crossed with domestic cattle, produced the Brahman which is the origin of most Brazilian herds. Cerrado cattle with Zebu blood average over 425 kg live weight and 245 kg dressed weight. The Zebu-type is now being compared with Holsteins (Silva et al., 1971) and with Holstein-Zebu hybrids (Milagres et al., 1971) for weight gains during the dry and wet seasons.

The resilience of Gir strains is offset by its slower breeding rate; spermatogenesis in the Gir starts much later than in the taurine breeds (Godinho, 1970). On open range, the Zebu traits are an advantage, but on improved pasture an increased taurine inheritance is more successful and this trend will increase as pastures are improved.

In addition to being sparse in general, cerrado forage is low in protein, unpalatable in the dry season and lacks some minerals, notably phosphate. Research on blood serum analyses suggests that phosphate and possibly magnesium may be limiting to growth on occasions (Lopes et al., 1972). Although Brazil is the leading South American milk producer (over 7 million m^3 y^{-1}), the daily yield is a meagre 1—1.4 l per cow in production or, 700 to 1000 kg y^{-1}, less than one-third a "normal" European average.

In spite of a resistance to pests, most cerrado cattle are parasitized, particularly by helminths which are listed by Freitas and Costa (1970). Nearly all cerrado cattle are infested with *Cooperia*, and over 70% by *Haemoncus*, *Bunostomum*, *Oesophagostomus* and *Trichuris* according to Costa et al. (1970, 1971). The occurrence of infective larvae is greatest after the wet season (Costa Val et al., 1971, 1972; Guimarães, 1972).

Experiments in Mato Grosso showed that providing salt licks of minerals including sodium chloride and bone meal increased the gain in weight of cattle. Salt licks provided with hormonal additives did not lead to significantly different results on improved cerrado pasture.

Brazil is thought to have the most horses in the world — of the order of 10 million head, many of them pedigreed blood-stock from the south of Brazil. Cerrado horses are strictly work-horses, tough but with little breeding.

Management of herds. There is no migration from pasture to pasture between the wet and the dry season. In the dry season, cattle graze on the edges of streams or in marshes. As the dry season progresses and fires become more frequent the cattle eat the tender young shoots which sprout a few days or weeks after a fire and which are relatively high in protein. Many fires are set on purpose to provide this source of food during the most critical time of the year for cattle. Over the long term, however, burning must be prejudicial to the ecosystem.

Cattle stocking rates that do not progressively degrade the pasture are as follows: wild pasture, 5—8 ha per head; molassas grass pasture, about 4 ha per head; *Brachiaria* pasture, about 2 ha per head. Best stocking rates for sheep and goats in cerrado vegetation are not known since these animals are rarely raised, nor is the best rate of mixed stocking known.

Most cattle are driven to the slaughter-house which may be several weeks away. Thin cattle travel these large distances surprisingly well. Occasionally the troop manager rents a pasture for the herd. Increasingly, the herd is put on a fattening pasture near the abattoir for some weeks before slaughter. Most of the large abattoirs are south of the main cerrado area, where the soils are richer and the dry season not as harsh. Increasingly too, as roads are built and ancient cattle tracks are improved, the cattle are transported to market in large trucks holding a dozen head or more. Slaughter periodicity is to a certain extent controlled by the government as part of a price-regulating mechanism.

Much beef is consumed locally by the rural dwellers but this is not a critical aspect. The value of a carcass is so very much higher at the cold-storage plant, compared with the price on the hoof at origin, largely because of the high costs and crippling losses by attrition en route.

The distribution of slaughter-houses reflects the demography. Cerrado cattle are slaughtered mainly in Anapolis and Goiania in Goiás, northern São Paulo, western Minas Gerais (such as Ituitaba and Uberaba) and the Belo Horizonte area. Principal cold-storage plants are modern and efficient, all contain freezing facilities and often ancillary product factories are associated with them. The cold-storage plant is always well connected with a marketing network, as they are located at a railhead or on a fast highway. It is likely that cerrado beef supplies the Brazilian consumer. Most export beef is probably of higher quality from more temperate southern Brazil, Paraná, Rio Grande do Sul and Santa Catarina where the pasture is richer and the dry season attenuated.

From the above it is clear that the cerrado cattle-raising process becomes more capital and hand-labour intensive, and more technological with increasing distance from the main centres. Although management is still somewhat primitive, slaughter and distribution have been modernized. In general, most of the beef entering this process is raised on vast latifundia ranches (fazendas) owned by rich absentee landlords or by large agro-industrial corporations. Small landholders do not supply cattle to this system, partly because they do not raise many cattle and partly because the large ranches have enough of their own where they are so cheap. The only small landholders contributing cattle are those holding rich pasture on the southern edge of the cerrado near a cold-storage plant. Such places can be used profitably for fattening.

Ranch-hands are employees of the agro-industrial corporation or of the latifundia. They may be lent a dwelling and given produce in lieu of the minimum wage. Since cattle have recently become enormously profitable, latifundias have expanded with a concomitant demand for labourers on these remote unpopulated ranches. This has led to a human migration principally from the impoverished and economically-depressed areas such as the north-eastern drought zone and interior Minas Gerais.

Cerrado grazing occurs almost entirely on open range. The distances involved are huge and prohibitively expensive to fence, in any case, annual fires would damage such fences. Cerrado range must remain open to permit the cattle to reach water and shade when necessary, to allow access to different grazing as each type of cerrado becomes palatable, and to enable cattle to follow the nutritious, succulent sprouts appearing after the frequent but irregular fires. There is no seasonal migration of cerrado cattle and there is no human nomadism in cerrado. Most cerrado grazing occurs in the open areas and in the abundant marshy areas, and to a much lesser extent in the more densely wooded cerrado. Interestingly, the densely wooded cerrado and cerradão, and the dry mesophytic forest occur on richer soils more suitable for agriculture than the soils of the open cerrado. Wells are dug in cerrado for human use; rarely is it necessary or practicable to do so for cattle.

Pasture improvement

Some ranchers feel that cerrado pasture is improved by felling trees and by clearing shrubs, but this is not proven. Woodland areas, cerradão and dry forest can be converted into useful pasture. The usual practice is to clear-cut, burn, then seed with various grasses, although it may well be preferable to thin the trees to the point at which grass will thrive. Such a thinned system is likely to be more sustainable than the clear-cut system in the long term. Most cerrado trees are so well adapted to the inhospitable cerrado dry season that they remain green throughout. Native trees may be important in the maintenance of soil fertility by pumping nutrients from deep within the soil profile and depositing them on the soil surface as fallen leaves where their decomposition products become more accessible to the ground layer.

Pasture in cerrado has always been made up of the native grasses and legumes. Some agronomists and ranchers in the southern part of the cerrado region have been seeding pastures (made by completely clearing cerrado) to molassas grass (*Melinus minutiflora*, "capim gordura") and to *Hyparrhenia rufa*, "capim jaraguá". The tendency now is to plant *Brachiaria* species, especially *B. decumbens*, *B. radicans* and *B. humidicola*. The first species, most planted and excellent nutritionally, has the disadvantage of often harbouring a *Phormopsis* fungus which causes photosensibility in young cattle. If the animals are shifted to other pasture species the condition soon disappears. Also, an orthopteran insect-pest, "cigarrinha", severely attacks *B. decumbens*.

Quite distinct from the above is the improvement of pasture degraded by the invasion of shrubs. This process is often related to abuse such as overgrazing and so far is considered serious only near population centres such as upstate São Paulo where it has become economic to improve pastures. Quinn et al. (1956), Quinn and Griffing (1958), Fisher and Quinn (1959) and Quinn (1961) report that such shrubs as "leiteiro" (*Tabernaemontana fuschsiifolia*, Apocynaceae), "amendoim" (*Pterogyne nitens*, Leguminosae), "monjoleiro" (*Acacia polyphylla*, Leguminosae) and "unha de gato" (*Acacia paniculata*, Leguminosae) are the most serious pests. The authors recommend aerial and ground spraying of Karmex (Krovar, Diuron) Silvex (Kuron, Weedone-TP) and other hormones. Recently the notorious "Agent Orange" (2,4-D—2,4, 5-T) Vietnam surplus, declared too dangerous for use in the United States, has been used in Brazil.

Poisonous pasture plants. The main poisonous plants of Brazilian pastures are reasonably well known, thanks largely to Hoehne's (1969) book on medicinal and poisonous plants. Ferreira (1971) has compiled a list from the Distrito Federal, which supports cerrado almost entirely. Some of the more serious poisonous plants are listed in Table 8.2. This list contains plants which might be eaten by cattle living in the cerrado although some of the plants are not strictly members of the cerrado flora and some, like mata-

TABLE 8.2

Some poisonous plants of the cerrado region

Species	Family	Common name	Vegetation
Cassia bicapsularis L.	Leg. Caes.	Mata-pasto	Ruderal, pasture, forest, cerrado
Cassia cathartica Mart.	Leg. Caes.	Sena do campo	Ruderal, pasture, cerrado, forest edges, campo rupestre
Cassia obtusifolia L.	Leg. Caes.	Mata-pasto	Pasture, ruderal (incl. disturbed cerrado)
Cassia occidentalis L.	Leg. Caes.	Fedegoso	Ruderal, pasture, cerrado
Cassia pendula HBK. ex Willd. var. *pendula*	Leg. Caes.	Mata-pasto	Forest, cerrado, ruderal, pasture
Cassia pentagonia Mill.	Leg. Caes.	Fedegoso	Ruderal, pasture, forest
Cassia uniflora Mill.	Leg. Caes.	Mata-pasto	Ruderal, pasture
Crotalaria anagyroides HBK.	Leg. Pap.	Cascalheiro, Cascavél	Ruderal, disturbed forest, pasture
Crotalaria velutina Benth.	Leg. Pap.	Cascalheiro, Cascavél	Marsh, forest, cerrado
Dimorphandra mollis Benth.	Leg. Caes.	Faveiro, Faveira, Barbatimão da folha miúda	Cerrado
Diodia spp.	Rubiaciae		
Enterolobium contortisiliquum (Vell.) Morong	Leg. Mim.	Tamboril, Timbaúba Orelha de negro	Forest, pasture
Enterolobium gummiferum (Mart.) Macbr.	Leg. Mim.	Boizinho, Tamboril do campo, Orelha de onca.	Cerrado
Lupinus crotalarioides Mart.	Leg. Pap.		Cerrado
Lupinus subsessilis Mart.	Leg. Pap.		Cerrado
Palicourea coracea (Cham.) Schum.	Rubiaceae		Cerrado
Palicourea marcgravii St. Hil.	Rubiaceae	Erva de rato	Forest edges, old fields, pasture
Palicourea rigida HBK.	Rubiaceae	Gritadeira, Douradinha	Cerrado
Palicourea squarrosa (M. Arg.) Standl.	Rubiaceae	Roxinha escurinha	Cerrado
Pteridium aquilinum (L.) Kuhn	Polypodiaceae	Samambaia	Forest edges; rare in cerrado, pasture
Pterogyne nitens Tul.	Leg. Caes.	Amendoim	Forest, pasture
Rhynchosia phaseoloides DC.	Leg. Pap.	Olho de cabra	Cerrado, forest, pasture
Solanum spp.	Solanaceae	Joá	Forest, cerrado, waste ground, pasture
Tabernaemontana fuchsiifolia and other spp.	Apocynaceae	Leiteiro	Cerrado, forest, pasture

pasto and fedegoso are rather avoided by grazing animals. Bracken (*Pteridium*), given on the list, is very rare in cerrado (it does occur in Brasília cerrados), but is common on the edges of gallery forests, in marshes and in ditches, especially where fire occurs frequently. "Pasture" refers to pasture derived by clearing forest land. More recently Andrade et al. (1971) reported nitrate poisoning of cattle after grazing on Tanner grass (*Brachiaria decumbens*, *B. ruzizzensis* and *B. brizantha*).

Size of properties. Most cerrado is owned by very few landlords in the form of enormous latifundias. The data provided by Stefan (1960) can be reworked to show that nearly half of the land consists of ranches over 1000 ha in size. Large agro-industrial groups are now trying to establish that their ranches must be at least 15 000 ha and already there are enormous ranches extending over 600 000 ha in the area. It is likely that land ownership in the cerrado is even less equitable than in the rest of Brazil, where 23% of the properties occupy 80% of the area. Indeed, there are 175 properties of over 100 000 ha each — a total of 18 million ha, whereas 71% of the proprietors own minifundias totalling scarcely 13% of the land. Stefan (1960) provides data on the number of farms in various size classes in four areas in the "Tristate" region and concludes that small farms predominate. This is true in that there are more small farms than large farms. However, from his data, one can calculate that most of the land is owned by very few people. Assuming the average area of each property is the arithmetic mean within each of Stefan's size classes, the results are presented in Table 8.3.

Recommended land-use. World consumption of beef will probably be curtailed in a generation or so and be replaced by an increased consumption of vegetable products such as grains, pulses, fruits and vegetables. Cattle rearing is not helping the people of the cerrado as much as small fixed farms would. It can be concluded for the long term therefore, that the people of

TABLE 8.3

Sizes of cerrado properties (after Stefan, 1960)

Size classes (ha)	Number of properties	Estimated total area (ha) [a]
0— 10	3669	183 000
10— 50	11 888	297 000
51— 100	5465	410 000
101— 200	4629	695 000
201— 500	4626	1 620 000
501—1000	1778	1 330 000
over 1000	1639	2 460 000

[a] Mean size of property in each class is assumed to be the class mid-point.

the areas would be more profitably engaged in their own smaller holdings. It is unnecessary to delay this process until world environmental and economic conditions dictate it. For the short term and in small areas, pasture improvement by encouraging browse trees of legumes and phreatophytes and by decreasing the canopy rather than clear-cutting, will be worthwhile in the cerrado.

REFERENCES

Andrade, S.O., Retz, L. and Marmo, O., 1971. Estudos sobre *Brachiaria* sp. (Tanner grass): III. Ocorrências de intoxicações de bovinos durante um ano (1970—1971) e níveis de nitrato em amostras da gramínea. Arq. Inst. Biol., São Paulo, 38(4): 239—252.

Cabrera, A.L. and Willink, A., 1973. Biogeografia de America Latina. Serie de Biologia, Monografia no. 13, OEA, Washington.

Costa, H.M.A., Freitas, M.G. and Guimarães, M.P., 1970. Prevalência e intensidade de infestação por helmintos de bovinos procedentes da área de Três Corações. Arq. Esc. Vet., Univ. Fed. Minas Gerais, 22: 95—101.

Costa, H.M.A., Costa, J.O., Guimarães, M.P. and Freitas, M.G., 1971. Helmintos parasitos de bezerros procedentes da bacia leiteira de Ibiá, Minas Gerais. Arq. Esc. Vet., Univ. Fed. Minas Gerais, 23: 221—228.

Costa Val, V.P., Pessoa, J.M. and Ferreira Neto, J.M., 1971. Variações no hemograma de bezerros durante os primeiros estágios de vida após o nascimento. Arq. Esc. Vet., Univ. Fed. Minas Gerais, 23: 291—296.

Costa Val, V.P., Pessoa, J.M. and Ferreira Neto, J.M., 1972. Influência de minerais sobre o hemograma de bovinos em crescimento. Arq. Esc. Vet., Univ. Fed. Minas Gerais, 24(2): 175—177.

Eiten, G., 1968. Vegetation forms. A classification of stands of vegetation based on structure, growth form of the components and vegetative periodicity. Inst. Bot. São Paulo, Bol. 4, iv + 88 pp.

Eiten, G., 1972. The cerrado vegetation of Brazil. Bot. Rev., 38(2): 201—341.

Ferreira, M.B., 1971. As plantas tóxicas do Distrito Federal. Cerrado (Brasília), 3(14): 26—30.

Fisher, G.E. and Quinn, L.R., 1959. Control of three major brush species in grazing lands in the United States, Cuba and Brazil. J. Range Manage., 12: 244—248.

Freitas, M.G. and Costa, H.M.A., 1970. Lista de helmintos parasitos dos animais domésticos do Brasil. Arq. Esc. Vet., Univ. Fed. Minas Gerais, 22: 33—94.

Godinho, H.P., 1970. Puberdade em bovinos Gir estimada pela análise do semen. Arq. Esc. Vet., Univ. Fed. Minas Gerais, 22: 165—169.

Goodland, R., 1969. An ecological study of the cerrado vegetation of south central Brazil. Ph.D. thesis, Bot. Dep., McGill University, Montreal, 224 pp.

Goodland, R., 1970. Plants of the cerrado vegetation of Brazil. Phytologia, 20: 57—78.

Guimarães, M.P., 1972. Variação estacional de larvas infetantes de nematóides parasitos de bovinos em pastagem de cerrado de Sete Lagoas, M.G. Arq. Esc. Vet., Univ. Fed. Minas Gerais, 24(1): 97—113.

Hoehne, F.C., 1969. Plantas e substancias tóxicas e medicinais. Graphicars, São Paulo, 365 pp.

Lopes, H.O.S., Ferreira Neto, J.M. and Veloso, A.F., 1972. Estudos dos teores de cálcio, fósforo, manganês e atividade da fosfatase alcalina em bovinos criados no cerrado. Arq. Esc. Vet., Univ. Fed. Minas Gerais, 24(1): 33—43.

Melo Carvalho, J.C., Coimbro Filho, A.F., Sick, H., Narchi, W., Bokermann, W.C.A., Britski, H.A., Otero, L.S., Pádua, L.F.M. and Pádua, M.T.J., 1978. Atlas da fauna brasileira. MA/IBDF-MEC/ FENAME, São Paulo, 128 pp.

Milagres, J.C., Tôrres, J.R., Carneiro, G.G. and Gomes, F.R., 1971. Diferenças de reações entre novilhas zebus e mestiças holandesas a condições climáticas de Leopoldina, Minas Gerais. Experientiae (Viçosa, Minas Gerais), 12(8): 227—280.

Quinn, L.R., 1961. O controle de arbustos nas pastagens do Brasil. IBEC Res. Inst., Nota Técnica, 6: 1—19.

Quinn, L.R. and Griffing, J.B., 1958. O controle do "leiteiro" e do "amendoim" em pastagens. IBEC Res. Inst., Folheto de Divulgação Ampla.

Quinn, L.R., Swierczynski, K.L., Schilmann, W.L. and Gullöve, F.H., 1956. Experimental program on brush control in Brazilian pastures. IBEC Res. Inst. (Bull.), 10: 1—35.

Sick, H., 1965. A fauna do cerrado. Arq. Zool., 12: 71—93.

Silva, H.C.M., Carneiro, G.G., Rosa, F.F. and Tôrres, J.R., 1971. Diferenças de pêso e ganho em pêso de novilhas zebus e holandeas, durante as estações seca e chuvosa, na área de cerrado em Minas Gerais. Arq. Esc. Vet., Univ. Fed. Minas Gerais, 23: 153—157.

Stefan, E.R., 1960. Agricultura. In: M.V. Galvão (Editor), Grande Região Centro Oeste. Conselho Nac. Geogr., Rio de Janeiro, pp. 223—248.
Vanzolini, P.E., 1963. Problemas faunisticos do cerrado. In: Simpósio sôbre o cerrado. Editôra da Universidade de São Paulo, São Paulo, pp. 305—321.
Von Ihering, R., 1968. Dicionário dos animais do Brasil. Editôra Universidade de Brasília, São Paulo, 790 pp.

Chapter 9

LAND-USE OF SEMI-DESERTS IN THE U.S.S.R.

M.P. PETROV †

Leningrad State University, Department of Geography, Leningrad (U.S.S.R.)

INTRODUCTION

This review includes those territories lying between real deserts, where irrigated agriculture is conducted, and steppe zones where stable yields of field crops can be achieved under normal atmospheric precipitation. According to the landscape classification used in the U.S.S.R., the zone thus described is referred to as semi-desert (Anonymous, 1968). Semi-desert is understood by Soviet geographers to be a type of landscape formed under the condition of an arid climate and characterized by a vegetation cover combining fragments of steppe and desert landscapes. Steppe associations of a semi-desert are dominated by sod cereal herbs, and the desert associations by wormwoods (*Artemisia* spp.) [1], saltworts (*Salsola* spp.) and other types of miscellaneous herbs. Evaporation is, as a rule, three to six times greater than the total annual precipitation, the latter usually not exceeding 300 mm. Soils are light-chestnut, sub-alkaline and weakly carbonate. In general these regions are subject to drought and do not produce stable crop yields without irrigation. The preceding definition of the term "semi-desert" corresponds to the meaning of the term "semi-arid" in Western European geographical literature.

The geographic position of the semi-desert zone corresponds roughly to the regions between the July isotherms of 20 and 25° C. However, this climatic boundary is only approximate since meteorological conditions vary considerably from year to year. Furthermore, the boundaries between the semi-desert, steppe and desert zones meander north and south, depending on the geological—geomorphological conditions. In individual places the boundary coincides well with that between light-chestnut desert—steppe and brown desert soils. Deviation of the boundary to the north is related to saline and carbonate substrata in the hilly lands and on plains. Southward curves are due to high massifs and the distribution of soils of light mechanical composition on the plains.

The displacement of landscape boundaries in the semi-deserts of Kazakhstan, depending on the litho-edaphic conditions, is especially well demonstrated in the instance of sand-massif landscapes. The sands are known to

[1] Nomenclature is according to "Flora of the U.S.S.R", 1934—1964.

have hydrophysical properties favourable for the growth of plants. They retain moisture well and are heat-conductive. These properties of the sands facilitate the expansion of moisture-loving plants further south and of heat-loving ones to the north. An example of this is the pine forests in the Naryn sands of northern Kazakhstan which are located several hundred kilometres south of the principal pine-forest area of the Russian plain. It is also known that sands on the southern slopes of hills in the zone of taiga are populated with steppe plants.

Thus the boundary between the steppe and desert zones is very sinuous and complicated in its structure. It can be gradual (diffuse) or distinct, almost linear, because of the contact of territory types differing in their geomorphology, or of an island type. In the case of a gradual transition from steppes to deserts there appears a transitory semi-desert belt with a combination of steppe and desert associations, peculiar in their floristic composition, which do not occur in the adjacent botanical-geographic regions.

Therefore it would be more correct to draw the boundary of the semi-desert zone by the character of vegetation. Landscapes with a prevalence of soddy cereal herbs of the genera *Stipa*, *Festuca*, *Agropyron* should be regarded as steppe associations. In the desert—steppe transition zone these herbs are joined by wormwood of the subgenus *Seriphidium*.

Typically desert associations are groups with a prevalence of xerophytic semi-shrubs (suffrutescent plants) — *Salsola arbusculyformis*, *Artemesia terraealbae* and *A. semiarida*. In drawing the boundary, attention should mainly be paid to the vegetation on watersheds, and should they be absent, to the vegetation on trains and lower parts of the slopes of hillocks.

NATURAL FEATURES OF THE SEMI-DESERTS OF THE U.S.S.R.

Physical features

The semi-desert zone of the U.S.S.R. occupies over 100 million ha, extending from the eastern piedmonts of the Great Caucasian Range, east across all of Kazakhstan to the piedmonts of Altai for almost 3000 km, and attaining up to 300 km in width (see Fig. 9.1).

In the west it includes a part of the Caspian lowland. A little further east it lies within the near-Emba and sub-Ural plateaus; still further east it includes the Turgai plateau, then the Kazakh hillocks, and in the form of a narrow belt passes into the Ziasan trough.

The semi-deserts of the European part of the U.S.S.R. are separated from those of Kazakhstan by the southern termination of the Ural mountains (the Mogodzhar mountains), where the landscapes are transitionary from semi-deserts to dry steppes. Here the semi-desert zone has a break (see Fig. 9.1).

Natural and climatic conditions are not uniform throughout the semi-desert zone. In the western half there are clearly distinguishable bands of

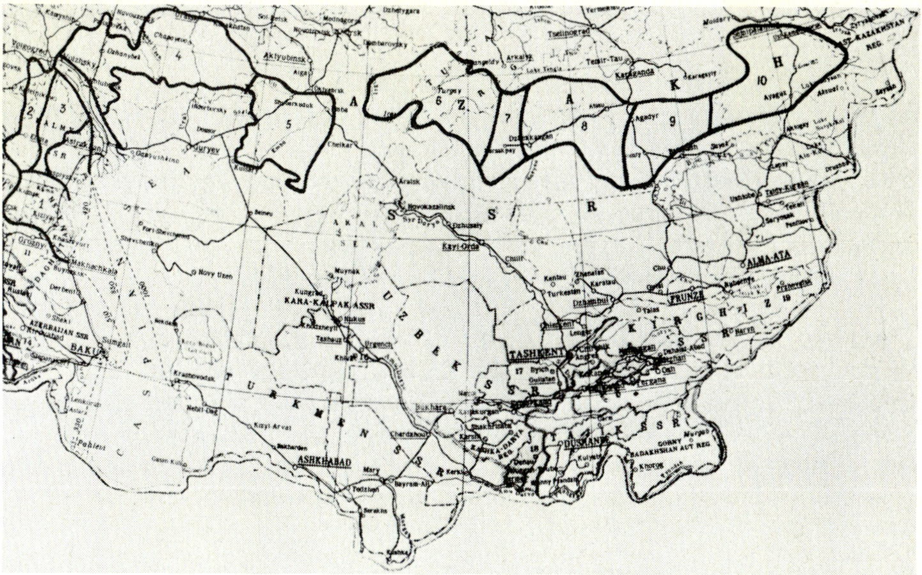

Fig. 9.1. Map of semi-desert zone of the U.S.S.R. (from: Physico-geographical regionalization of the U.S.S.R., 1968).

dry steppes, semi-deserts and island massifs of sandy desert. In the eastern half, just as clearly distinguished, are steppe and semi-desert landscapes of plains, low mountains and the mountains of central Kazakhstan.

The southern boundary of the semi-desert zone extends along approximately latitude 48° N, curving slightly northwards on the Turgai plateau and extending southwards within the Kazakh hillocks to almost latitude 47° N. The northern boundary, within the Caspian lowland, is at about 51° N. Within the Kazakh hillocks it extends south to about latitude 49° N and moves north along the pre-altai plains to almost latitude 52° N. The spurs of the southern Urals (Mugodzhar mountains) divide the semi-desert zone into the Eastern European and Kazakhstan provinces.

Administratively this territory is located mainly in the Kazakh S.S.R. and partly in the south-east of the European part of the U.S.S.R., in the Daghestan Autonomous Republic, the Volgograd and Astrakhan regions.

The geology and gemorphology of the semi-desert zone is highly varied. In the west semi-deserts occupy the northern part of the Caspian lowland, composed of Quaternary marine depositions. Beyond the Ural mountains, in Kazkhstan, they occur in the Irghiz—Turgai depressions and along the southern margin of the Turgai plateau, covered with deluvial loams. Sand massifs, for example those of Uil, Naryn and Irghiz, constitute large parts of this plain. The semi-deserts then grade into the southern parts of the old Kazakh hillocks, which are a peneplaned Paleozoic uplift, and end in the piedmont plains of Altai.

Climate

The climate of the semi-desert zone of the U.S.S.R. is markedly continental, with summers prone to drought and severe winters with little snow. This is well demonstrated in the annual temperature range of about 88° C between the absolute maximum and minimum air temperatures. The mean monthly temperature of the coldest month (January) varies from —3.5° C in the Caspian plain to —18° C in the east of Kazakhstan. At the eastern end of the zone the absolute minimum drops to —52° C (Table 9.1).

Summer in the semi-desert is fairly hot. The mean monthly air temperature in the hottest month (July) is 25° C, varying from 22° C in the west to 25° C in eastern Kazakhstan. In the hottest years the absolute maximum reaches 45° C (Table 9.1). The hot-weather period (25° C daytime air temperature) begins on the plains, on average, during the first third of June and ends during the last third of August. Thus the normal hot summer period varies from 2.5 to 3 months. Air temperatures start to fall in September.

TABLE 9.1

Principal climatic features of the semi-desert zone of the U.S.S.R. (data from Anonymous, 1966a,b,c)

Stations	Elevation above sea-level (m)	Air temp. (°C)				
		Jan. mean	July mean	Annual mean	Absolute max.	Absolute min.
East-European semi-deserts						
Kizlar	—5	—1.8	24.0	11.1	40.0	—32.0
Bazhigan	16	—3.5	24.6	10.4	44.0	—35.0
Elista	178	—6.7	24.2	8.6	43.0	—34.0
Volgograd	23	—9.2	24.2	7.6	42.0	—36.0
Baskunchak	34	—9.8	25.2	7.8	45.0	—37.0
Elton	5	—10.7	24.8	7.2	45.0	—36.0
North-Kazakhstan semi-deserts						
Uralsk	37	—14.0	22.9	4.8	42.0	—43.0
Dzhanybek	28	—12.3	23.2	6.1	44.0	—40.0
Kalmykovo	1	—13.0	25.0	6.4	44.0	—41.0
Furmanovo	8	—12.8	24.1	5.9	44.0	—40.0
Uil	88	—13.4	24.5	6.0	44.0	—42.0
Irghiz	114	—15.5	25.5	5.3	43.0	—43.0
Karabutak	229	—17.5	22.2	2.9	43.0	—44.0
Turgai	124	—17.0	24.1	4.1	43.0	—44.0
Dzhezkazgan	345	—16.1	22.5	5.0	43.0	—50.0
Kzyl-Zhar	361	—16.4	23.0	3.5	42.0	—52.0
Chubartau	646	—15.6	21.4	3.0	40.0	—49.0
Ayaguz	653	—18.1	20.7	2.1	41.0	—44.0

In the western half of the zone winter normally begins in November, while in the eastern half it begins late in October. It lasts on average for 5 months. Snow cover is thin and over most of the territory remains for 120—140 days. In the southern part, though, it remains for only 90—110 days. Mean annual precipitation varies from 306 mm in the west to 150 mm in the east (Table 9.1).

According to Meigs (1953), the semi-deserts of the U.S.S.R. are classed as cold (SbO_3). The total annual precipitation varies considerably in different years, by up to 500%, a feature which may be considered characteristic of the semi-deserts of the U.S.S.R. (Table 9.2). The seasonal distribution of precipitation shows no definite maximum, and part of it occurs in the form of snow.

Snow cover throughout the semi-desert zone varies greatly according to relief. In the mountainous and hilly regions it ranges in depth from 10 to 40 cm. In the plains of the western half of the zone it varies between 15 and 25 cm and in the south-central part between 15 and 20 cm. In snowy winters it may be as much as 55 cm in depth, in winters with little snow it is absent in the south and may be only 10—25 cm in depth in the north. Snow cover plays an important role in re-charging moisture reserves in spring.

Duration of frost-free period (days)			Annual precipitation (mm)	Snow cover average depth (cm)
Mean	Min.	Max.		
196	149	254	336	9
178	—	—	289	—
178	148	229	315	—
180	155	214	299	—
173	121	199	243	11
171	140	200	292	14
144	114	177	282	26
161	—	—	290	—
161	—	—	190	17
156	122	191	233	17
158	120	183	212	22
159	124	183	161	16
126	97	151	210	—
156	123	180	202	19
128	93	171	150	—
125	54	—	173	—
121	92	157	221	—
106	81	—	254	30

TABLE 9.2

Variation in annual precipitation in the semi-desert zone of the U.S.S.R. (in mm)

Locations	Mean	Max.	Min.
Steppe zone			
Volgograd	318	407	126
Urda	229	351	117
Uralsk	282	428	176
Orenburg	361	731	185
Aktubinsk	228	397	183
Atbasar	282	485	159
Semi-desert and desert zones			
Astrakhan	175	213	120
Guryev	172	268	51
Aral Sea	123	200	42
Dzhezkazgan	150	160	52
Balkhash	127	140	59

Surface water is very scarce and there are very few permanent rivers. The rivers of the Caspian lowland — the Bolshoi and Malyi Usen, the Kushum and other smaller rivers — terminate in endorheic depressions. In summer they break up into a series of lakes with brackish water. Only the Ural river, running from the Ural Mountains, crosses the semi-deserts to flow into the Caspian Sea. Rivers of the pre-Uralian plains, the Emba, Sagiz and Wil rivers, and the rivers of the Tenghiz Lake basin and the Sarysu, are filled with water only during the spring thaw. They discharge about 80% of their total runoff during the short spring. In summer small rivers dry up (Anonymous, 1967).

Ground water occurs throughout but, except for lake depressions, is at a considerable depth and has no influence on the growth of vegetation. However, this does not apply to the sand massifs where ground water is comparatively close to the surface and is accessible to root systems of the woody vegetation. In spite of the limited amount of precipitation, the semi-desert soils are sufficiently well wetted during the spring thaw to allow for the development of a fairly rich vegetation cover, based on this reserve of moisture.

To illustrate the peculiarities of the dynamics of the water regime of light-chestnut soils in semi-deserts, Fig. 9.2 gives the chrono-isopleths for such soils in the district of the Dzhanibeck experimental station (western Kazakhstan).

Soils

According to the classification used in the U.S.S.R., the soils of the semi-desert zone are described as light-chestnut and alkaline, often in conjunction with alkali soils (Lobova, 1960). In the south some areas have brown

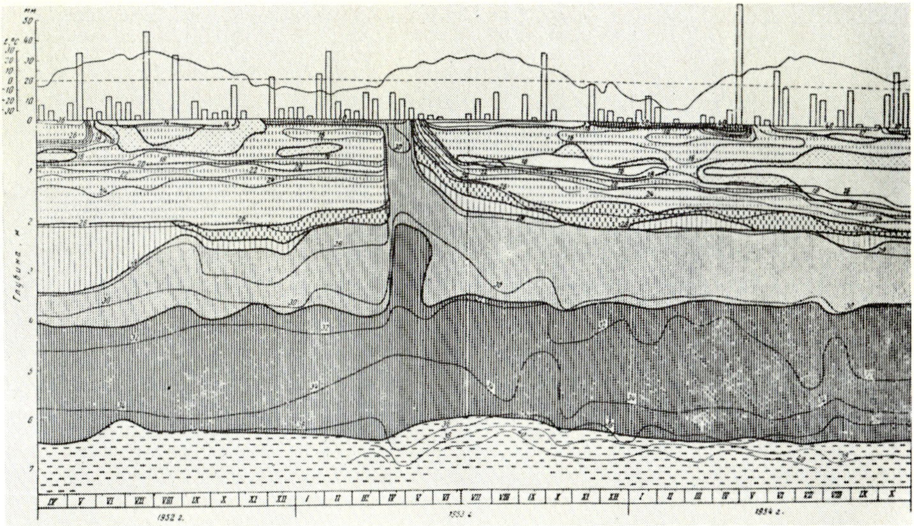

Fig. 9.2. Water regime of light-chestnut soil in the semi-desert zone of Kazakhstan. 1, less than wilting coefficient; 2, wilting coefficient; 3, from wilting coefficient till moisture of capillarity gap; 4, moisture of capillarity gap; 5, from moisture of capillarity gap till water capacity; 6, minimum water capacity; 7, from minimum water capacity till total moisture capacity; 8, total moisture capacity. (After Rode, 1963).

desert—steppe soils, also more or less sub-alkaline. In depressions, where ground water is close to the surface, solonchaks and solonchak-like soils are common.

Soil-forming deposits over the greater part of the semi-desert zone are represented by stony rock-debris loams. Over the remaining territory there are widespread sandy loams, loams and in places sands.

A characteristic feature of the above-mentioned soils is their low humus content (0.5—2.5%) and relatively excessive salinity, particularly in the form of carbonate. The soils of sandy massifs are more leached by atmospheric precipitation and are not so saline. The highest salinity occurs in the solonchaks (Table 9.3).

The considerable widespread salinity of soils and subsoil under the conditions of a strained water balance is manifested in the semi-desert in the form of the overall distribution of the combinations of zonal soils with those of the black alkali and black alkali—solonchak series.

A distinctive feature of the semi-desert Kazakh hillocks is the poorly or incompletely formed rock-debris stony soil, which extends either as vast massifs in the regions of hillocky-relief, or in combination with soils of other types. Their reduced profile with weakly differentiated horizons is replaced at the shallow depth of 40—50 cm by coarse fragments composing the eluvium of dense rocks of the Paleozoic base.

On the whole, however, all soils, except for the solonchaks, are suffi-

TABLE 9.3

Chemical composition of light-chestnut soils in the semi-desert zone of Kazakhstan (after Anonymous, 1968)

Location	Sample depth (cm)	Humus (%)	Solid residue (%)	Alkalinity							pH
				HCO_3^-	CO_3^{2-}	Cl^-	SO_4^{2-}	Ca^{2+}	Mg^{2+}	$Na^+ + K^+$	
Light-chestnut soil. Central Kazakhstan	0—10	1.8	0.022	0.012	Nil	0.002	0.002	0.003	Nil	0.003	7.2
	15—25	1.4	0.038	0.025	Nil	0.001	0.002	0.007	Traces	0.003	7.5
	30—40	—	0.049	0.033	Nil	0.001	0.003	0.010	0.001	0.001	8.2
	50—60	—	0.048	0.033	Nil	0.001	0.002	0.008	0.002	0.002	8.2
Light-chestnut soil. Central Kazakhstan	0—10	1.2	0.041	0.023	Nil	0.001	0.006	0.004	0.001	0.006	7.8
	10—20	1.1	0.026	0.015	Nil	Nil	0.004	0.006	0.001	Nil	8.0
	30—40	1.1	0.048	0.028	Nil	Nil	0.008	0.010	0.001	0.001	8.2
Light-chestnut solonetz soil. Central Kazakhstan	0—10	2.1	0.070	0.032	Nil	0.001	0.019	0.015	0.002	0.001	8.0
	16—26	1.7	0.080	0.042	Nil	0.001	0.016	0.008	0.002	0.011	8.5
	30—40	1.2	0.083	0.056	0.001	0.003	0.002	0.004	0.002	0.016	8.8
	50—60	—	0.158	0.079	0.005	0.017	0.016	0.005	0.002	0.039	8.5
Light-chestnut solonetz soil. Central Kazakhstan	0—10	1.05	0.034	0.013	Nil	0.007	0.003	0.001	Nil	0.010	7.8
	15—25	0.70	0.050	0.021	Nil	0.011	0.002	0.001	Nil	0.015	8.0
	25—35	0.91	0.437	0.048	Nil	0.219	0.008	0.006	0.003	0.154	8.4
Brown soil. Central Kazakhstan	0—10	1.1	0.028	0.019	0.001	0.001	0.001	0.002	0.001	0.004	—
	18—28	0.8	0.032	0.020	Nil	0.003	Traces	0.005	Traces	0.004	—
	38—48	0.5	0.043	0.030	Nil	Traces	0.003	0.006	0.002	0.002	—
Solonchak. Central Kazakhstan	0— 1	—	42.899	0.005	Nil	24.660	1.564	0.399	0.018	6.253	7.5
	1—10	—	7.256	0.010	Nil	3.366	1.184	0.304	0.014	2.378	8.0
	15—25	—	3.919	0.017	Traces	2.142	0.274	0.051	0.037	1.398	8.2
	35—45	—	4.878	0.012	Nil	1.819	1.295	0.304	0.008	1.440	8.2

ciently fertile and, depending on the availability of moisture, have a high potential biological productivity. Although the fertility of the light-chestnut soils makes them suitable for agriculture, it is difficult to reclaim them without irrigation. It would be necessary to have specially selected drought-resistant species and to use special agro-techniques. The cost of grain would be higher than that in the steppe zone on dark-chestnut soils.

Vegetation

The vegetation cover is fairly monotypic throughout, and consists of wormwood (*Artemesia*)—sod-grass (*Festuca, Stipa*) formations. Floristic composition over the vast area of the semi-deserts is highly varied, but the composition of life forms is basically the same. The most prevalent forms are highly drought-resistant euxerophytic and partially euryxerophytic grasses, such as *Festuca sulcata, Stipa capillata, S. sareptana, Agropyrum desertorum, A. sibiricum* and others, with a considerable amount of euxerophytic semi-shrubs, mainly wormwoods of the *Seriphicium* subgenus, as well as *Pyrethrum achilleifolium, Kochia prostrata* and some others. A fairly large amount of ephemeroids is also characteristic of the semi-desert flora. Especially abundant is *Poa bulbosa* var. *vivipara*, and to a lesser extent tulips and ephemers (winter crop annuals). The ephemeroids and ephemers are abundant in the western regions of the desert steppes. In these steppes there is also a variety of soil-surface lichens, both foliate migrating (*Parmelia vagans, P. ryssoles* and others) and crustose forms (*Psora* and other genera), blue-green algae (*Stratonostoc commune* and others), and sometimes mosses (*Tortula* species). The micro-floral crusts are significant for the water regime of semi-desert soils, since these plants readily absorb not only the moisture of atmospheric precipitation but also that of air.

The wormwood—sod-grass semi-deserts are usually of a complex nature. Soviet geographers understand a geographic complex to be a regular combination of geographical components (relief, climate, surface water, soils vegetation, fauna) joined in a complicated interaction and interdependence and forming a single inseparable system in which individual areas of different vegetation associations, or their fragments of a micro-relief, alternate on a genetically mono-typical territory, and are the stages of a single successional series. The character of complexes is highly varied. It has been mentioned that these semi-deserts are confined to sub-alkaline light-chestnut soils. Depressions with chernozem-like soils are occupied by sod-grass steppes dominated by *Festuca, Stipa* and *Agropyrum* with a smaller or larger addition of different steppe herbs, and sometimes steppe shrubs, mostly *Spiraea hypercifolia*. Associations of desert character, with the prevalence of semi-shrub wormwood are confined to areas of crusty and sometimes medium black alkali soils. Whole saltworts (*Salsola*) occur on the solonchak—alkali soils.

Over the enormous distance from the Black Sea coast to the Zaisan and

Alakol troughs the floristic composition of the wormwoods, and partly of the steppe sod-grasses, changes. In this connection several varieties of desert steppes can be distinguished: the Black-sea (with *Artemesia taurica*), Caspian-Kazakhstan (with *A. lecheana*), western-Kazakhstan (with *A. lessingiana*), and eastern-Kazakhstan (with *A. sublessingiana*) dominated by species of *Artemesia*, *Festuca* and *Stipa* and the Caspian-Kazakhstan steppes, with *Artemesia arenaria*, *A. astrachanica*, *Eurotia eversmanniana* and *E. certoides.*

The desert wormwood—sod-grass steppes are used mainly as pastures and only in years with particularly high production (with a wet spring and summer) is hay made. The band of these steppes lies in the zone of non-irrigated farming, but does not give stable yields because of the deficiency of precipitation. Individual massifs of less alkaline soils are, however, ploughed.

Fauna

The fauna of the semi-desert zone is not rich and consists of both steppe and desert animals. Most numerous among the mammals are various burrowing rodents: *Lagurus lagurus*, the small gopher (*Citellus pygamaens*), ground-squirrels (*Citellus fulvus*), jaculus (*Allactaga jaculus*), jaculus-emuranchik (*Scirtopoda telum*), pischukha (*Ochotona pusilla*) and the hamster (*Cricetulus eversmanni*). Rodents form extensive colonies and are one of the chief preys of predaceous birds, such as the steppe eagle, kurgannik (*Buteo fezox*) and others.

Small birds are represented by large numbers of small larks (*Melanocorypha calandra*) and field larks (*Alauda arvensis*), kamenka (*Saxicola oenantha*), Caspian plovers (*Aegialitis alexandrina*) and others.

In the southern districts of the semi-desert there are numerous herds of saigak antelopes (*Saiga tatarica*). In low-mountain massifs, such as the Ulutau and Kyzyltas mountains, arkhars (*Ovis polii*), occasional roes and a steppe cat (*Felis ornata*) still survive. More frequent are badgers, weasels, foxes and hares.

In terms of physical geography the semi-desert zone of the U.S.S.R. is divided into the Eastern European and Kazakhstan regions, each with five natural districts (see map). Those of the Eastern European region are the Tersko-Kumskoi, Yergeninskii, Sarpinskii, Volgo-Uralskii and Preduralskii districts. In northern Kazakhstan they are the Turgaiskii, Ulutaiskii, Sarysu-Tenghizskii, Kysyltaiskii and Chegisskii.

UTILIZATION OF THE NATURAL RESOURCES

Animal husbandry and pasture economy

The use of semi-desert rangelands for animal husbandry is of great importance in the people's economy. The cost of the products of rangeland animal husbandry in deserts and semi-deserts is low in comparison with other natural zones and the study, efficient utilization and improvement of semi-desert rangelands in the U.S.S.R. therefore receive considerable attention.

Considered from a range management point of view, the semi-desert rangelands of the U.S.S.R. are quite varied. On a life-form basis they are classed as semi-shrub-herbaceous vegetation, but according to their species composition, they can be divided into the following types.

(1) White wormwood pastures (*Artemesia lercheana*). Besides wormwoods, the valuable forage couch-grass also occurs, while in spring there is a rich herbaceous cover of ephemers and ephemeroids (annual and perennial plants with short growing seasons). The white wormwood is the most widespread species. It is a low xerophytic semi-shrub, 20—40 cm high. It is palatable throughout the year, especially in autumn. Its chemical composition classes it as a high-grade forage plant. Production varies from 400 to 650 kg ha^{-1}, depending on the climatic conditions of the year.

(2) Black wormwood pastures (*A. pauciflora*) are also widespread. It is a low xerophytic semi-shrub growing preferentially on alkaline and sub-alkaline brown and light-chestnut soils. It is most palatable in autumn, especially to sheep. It is one of the best forage plants of the semi-desert zone and has a high rating as forage. Average production is 300—400 kg ha^{-1}.

(3) Gray wormwood pastures (*A. terrae albae*) predominate in the northern band of the semi-desert zone. It is a low xerophytic semi-shrub, palatable to sheep, especially in winter, and has a high nutritive value.

(4) *Eurotia* pastures. *Eurotia ceratoides* is a xerophytic semi-shrub up to 40 cm high, and is a fairly good forage plant. It is palatable all year, especially to sheep. This plant prefers rock debris or sandy loam soils. Its production varies from 200 to 500 kg ha^{-1}.

(5) *Atriplex cana* pastures are widespread on solonchak-type soils. It is a haloxerophytic semi-shrub and is palatable only in winter and autumn (Bykov, 1969). All these pastures form different complexes with saltworts (*Salsola*) on sub-alkaline soils or with herbs on lighter soils. In years with considerable precipitation the wormwoods and other plants grow to such luxuriance that they are mown for hay. In such years insurance reserves for several years are built up.

(6) Herbaceous-wormwood pastures are widespread on light-chestnut soils throughout the semi-desert zone (Fig. 9.3). The principal species of herbs are feather grasses (*Stipa capillata*, *S. sareptana*, *S. lessingiana*, *S. kirghisorum*, *S. sublessingiana*), *Agropyrum sibiricum*, *A. desertorum* and *Festuca sulcata*.

(7) Summer cypress pastures with the predominance of the xerophytic semi-shrub *Kochia prostrata* are fairly widespread. They yield on average about 400 kg ha^{-1}.

Year-round grazing of livestock is traditional and has been practised since historical times. It is still practised today but is being constantly expanded and improved. Considerable success has been attained in the study and

Fig. 9.3. *Agropyrum desertorum* + *Artemisia* pasture in the semi-desert zone of the Kazakhstan (Photograph: M. Petrov).

reclamation of the ranges. Large-scale mapping has been carried out for the pasture areas, their production and carrying capacity have been estimated, and the nutritive value and palatability of rangeland plants and sheep forage have been evaluated. Adequate reference books with this information have been compiled (Larin, 1951/1956, 1969).

A scientific basis for pasture turnover calculations has been worked out and methods have been recommended to livestock farms for developing management plans to ensure optimum use. Much thought has been given to developing methods of radical improvement of pastures. Some of the measures in this respect are listed below.

(1) Large-scale mapping of pastures. Such maps usually show the outlines of pastures (ordinarily combinations of complexes of several types of pastures). A pasture is named after prevalent forage or forage groups. The maps show pasture areas, average yields of forage and the season when pastures are used.

(2) A scheme of the locations of pasture regions. The scheme produces a general assessment of the pasture territory of a farm. Large natural-economic and pasture-hayfield regions correspond to the natural features of the terri-

tory, their use (with respect to vegetation), water supply, soils and measures on their improvement.

(3) A plan of routes over which animals are driven, showing places of shearing, insemination, watering and stopovers. The plan is compiled on the basis of a general geographic and pasture map of a farm and of pasture lands allotted to the farm.

(4) A scheme of hayfields, compiled annually in accordance with predictions of yield and time of moving.

(5) Lists of forage plants, weeds, poisonous and obnoxious plants, as well as medicinal, edible and other useful plants.

(6) Allocation of areas where seeds should be collected, for use in the improvement and stabilization of pastures. Grazing in these areas should be limited in spring and summer (and sometimes in autumn too).

(7) Large-scale land-use maps of pastures for individual flocks, showing grazing boundaries and types of pastures in more detail than on the pasture maps. These land-use maps show roads, paths, wells, animal shelters, places considerably trampled, weedy sites, forest sites where grazing is limited and places where fodder can be prepared.

(8) A scheme of probable areas for fenced pastures (also for individual flocks).

(9) Schedule of grazing on distant pastures. This schedule is compiled for each season, account being taken of the reserves of forage in each pasture area.

(10) A scheme of pasture turnover made on the basis of a pasture map, with account taken of natural conditions, load on the pasture depending on the degree of trampling, weediness, grade and season of forage maturity.

(11) A plan for the improvement of pastures and hayfields.

(12) A table of forage reserves and forage stores on pastures for a farm on the whole and for pasture areas. It is compiled by specialists with the aid of livestock breeders. Calculations can be made on the basis of the available data on yields and phenology of plants. These data are supplied by agrometeorological stations. Annual records should be kept for pastures because of yield variations in different years and seasons, depending on weather conditions (Kurochkina and Osmanova, 1973).

In recent years botanists and agrometeorologists have together been developing methods for forecasting pasture production in semi-arid regions and these have given rise to practical suggestions in the form of instructions and recommendations (e.g. Fedoseyev, 1969).

Agroclimatic reference books are available which describe the peculiarities of climatic conditions as they affect livestock management in all the rangeland regions of the semi-desert zone. These reference books contain the following information: the agrometeorological conditions required for renewed growth of annuals and of ephemers, the climatic conditions under which rangeland plants are scorched in summer, as well as those under which they resume growth again in autumn, the meteorological conditions required for winter pasturing of sheep and goats, the dates when animals should begin grazing on green herbage, meteorological conditions at the lambing period, and the conditions for spring shearing of sheep and for summer grazing. The dates of the early-spring and spring lambing of sheep are adjusted to the time when it is possible to make the maximum use of natural pasture forage.

Animals graze almost year-round, but livestock farms have reserves of fodder for the periods of heavy snowfall, when animals are transferred to semi-stall maintenance.

The principal domestic animals raised on the semi-desert rangelands of the U.S.S.R. are sheep and some goats. The flocks are large and the number of animals runs into millions.

The strains of sheep in the European part are Soviet merino, Stavropol, Caucasian and Grozny fine-fleeced sheep (Fig. 9.4). A sheep yields on average 4—5 kg wool per annum. The semi-deserts of western Kazakhstan are the principal breeding zone of fat-tail, meat-and-fat sheep which also yield coarse and semi-coarse wool, and high-grade sheepskins — valuable raw materials for the production of cloth, carpets and clothing. In these regions the Edilbai sheep is the best improved strain of meat-and-fat sheep. It is the largest of all the fat-tail sheep in the semi-desert zone, and along with its high meat and fat production it has the best wool of all these breeds of

Fig. 9.4. Sheep grazing the pasture of semi-desert zone of Kazakhstan (courtesy of TASS).

sheep. The average live weight of ewes is around 75 kg and that of rams is up to 100—120 kg.

Camel-breeding is not so widely developed as sheep-breeding, but it still plays an important role in livestock-raising in the Republic. Camels are used as transport animals for local needs, and they are also used for meat, pelts and wool. Their milk is used for making a refreshing and health-giving drink, locally known as "chad". The species of Kazakhstan camels is the two-humped *Camelus bactrianus* (Fig. 9.5). Their numbers run into tens of thousands.

The system of pasture economy management has been built up on the basis of a planned socialistic principle. All the pasture areas are state property. In these territories there are hundreds of Soviet and collective livestock-raising farms, each of which is allotted a definite pasture area, depending on the number of its animals. The grazing system is drawn up according to the peculiarities of the pasture areas of each farm. The usual practice is free grazing and pasture turnover.

The annual cycle of sheep maintenance is primarily a pasture one. The animals graze on the rangelands practically all year round. In winter they receive additional fodder or stall maintenance only on days of inclement weather. Weather has a double influence on animals in pasture maintenance; it determines the condition of the pasture vegetation and directly affects the animals themselves.

Fig. 9.5. Bactrian camels (courtesy of TASS).

The prime conditions for successful animal maintenance are sufficient pasture forage and sufficient watering points.

Weather conditions can be favourable or unfavourable for animals, depending on the species, strain, age, condition, sufficient food supply, the maintenance system, degree of acclimatization and other factors.

During wintertime unfavourable conditions for sheep are snow cover, ice crusts, strong winds in combination with sub-zero temperatures and snow storms. A snow cover, depending on its depth and density, layers of ice and frozen snow make grazing very difficult and sometimes impossible, since the animals cannot get to the plants. Low air temperatures in combination with wind cause great losses of heat and lead to catarrhal diseases. Snow storms and heavy drifting snow prevent grazing and cause catarrhal diseases and frost-bites. Summer grazing and fattening are of considerable importance as they determine the condition of the animals for the winter.

A peculiarity of natural forage lands of the semi-desert zone is the distinctly seasonal use by animals of a considerable proportion of the pastures. Only a part of them can be more or less fully used for grazing in the spring—summer and autumn periods. Thus, for instance, all wormwood—saltwort pastures can be used only in early spring, and again in autumn and winter, since by the latter time the plants are dry and their salt content has been leached by rains.

Efficient organization of livestock-raising on these rangelands calls for the use of individual seasonal pastures during the years when they attain maximum production. Historically, this led to the organization of livestock-raising in semi-deserts based on the driving of animals from their pastures to others, often over great distances.

In the livestock-raising regions of the central and south-western Kazakhstan, dry and semi-desert steppes, used mainly for spring, summer and autumn pasturing, predominate. The southern slopes of hills and sand massifs are used as winter pastures. Therefore, the predominant form of animal husbandry is to drive them to winter pastures in the sands of Sami, Uimgut and other places, over distances of 300 km and more. The winter period is the most difficult. The problem is actually one of minimizing the number of deaths. During the winter months sheep lose considerable weight owing to the shortage of forage, but in the first spring months weight is gradually restored, while in summer and autumn sheep are well fattened.

During the years of very low precipitation a dearth of forage can result, and it is therefore important for livestock-raising in the semi-desert zone of the U.S.S.R. to build up insurance fodder reserves during the wetter years for additional feeding of animals in periods when grazing is not possible, or when the quality of herbage does not meet the requirements of the sheep. At times when grazing is for some reason impossible, straw of various cereals and hay are used for additional feeding.

Grazing is a dominant factor affecting the dynamics of rangeland vegetation in the semi-deserts. Moderate grazing, when the animals consume about 60—70% of the annual production, has a beneficial effect on the growth of vegetation (Nikolayev, 1977). This accounts for the maintenance of the species composition of semi-desert vegetation in spite of its long-term use by pastoralists.

Over-grazing, however, results in due course in the destruction of the basic types of vegetation. This is accompanied by the replacement of palatable plants by unpalatable weeds. In Russian literature this process has been named "pasture digression" (Vysotskiy, 1915). This process generally includes not only the replacement of species, but also radical changes in other ecosystem processes, as it has many side-effects (see Fig. 9.6).

Some plants react negatively to excessive compaction of soil and are characteristic of little-used pastures. Moderate loosening of soil by the hooves of animals promotes better infiltration. Moreover, partial browsing of the shoots of perennial plants frequently increases their productivity. Finally, pasturing facilitates the burial of seeds, which aids renewal of the vegetation.

Excessive utilization of vegetation and trampling of the soil surface transforms good pasture into areas of low production. The consequences of this

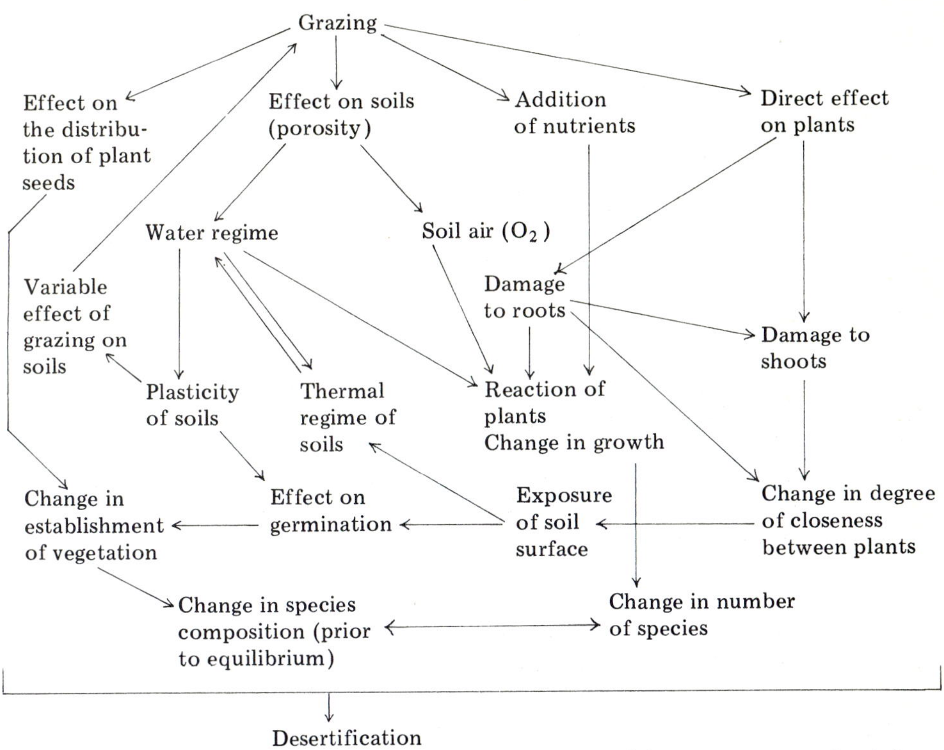

Fig. 9.6. Pasture degradation in semi-deserts and deserts resulting from over-grazing (after Shvan-Guriisky et al., 1974).

process can be seen near wells which are usually surrounded by a ring of barkhan sands, and on the outskirts of settlements where pastures have been completely trampled (Fig. 9.7). However, the transformation of sand desert pastures into shifting barkhan sands occurs over comparatively small areas. The degradation of pastures in deserts and semi-deserts is more commonly the result of over-grazing.

It should be noted that in general the potential for increasing the carrying capacity in deserts and semi-deserts is greatly limited by insufficient precipitation. Phytomelioration can help to raise this potential by no more than around 50%. Consequently, increasing the productivity of livestock in arid regions requires particular attention to the following two points: (a) improvement of the strain of animals and combating their diseases, and (b) growing plants for fodder on irrigated lands for additional feeding of livestock in drought years (Nikolayev, 1977).

(1) Water supply in rangelands. One of the most important conditions for the successful pasture maintenance of animals is the availability of a sufficient number of watering places. Much attention is given to this problem and it has been solved in different ways. The most common and traditional

Fig. 9.7. Overgrazed pasture near well. Western Kazakhstan (Photograph: M. Petrov).

way is the supply from wells, of which there are thousands in the semi-desert zone. Their water is usually of good quality, with little hydrocarbon salination, although the water is sometimes slightly saline due to chlorides and sulphates.

In some regions there are still considerable areas of pastures that are little used because of a lack of water. This difficulty can, however, be fairly easily overcome. Hydrogeological explorations in the semi-desert zone of the U.S.S.R. have shown a widespread availability of considerable reserves of ground water. Its utilization by means of wells and artesian boreholes will be achieved without any particular engineering difficulties. There is enough water both for animals and people.

In recent decades extensive artesian basins have been found in the semi-desert regions of the U.S.S.R., so that in the case of insufficient or unstable ground water it is possible to ensure water supplies by means of artesian pressure waters. These waters are used each year in constantly increasing volume to supply water to tens of millions of hectares of new pastures. The quality is suitable not only for animals but also for supply to human settlements, and there is sufficient water for the irrigation of small flower and vegetable gardens.

Extensive use is also made of river water. Even though there are virtually no permanent rivers in the semi-desert, spring-season and rain waters sometimes fill dry river-beds and provide a temporary flow (e.g. the Bolshoi and Maliy Uzen-Irghiz, Turgay and other small rivers). Dams have been built on such rivers for collecting the water in reservoirs. Finally, water for livestock is supplied from such large rivers as the Volga and Ural, via canals and aqueducts.

(2) Improvement of pastures. In recent decades the number of livestock in some parts of the semi-deserts of the U.S.S.R. has increased to an amount exceeding the carrying capacity of the pastures. Over-grazing and degradation of pastures has resulted in some places, and thus new requirements have now been set for the use of these pastures. They have been necessitated by the need for reinforcing the reserve forage base, for expansion of the pasture areas and for their improvement so that the number of livestock could be further increased. Specialists in pasture management were required to draw up measures for the improvement (phytomelioration) of the pasture lands. The work is conducted on a large scale. For the improvement of pastures local forage plants are used, e.g. *Kochia prostrata*, *Agropyrum desertorum*, *A. sibirica*, *Artemisia arenaria*, *A. astrachanica*, *A. glauca* and others. Seeds of these plants are produced on specialized farms.

Forage grass cultivation requires preparation of the soil by ploughing bands of virgin land. This increases the pasture carrying capacity by 60—70% (Nikolayev, 1972). Moreover, in years of abundant atmospheric precipitation when pasture vegetation is especially luxuriant, livestock raisers mow grass for hay. In addition fodder herbs are grown in depressions which receive additional moisture from local runoff. This herbage is also mown for the reserve fodder banks.

As a result of all these measures, pasture maintenance of livestock in the semi-desert zone of the U.S.S.R. has developed new management techniques differing considerably from the traditional nomadic livestock raising. In Soviet literature it has been given the name of "livestock raising on distant pastures". Under this system the bulk of the population lives in the agricultural regions, and only small teams of herdsmen are sent to seasonal pastures with the animals. Thus, agriculture is now combined with pasture maintenance of livestock, and almost all of the population is permanently settled.

Livestock raising in the semi-desert zone in this improved form will retain its importance for many future decades, since the possibility of year-round grazing ensures maximum revenues at minimum labour expenses.

Agriculture

The problems of irrigated agriculture in the semi-desert zone of the U.S.S.R. are not discussed here, as they do not fall within the scope of this chapter.

Specific natural characteristics of the semi-desert zone of the U.S.S.R. affecting yields of cultivated crops include a sharply continental climate, insufficient atmospheric precipitation and strong dry winds resulting in the deflation of soils and thin snow cover. The following types of agriculture are practised: (a) dry-land agriculture which relies solely on atmospheric precipitation; (b) irrigated agriculture using a permanent system of irrigation channels; (c) agriculture irrigated by well and artesian water. Areas of irrigated agriculture are very limited, but they produce high and stable crop yields.

The assortment of cereals and other herbs cultivated in the semi-desert zone is small: winter rye, spring wheat, barley, millet and forage cultures for green fodder (corn, annual and perennial herbs).

Dry-land agriculture is widespread over large areas, particularly in depressions which have darker soils and receive additional moisture from the inflow of thaw water. Crop yields in these areas are comparatively small and subject to constant fluctuations depending on the weather conditions of the year. They suffer especially from dry winds and droughts.

Climatologists distinguish five types of drought: early spring, spring—summer, summer—autumn, combined and stable-droughts. They usually result from disturbance of normal atmospheric circulation. A prevalence of meridional circulation over latitudinal circulation over the U.S.S.R. always results in droughts. Consequently, droughts have no distinct periodicity and drought years are often separated by considerable periods (see Table 9.4).

Unfortunately there are no reliable criteria for forecasting droughts, depending as they do on the dynamics of air masses.

TABLE 9.4

Frequency of droughts in the U.S.S.R. (after Alpatyev and Ivanova, 1958)

	Total number of years	Drought years
Saratov	60	33
Volgograd	56	39
Uralsk	58	45
Orenburg	61	28

Successful development of agriculture in the semi-desert zone is greatly hindered by wind erosion. In these regions wind erosion processes are widespread and become catastrophic in very dry years. A complex of measures on wind erosion control is represented in Fig. 9.8 which shows that control of wind erosion must be aimed at (a) reducing the velocity of wind near the soil surface, (b) soil protection (together with sown seeds, sprouts and young plants of agricultural crops), and (c) increasing soil resistance to deflation by numerous measures, e.g. the application of different farming practices.

Protective afforestation in the semi-desert zone

Protective afforestation in the semi-deserts of the U.S.S.R. is introduced for the following purposes: creation of protective forest belts for agriculture, creation of protective shelters for livestock against strong winds in winter and hot weather in summer, landscaping of settlements, and growing timber for local construction (Fig. 9.9).

Protective afforestation in the dry agricultural regions helps the accumulation of snow in winter, which leads to improved soil moisture.

Forest belts on pastures raise forage production by 50—100% (Nikitin, 1971). Richer vegetation is formed under the shelter of these belts, consisting of cereal herbs such as fescue and *Koeleria*.

To achieve stable, unirrigated plantations in the semi-desert zone requires that trees and shrubs be drought-, heat-, salt- and frost-resistant. A comparatively small number of species meet these requirements. Among the trees which do, are the small-leaf elm (*Ulmus pumile*), oleaster (*Elaeagnus angustifolia*) and white acacia (*Robinia pseudoacacia*). Among the shrubs are the golden currant (*Ribes aurea*), honeysuckle (*Lonicera tatarica*) and tamarix (*Tamarix ramoissima*). On sandy massifs where ground water is at a fairly shallow depth, more moisture-loving trees are grown, such as pine (*Pinus sylvestris*), birch (*Betula alba*) and black poplar (*Populus nigra*).

Of these species the most promising one for the drought conditions of the semi-deserts is the small-leaf elm (*Ulmus pumila*) and it is consequently the principal tree in protective belts. It is a small tree, 6—7 m high, with a

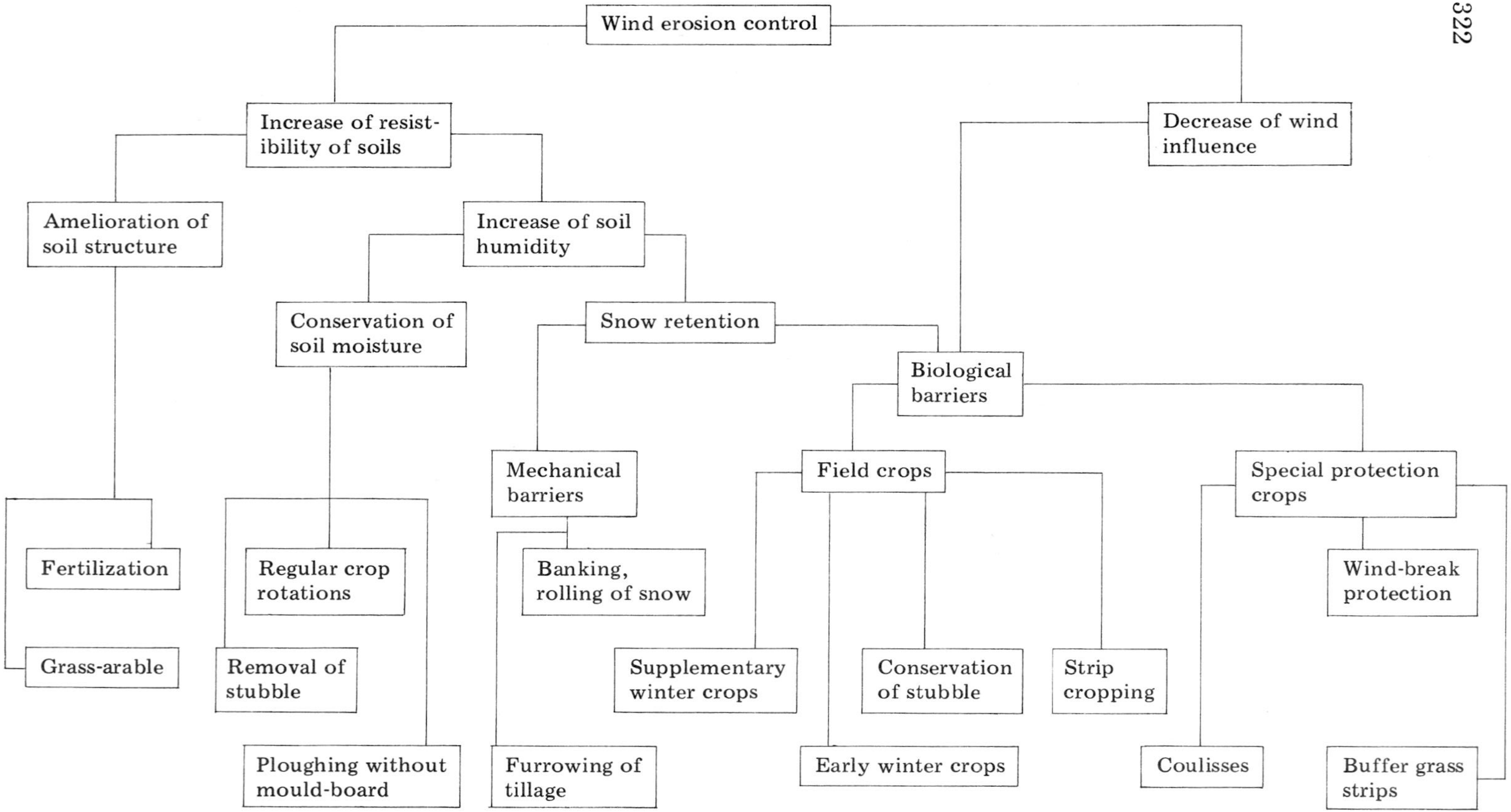

Fig. 9.8. Wind erosion control (after Armand, 1963).

Fig. 9.9. Shelter belt on pasture of the semi-desert zone of Kazakhstan (Photograph: M. Petrov).

shallow root system. It is grown from both seeds and root grafts.

Oleaster (*Elaeagnus angustifolia*) is also a small tree, 6—7 m high, with a variable root system which is either of a surface or deep-reaching type, depending on the depth of ground water. It is well capable of renewal by root and surface shoots. It multiplies by seeds, shoots and grafts.

White acacia (*Robinia pseudoacacia*) is a slightly taller tree, 7—8 m high with a surface-type root system. It multiplies by seeds, stump and root shoots. It has, however, little resistance to frost.

A characteristic feature of the shrubs grown in protective plantings in the semi-desert is the great variability of their root systems. This makes it possible for them to utilize additional soil volumes by means of their abundant adventitious roots which absorb even hardly accessible moisture. Their heat-resistance allows them to stand tissue heating up to 45—50°C. They are also salt-resistant and withstand seasonal salinations of up to 2—3% in the soil solution. The most salt-resistant shrubs are tamarix, which can stand salinations of up to 15—20%.

The growth and development of shrubs is characterized by high growth rates in their first years, especially of their root systems. They enter the reproductive phase early, and thereupon their growth slows down. They are not long-lived.

At the Dzhanibek station of the Academy of Sciences of the U.S.S.R. methods have been developed for growing protective forest plantings in light-chestnut soils of the semi-desert sub-alkaline complex. Schemes have been determined for efficient construction of protective belts for different purposes: for sheltering crop areas, roads and for landscaping. Recommendations are made for the selection of tree and shrub species for these purposes. For instance, a scheme has been drawn up for single-row planting of the small-leaf elm, this ensuring accumulation of up to 100 mm of snow per year and thus supplying moisture to agricultural fields. These recommendations have become the basis of agricultural reclamation of the semi-deserts of the northern Caspian region, without requiring irrigation (Nikitin, 1971).

Game and hunting

The vast territory occupied by the semi-deserts of the U.S.S.R. provides favourable living conditions for various animals. Saigak antelopes (*Saiga tatarica*) occur in large numbers. Common predators are the karaganka fox (*Vulpes vulpes karagan*) and the wolf (*Canis lupus*).

Many of the species are hunted as game. Of greatest economic importance

Fig. 9.10. Antelope-saigak (courtesy of TASS).

is saigak hunting. It is very widespread in the semi-deserts and its numbers are estimated at about 20 million animals, of which up to 200 000 are shot annually on a planned basis. The saigak antelope is a fairly large animal, weighing up to 40 kg and its meat is of high quality. Its hide is used for leather (Fig. 9.10).

Second in economic importance is the gopher (*Citellus pygmaens*) which is used for its pelt. The number of these animals is enormous, and hundreds of thousands of pelts are procured annually. In Kazakhstan tens of thousands of pelts of the local karaganka fox are obtained each year. These are relatively small animals, up to 1 m long, with handsome grey-brown fur. Game-bird shooting is of considerable local importance.

Hunting is regarded as a special branch of the economy, and there are prepared plans for pelt deliveries, for censusing the number of animals, and for determining the regions where they are most abundant. At the same time measures are developed for increasing the number of these animals, protecting them from epizootics, primarily from plague.

The hunting season and shooting quotas have been decreed by the government to prevent a decline in the number of animals to provide for their safety at breeding time, and to maintain the high grade of pelts.

Mining, industry, urbanization and construction of transport facilities

Industrial development of natural resources in the semi-arid regions of the U.S.S.R. is comparatively new, and very promising, since economically valuable minerals of the regions are rich and varied. This has led to a reassessment of the importance of the semi-arid region territories in the economics of the country. They have moved from a low-productivity category into the high-productivity category.

In consequence, the economic activity of Man in the semi-deserts of the U.S.S.R. has greatly increased in recent years. It has been due to the discovery and successful exploitation of major deposits of economic minerals, such as the rich deposits of copper ore in the Dzhezhkazgan and Karsakpai regions of the northern Caspian region.

The semi-deserts have been crossed by new railroads, for example the Chu—Mointy—Karagandu—Tselinograd in central Kazakhstan, and the Kandagach—Emba in western Kazakhstan, and also by thousands of kilometres of new automobile roads and by oil and gas pipelines.

Finally, in the semi-deserts of western Kazakhstan a large canal is under construction from the Volga to the Ural river. In 1975 construction was completed of a canal from the Irtysh river to the city of Karaganda, a distance of 458 km.

These canals are of great importance for the water supply of the towns and industrial plants of the metallurgical, coal and chemical industries, located in the semi-desert zone. They will also supply water for irrigated agriculture, allowing the development of large new farms for growing vegetables, potatoes and livestock fodder. A part of the water will also be used for water supply to livestock ranges.

The human population has increased accordingly. New industrial plants with accompanying settlements have sprung up at Dzehezkazgan, Kounrad, Karaganda, Tselinograd and other centres, and the old towns and settlements have greatly expanded. The introduction of industrial methods for the development of the arid regions creates relationships between man and his environment entirely different from those involved in the agricultural use of the regions.

CONCLUSIONS

The development of natural resources of the semi-arid regions of the U.S.S.R. is based on a planned socialistic economy. It is conducted along two major lines — agriculture and the mining of economic minerals. The scale of work along these lines expands annually, leading to the growth of the population, the appearance of new settlements, and development of all forms of transportation. Livestock raising is the basis of the rural economy.

The increased rate of development of the semi-arid regions has led to a sharp increase in the pressure on these ecosystems. There can arise danger in the future of the development of the anthropogenic ecosystems with low biological productivity.

Adequate measures are being worked out for the prevention of such processes. More specifically, there are detailed recommendations as to the maximum pressure on the semi-desert ecosystems for each different type of economic development, such as livestock raising, agriculture, mining, etc.

Whenever measures are drawn up for the development of the natural resources of semi-arid regions, changes in their structure are scientifically predicted to prevent the development of natural phenomena detrimental to the economy. Prediction is done on the basis of the analysis of interrelations of the ecosystem's structure.

The predictions are both short-term, made for the period of construction, and long-term for the period of operation. In other words, the prediction should be dynamic and will require constant corrections.

Of particular importance is the long-term prediction of physiogeographical processes in places where human activity has a great influence on nature. The development of these predictions is very complicated, requiring consideration of a large number of factors and processes, and can be made only with the aid of mathematical and physical methods, including the modelling of natural processes.

Thus, within the entire semi-arid zone protective measures are implemented and reasonable methods for the utilization of the environment are introduced.

Humanity can and should penetrate the semi-arid zones of the globe for development of their natural resources, using the achievements of scientific and technological advancement. This will require, however, considerable changes in the traditional methods of the utilization of deserts. All this will permit the preservation of the dynamic equilibrium of the ecosystems in the semi-arid regions and raise their biological productivity.

REFERENCES

Alpatyev, A.M. and Ivanova, V.N., 1958. Description and geographical distribution of droughts. In: Droughts in the U.S.S.R., Their Origin, Frequency and Influence on Yield. Gidrometeoizdat, Leningrad, pp. 31—45.

Anonymous, 1966a. Handbook on the Climate of the U.S.S.R., Vol. 13. Part II. Air and Soil Temperatures. Part IV, 1968. Air Humidity, Precipitation, Snow Cover. Leningrad.

Anonymous, 1966b. Handbook on the Climate of the U.S.S.R. Vol. 15. Part II. Air Soil Temperatures. Leningrad. Part IV, 1969. Air Humidity, Precipitation, Snow Cover. Leningrad.

Anonymous, 1966c. Handbook on the Climate of the U.S.S.R., Vol. 18. Part II. Air and Soil Temperatures. Part IV, 1968. Air Humidity, Precipitation, Snow Cover. Leningrad.

Anonymous, 1967. Resources of Superficial Water in the U.S.S.R. The Main Hydrological Features, Vol. 13. Gidrometeoizdat, Leningrad.

Anonymous, 1968. Physico-geographical regionalization of the U.S.S.R.: Characteristics of Regional Units. Moscow.

Anonymous, 1968. Soils of Kazakhstan. Aktubinsk Region, Vol. II, Nauka, Alma-Ata.

Armand, D.L., 1963. Agricultural factors of the origin and suppression of storms. In: Dust Storms and Prevention of Them. Acad. Sci. U.S.S.R., Moscow, pp. 156—167.

Bykov, B.A., 1969. Pastures and Hayfields in Kazakhstan. Alma-Ata.

Fedoseyev, A.P., 1969. Scientific foundation of the agrometeorological forecasts of the desert pasture harvest. In: Problems of Desert Development, No. 4, pp. 3—13.

Flora of the U.S.S.R., 1934—1964. Vols. 1—30. Nauka, Moscow.

Kurochkina, L.Y. and Osmanova, L.T., 1973. Sand Desert Pastures of the Kazakhstan. Alma-Ata.

Larin, I.V. (Editor), 1951/1956. Fodder grasses of hay-fields and pastures of the U.S.S.R., Vols. I and II, 1951; Vol. 3, 1956. Moscow—Leningrad.

Larin, I.V. (Editor), 1969. Hayfields and Pastures. Moscow.

Lobova, E.V., 1960. The Soils of the Desert Zone of the U.S.S.R. Moscow.

Meigs, P., 1953. World distribution of arid and semi-arid homo-climates. In: Review of Research on Arid Zone Hydrology. Arid zone programme, Vol. I. UNESCO, Paris, pp. 203—210.

Nikitin, S.A., 1971. Shelter Belt Planting in Semi-Desert. Nauka, Moscow.

Nikolayev, V.N., 1972. Natural Forage Resources of the Turkmenistan. Ashkhabad.

Nikolayev, V.N., 1977. Desert Pastures, Their Forage Valuation and Taxation. Moscow.

Petrov, M.P., 1975. Boundaries of arid lands subbinding units. Problems of Desert Development. N 2, pp. 3—10.

Rode, A.A., 1963. Water regime and water balance of virgin soil of the semi-desert complex. In: Water Regime of Semi-Desert Soils. Acad. Sci., Moscow, pp. 5—83.

Shvan-Guriisky, I.P., Sinkovskaya, A.S. and Williams, J.K., 1974. Rational Use and Improvement Methods of the Arid Zones Pastures. Moscow.

Vysotskiy, G.N., 1915. Natural Site Conditions and Results of Afforestation on Ergeni District. Petrograd.

Chapter 10

LAND-USE OF SEMI-DESERTS IN CENTRAL ASIA

M.P. PETROV †

Leningrad State University, Department of Geography, Leningrad (U.S.S.R.)

NATURAL FEATURES OF THE SEMI-DESERTS

Physical features

Semi-desert occupies a considerable area in Central Asia, spreading mainly within the boundaries of the Mongolian People's Republic (MPR) and partly over the territory of the Chinese People's Republic (CPR), in Inner Mongolia. It extends in a fairly narrow belt from the Great Lakes Basin in the west, where it is 80—100 km wide, to the foothills of the Great Khingan in the east, increasing to 300 km in width. The total length of the zone is about 1600 km (see Fig. 10.1).

The semi-deserts change in relief from the inter-montane valleys and basins in their western part to the extensive Eastern-Mongolian plain with low island ridges and hills, composed of old rocks.

The Great Lakes Basin is not a monotonous plain. It contains uplands which divide it into three isolated depressions — the northern, central and southern depressions, with large lakes such as Ubsu-Nur, Hara-Us-Nur, Hara Nur and others. The lakes occur at elevations ranging from 759 to 1153 m above sea-level. The surface of the plains is composed of proluvial and lacustrine-river deposits of sandy loam and loam with frequently occurring pebble beds. Large expanses of the eastern part of the Basin are occupied by sand massifs such as Borig-Elesu, Altan-Elesu and others. In wintertime these territories are used as pastures.

South-east of the Great Lakes Basin, between the Gobi Altai and the Khangai ridges, there is an inter-montane depression called "The Lake Valley" containing a chain of numerous endorheic lakes occurring at elevations from 800 to 1300 m.

East of the Lake Valley the semi-desert zone becomes broader and enters the Eastern-Mongolian plain of Eastern Gobi. There it represents an extensive plain continuing east into China (Inner Mongolia). The topography of this plain is fairly complicated. The flat areas are interrupted by inselbergs and small hills separated by dry ravines. Sand mounds and hills, which occupy about 2.5% of the Gobi territory, also contribute to the ruggedness of the relief.

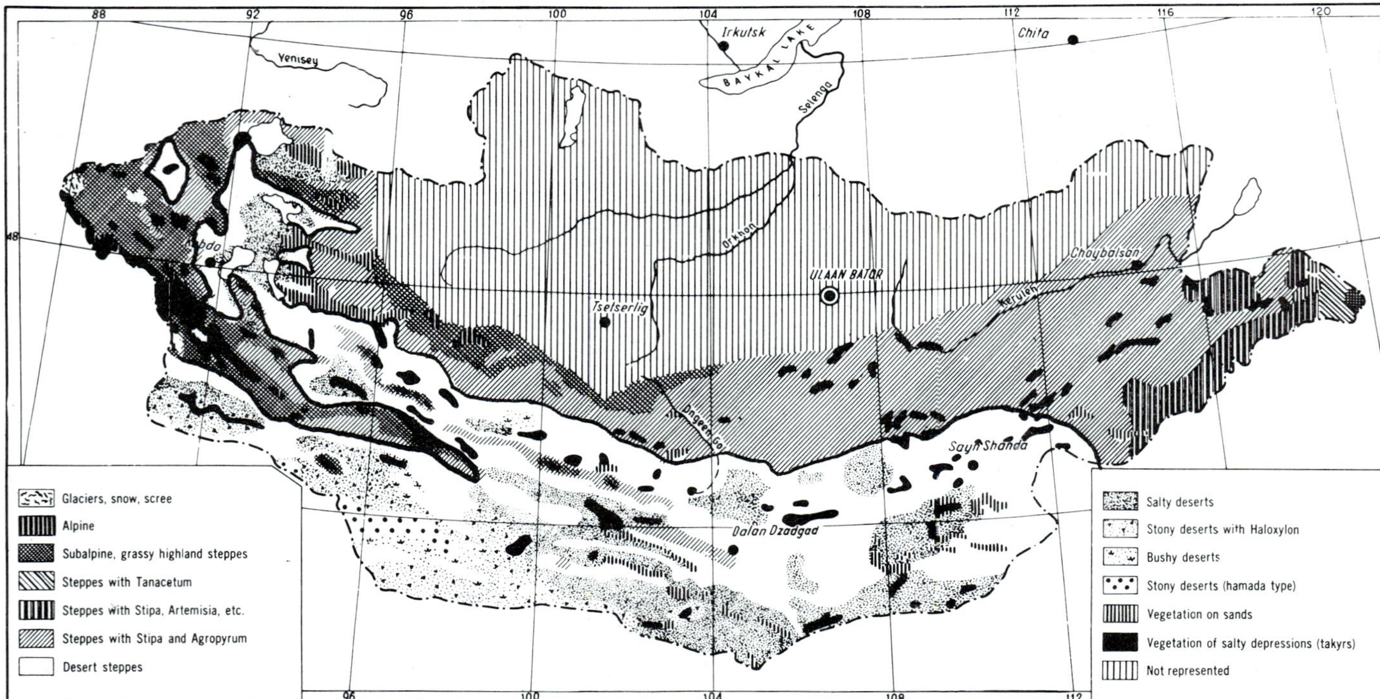

Fig. 10.1. Vegetation cover and landscapes in the Mongolian People's Republic. Black line, semi-desert zone. After Yunatov, 1950.

Insofar as the semi-desert zone of Eastern Asia is located mainly in the territory of the Mongolian People's Republic, with only a part of it extending into Inner Mongolia and the Chinese People's Republic, the data pertaining to it refer mostly to the territory of the MPR.

Climate

The climate of the semi-desert zone of Central Asia is markedly continental, owing to its location in the centre of the Asiatic continent, far from seas and oceans, and separated from them by immense mountain ranges. Severe winters are also due to the Mongolo-Siberian anticyclone, which leads to a wide range of annual and daily temperatures. Mean monthly air temperatures in January vary from −17°C to −27.9°C and in July from +19.6°C to +23.8°C (see Table 10.1 and Fig. 10.2).

Winter in Mongolia lasts for 3—4 months and is very severe, with absolute minimum temperatures of −48°C in the Great Lakes Basin and −36.5°C in the Mongolian part. The severe winter climate has an adverse effect on agriculture and creates considerable difficulties for animal husbandry. Livestock breeders have to devote considerable effort to the construction of warm shelters for animals, the preparation and distribution of additional fodder, and finding both local and distant migration routes in search of winter pastures.

Spring in the zone of Asiatic semi-deserts is, as a rule, cold, dry and very windy, resulting in soil deflation and desiccation. Summers are hot but more humid, since maximum precipitation occurs in July—August (50—60% of the annual total). This is due to the influence of summer monsoons. In general the desert zone gets a limited amount of precipitation — from 77 to 200 mm, mainly in the form of rains. Winter precipitation in the form of snow is negligible. The snow cover, when it forms, is thin. This is favourable

TABLE 10.1

Principal climatic characteristics of the semi-desert zone of Central Asia (from Anonymous, 1974a)

Station	Elevation above sea-level (m)	Air temperature (°C)					Total annual precipitation (mm)
		Monthly mean		Annual mean	Absolute maximum	Absolute minimum	
		Jan.	July				
Kobdo	1324	−27.9	+19.6	−1.7	+35.4	−48.0	125
Dzamyn-Ude	910	−18.2	+23.8	−3.7	—	—	77
Dalan-Dzadagad	1486	−17.0	+22.0	+2.9	+38.3	−36.5	128
Sasin-Shand	911	−20.4	+23.3	+2.2	+40.7	−414	115
Mandal-Gobi	1350	−20.3	+19.7	−0.3	+35.4	−36.8	196

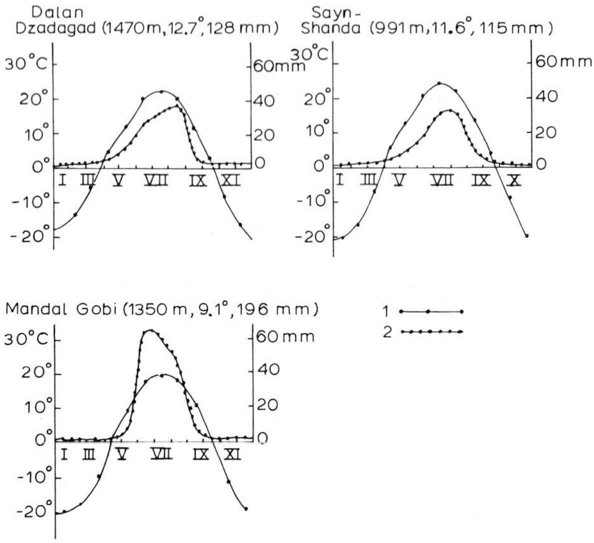

Fig. 10.2. Annual temperature (1) and precipitation (2) in the semi-desert zone of Central Asia.

for animal husbandry, since a thin snow cover does not obstruct grazing by animals. But, on the other hand, it has an adverse effect on agriculture because, unprotected by a snow cover, the ground becomes deeply frozen, which is unfavourable for winter crops. Moreover, the absence of snow facilitates considerable deflation of soil in spring.

According to Meigs (1953), the semi-desert territories of Central Asia are referred to as semi-arid regions with a cold winter and a summer maximum of precipitation (SbO_3). The frost-free period in the semi-desert zone is comparatively small, not exceeding 180 days.

In the western part of the semi-desert zone (the Great Lakes Basin and the Lake Valley) surface waters are widespread, occurring as small permanent rivers or intermittently flowing creeks, as well as large and small lakes. The eastern regions of the zone are almost devoid of surface waters, where they occur only as rare saline lakes and temporary streams.

The Gobi rivers are shallow, narrow, carry little water and run dry shortly after leaving the mountains or while still in the valleys. Their water is soaked up by alluvial deposits and evaporated, and the normally dry river beds are filled with water only during times of heavy rains. Such rivers occur mostly in the western part of the Gobi region, originating in the Gobi-Altai mountain chains or in their spurs. In the south-eastern region not even these sort of rivers occur. There are only rare spring streams which create small oases and a network of small dry river beds, which are wide, gently sloping channels with sandy bottoms, extending for only short distances and ending in the nearest drainless basins. Ground water supplies are consequently very important for animal husbandry, since they are the principal source of water. Thousands of wells and hundreds of artesian boreholes are used on the rangelands of the semi-desert zone (Murzayev, 1952).

Soils

Soils of the semi-desert zone of Central Asia are zonal types of light-chestnut and brown, weakly carbonate soils with little humus. Because of their mechanical composition the light-chestnut soils are mostly represented by varieties with considerable rock-debris content, and the loamy soils by much sand. The depressions contain many solonchaks, black alkali soil pockets and combinations of both.

Light-chestnut soils are characterized by little accumulation of humus (up to 1.8%), a shallow soil profile, and a concentration of $CaCO_3$ in the form of lime nodules or solid sinters, covering rock debris and stones at a depth of 40—70 cm. The $CaCO_3$ content is at a minimum in sandy soils and at a maximum in those with rock debris. The total amount of salination with easily soluble salts is negligible (see Table 10.2). On sandy massifs soils are weakly developed and are often greatly wind-blown.

The most suitable areas for agriculture are fairly flat, with brown silty soils of sufficiently high fertility. However, they can only be cultivated under irrigation.

Vegetation

The vegetation cover of the semi-desert zone is not very rich and is represented by the depauperate plant associations of desertified steppes. It is very sparsely distributed and consists mainly of sod-forming perennial cereal herbs: feather grasses (*Stipa glareosa, S. gobica* [1]), snake-weeds (*Cleistogenes mutica*), a small quantity of different herbs, mainly *Allium* species (e.g. *A. polyrrhisum*) and, in places, wormwood (*Artemisia frigida* and *Eurotia ceratoides*). The rare shrubs include *Caragana pygmea, Anabasis brevifolia* and *Hyppolitis achylleoides*. On more saline soils the halophylous shrubs *Anabasis, Salsola passerina* (Fig. 10.3), *Kalidium gracile, Reaumuria soongorica* and annual saltworts (*Salsola* spp.) occur. Unlike those in Kazakhstan, the semi-deserts of Central Asia have no ephemerals.

Richer vegetation grows on sand massifs: rare thickets of saxaul (*Haloxylon ammodendron*), *Hedysarum mongolicum, Nitraria schoberi, Caragana bungei* and *Artemisia arenaria* (Fig. 10.4). Of the perennial cereal herbs there occur individual plants of *Psammochloa villosa*. The annual *Agriphyl-*

---

[1] Nomenclature is according to V.I. Grubov, 1955. Synopsis of the Flora of the Mongolian People's Republic. Leningrad (in Russian)

TABLE 10.2

Chemical composition of light-chestnut soils from the semi-desert zone of Central Asia (% of dry soil) (after Bespalov, 1957)

Location	Sample depth (cm)	Humus (%)	Solid residue (Ash %)	Alkalinity		Cl⁻	SO₄²⁻	Ca²⁺
				CO_3^-	HCO_3^-	Cl^-	SO_4^{2-}	Ca^{2+}
Loamy soil with detritus and *Stipa*—*Artemisia* vegetation; 120—125 km NE from Aranbai-Here	0—15	1.80	0.520	—	0.040	0.002	0.004	0.005
	10—15	1.10	0.074	—	0.050	0.003	0.003	0.011
	35—40	0.40	0.068	—	0.038	0.003	0.004	0.008
	65—70	—	0.133	—	0.050	0.015	0.008	0.005
Loamy soil with detritus and *Artemisia*—*Stipa* vegetation; near Opchin-Gol river	0— 5	1.54	0.080	—	0.040	0.002	—	—
	10—15	1.08	0.050	—	0.044	0.002	—	—
	25—30	0.76	0.065	—	0.040	0.002	—	—
	65—70	—	0.178	—	0.024	0.017	—	—
Sandy-loam soil with *Artemisia*—*Allium* vegetation; 20—25 km SW from Sukhe-Bator	0— 5	1.85	0.085	—	0.014	0.017	0.003	0.005
	15—20	1.02	0.061	—	0.014	0.001	0.001	0.005
	35—40	0.63	0.043	—	0.010	0.001	—	0.003
	55—60	—	0.027	—	0.011	0.001	—	0.003
	75—80	—	0.032	—	0.011	0.001	0.001	0.002

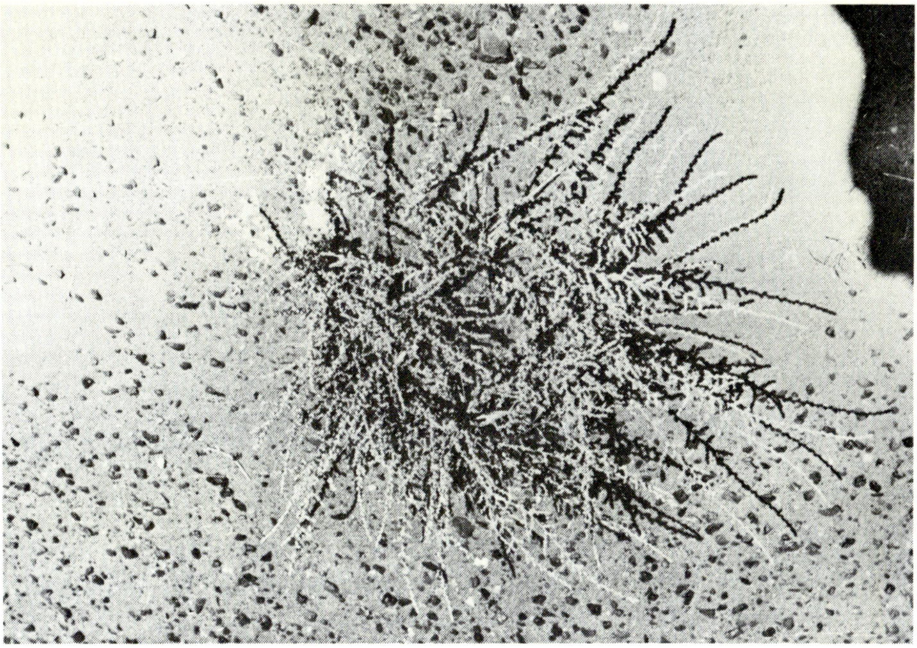

Fig. 10.3. Semi-shrub *Salsola passerina* — one of the main fodder plants of the semi-desert zone of Central Asia (Photo: Z. Bespalovoy).

Fig. 10.4. Sandy wormwood (*Artemisia arenaria*) (Photo: M. Petrov).

Fig. 10.5. Mongolian elm "haylas" (*Ulmus densa*) near well (Photo: M. Petrov).

lum gobicum is fairly common here. The local population uses the last two species for food, preparing them by grinding or frying (pulkhir).

Elm (*Ulmus pumila*) is the only tree occurring in the semi-desert zone. It grows in the bottoms of dry stream beds or in troughs where desalted waters are close to the surface (Fig. 10.5).

The semi-desert vegetation is extensively used for pasturing sheep, goats and camels. Crop yields in these ranges are not very high — from 200 to 300 kg ha^{-1}, but the plants are highly nutritious and provide forage for many thousands of animals.

Fauna

The fauna of the Central Asian semi-deserts is fairly varied. The dominant animals, as in all arid regions of the world, are rodents and reptiles. Amongst the large mammals there are widespread ungulate populations — the antelopes "dzeren" (*Gazella gutturosa*)[2] and "dzheiran" (*G. subgutturosa*) and the rarely encountered saiga antelope (*Saiga mongolica*, Fig. 10.6). Relatively common, for the semi-deserts, is the wild ass "kulan"

[2] Nomenclature according to: The Animal Kingdom of the U.S.S.R., Vol. II. Desert Zone. Leningrad, 1948 (in Russian).

Fig. 10.6. Antelope "dzeren" (*Gazella gutturosa*).

(*Equus hemionus*), occurring in herds of hundreds of animals. Predators in the semi-deserts of Central Asia are wolves (*Canis lupus*), a small korsak fox (*Vulpes vulpes caragana*) and the wild cat "manul" (*Felis manul*).

Rodents include hares (*Lepus tolai*), marmots (*Marmota sibirica*), gopher (*Citellus eversmanni*) gerbils (*Rhombomys optimus*, *Meriones meridianus*) and some widespread species of jerboa (*Dipus* spp.), hamsters (*Cricetulus* sp.) and other small rodents.

There is also a fairly large variety of birds in the semi-deserts. Flocks of *Syrrhaptes paradoxus*, *Podoces panderi*, *Galeria cristata* and others occur abundantly.

THE USE OF NATURAL RESOURCES

Animal husbandry and pasture economy

The population density is not very high in the Mongolian People's Republic (MPR) and in Inner Mongolia of the Chinese People's Republic (CPR) (Fig. 10.7). However, these territories have considerable economic value and are being actively developed.

The natural vegetation of the semi-deserts of Central Asia has been used for pasturing for many thousands of years, and is being used in much the same way today. Millions of sheep (Fig. 10.8) and goats, and tens of thousands of camels (Fig. 10.9), are pastured on extensive territories of the semi-

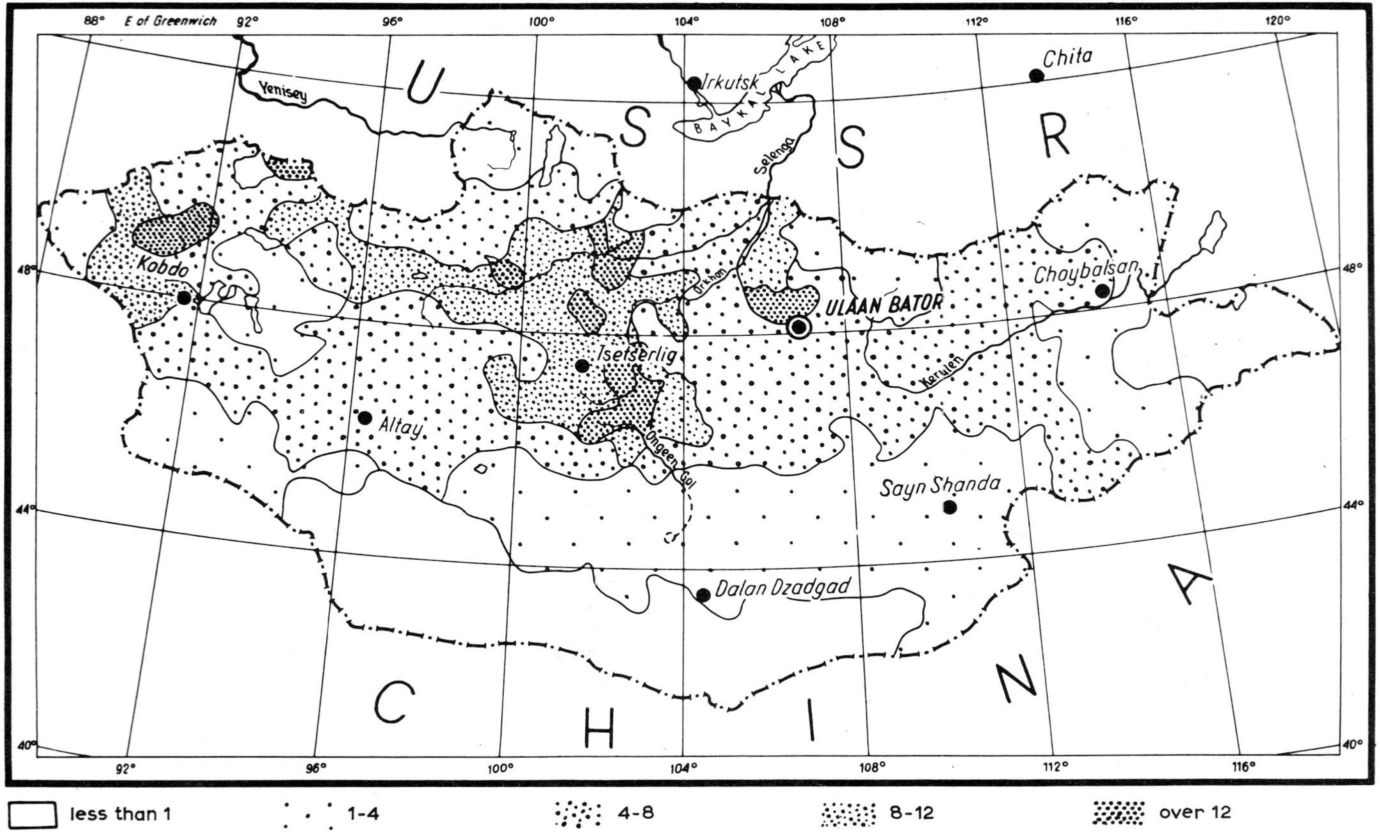

Fig. 10.7. Human population density in the Mongolian People's Republic. Number per 10 km^2 (after Badamajab et al., 1966).

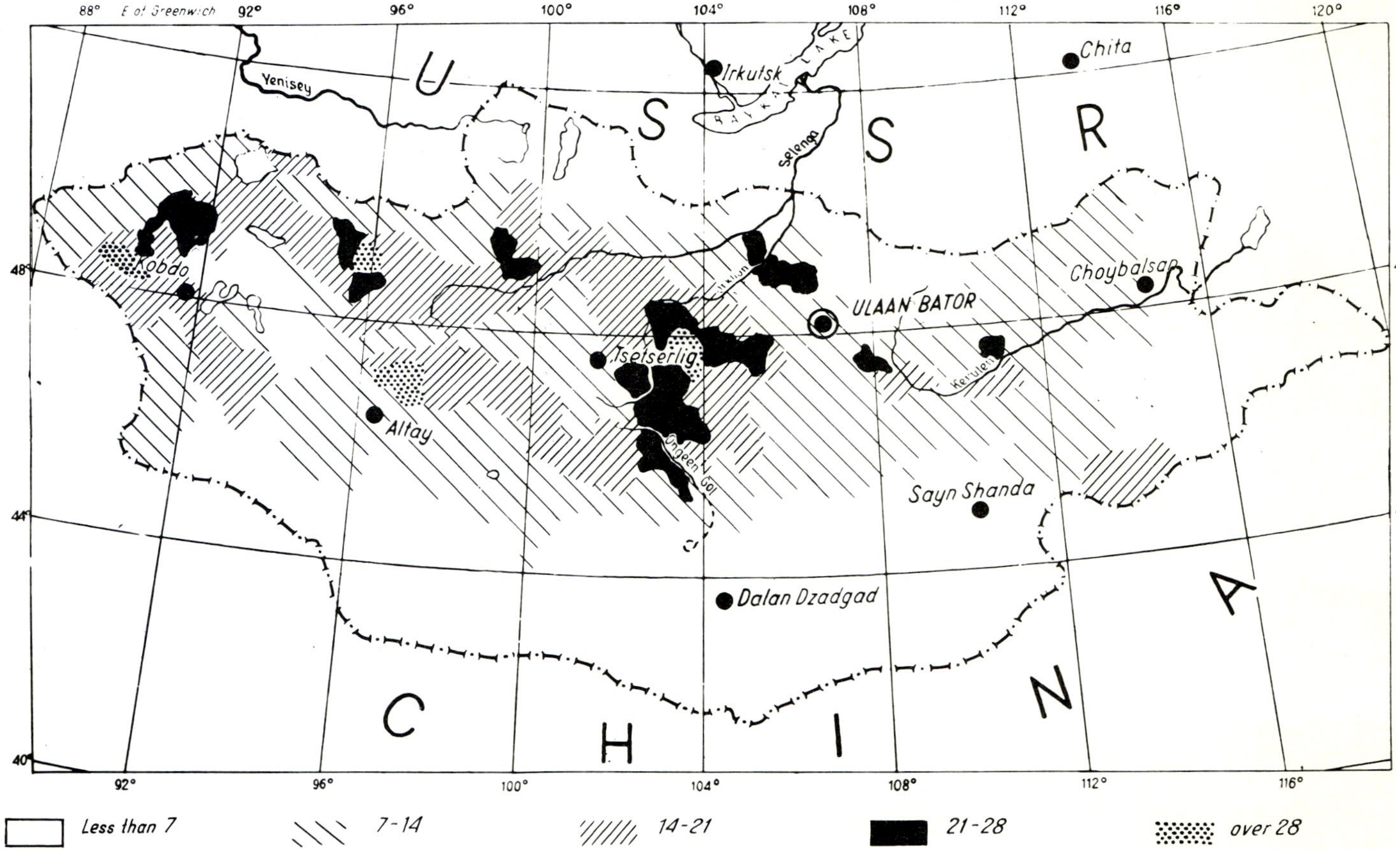

Fig. 10.8. Sheep distribution in the Mongolian People's Republic. Number per 10 km^2 (after Badamajab et al., 1966).

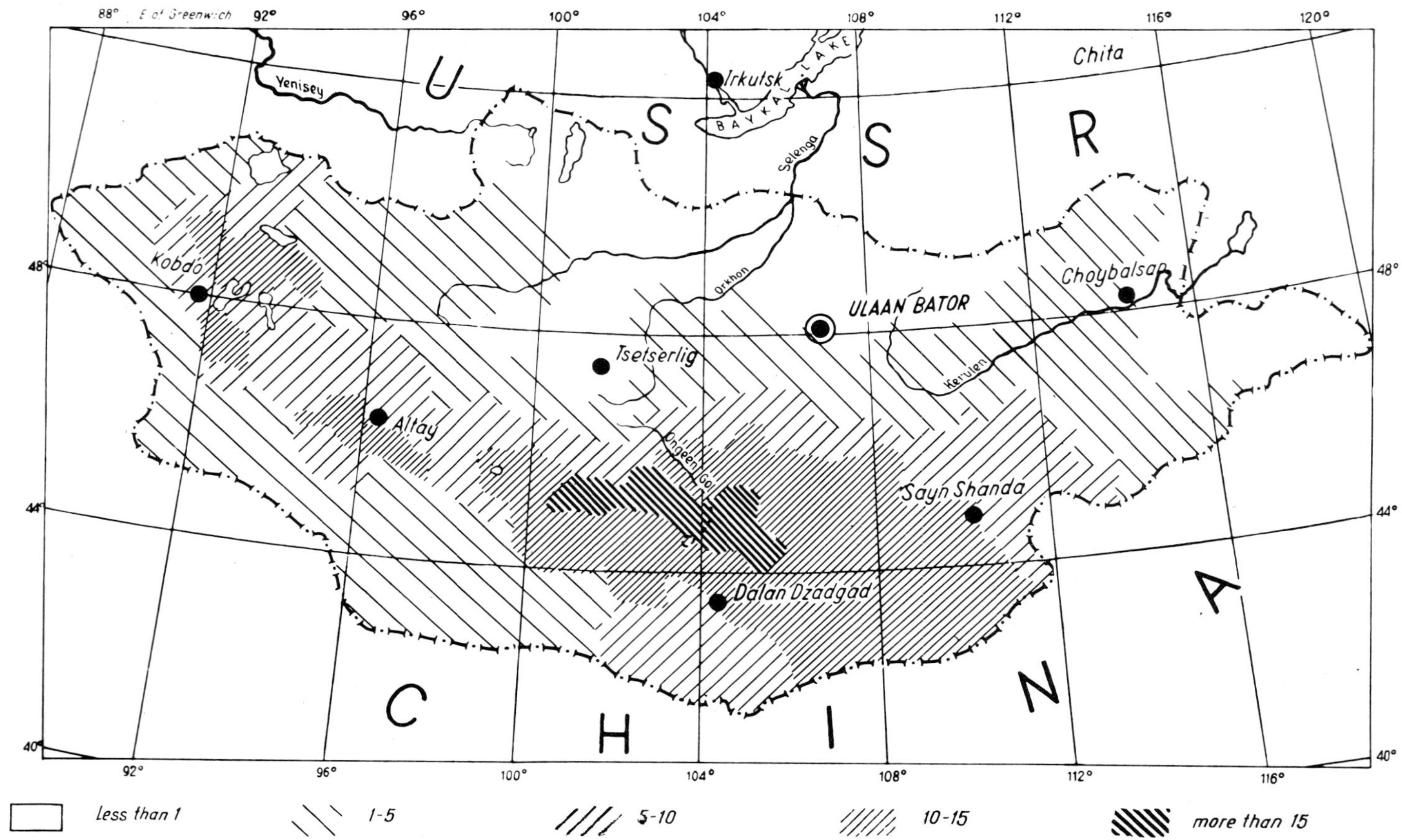

Fig. 10.9. Bactrian camel distribution in the Mongolian People's Republic. Number per 10 km^2 (after Badamajab et al., 1966).

desert rangelands of the MPR and Inner Mongolia of the CPR.

The principal breed of sheep is the local Mongolian short-fattailed white sheep with black spots, which is very well adapted to living under the rigorous conditions of the Central-Asian climate, and can graze on natural rangelands all year round (Fig. 10.10). Sheep-breeding is the major form of animal husbandry in the arid regions of the MPR and produces about 20% of all marketable products (Tables 10.3 and 10.4). There are about 15 million sheep, although this number is apt to vary owing to losses caused by inclement weather. (The 15 million applies to all of the arid zone of the MPR, including steppes and deserts, there being no detailed figures for individual districts of the Republic.)

Sheep wool is the major export commodity of the MPR (about 70% of the wool produced in the country). Mutton is widely used at home and is also exported. Sheep fat and suet are used extensively while sheep milk is used for cooking milk dishes which make up a considerable part of the Mongolian diet. Sheepskin is used for making clothes, and the wool for making felt, clothing and miscellaneous other articles (e.g. lining and shalbracks under saddles of pack animals). In recent years the strain of Mongolian coarse-wool sheep has been improved by cross-breeding with semi-fine-fleeced and fine-fleeced sheep imported from the U.S.S.R.

Goat-breeding is of equal importance to the economy of the MPR. There are about 4.5 million goats, the principal breed being the local populations

Fig. 10.10. Mongolian sheep (courtesy of TASS).

TABLE 10.3

Numbers of animals in the Mongolian People's Republic (in thousands of animals) (data from Anonymous, 1974b)

	1960	1965	1970	1973
Sheep	12 102	13 838	13 312	14 077
Goats	5000	4786	4204	4442
Camels	859	685	634	603

of light-grey, black and brown varieties. They are hardy animals which do not require choice pastures.

Camel-breeding is also an important part of animal husbandry in the MPR. There are about 600 000 camels which, in terms of camel numbers, places the MPR third in the world, after Egypt and the U.S.S.R. The main breed is the local two-humped bactrian camel (*Camelus bactrianus*), a large and hardy animal. The marketable products of camel-breeding are meat, hides, wool and milk. They are also used as transport animals.

All of these animals graze almost year-round, and the rangelands of the semi-desert zone of Central Asia constitute the region's most important natural resource. There are, however, relatively few vegetation types in the semi-desert zone of Central Asia. The principal ones are *Allium—Stipa*, *Cleistogenes—Stipa*, *Anabasis—Stipa*, *Salsola—Passerina—Kalidium*, and their different combinations (see Fig. 10.11).

The ranges can be subdivided into four principal types, according to their production: (a) ranges giving high yields of high-grade forage, with average yields of 540 kg ha^{-1} and making up 32.1% of the total range area, (b) medium-grade ranges yielding, on average, 420 kg ha^{-1} (51% of the area), (c) low-yield, low-grade ranges suitable mainly for pasturing camels, giving on average 210 kg ha^{-1} (11% of the area), and (d) land entirely unsuitable

TABLE 10.4

State Purchases of the Products of Animal Husbandry in the Mongolian People's Republic (data from Anonymous, 1974b)

	1960	1965	1970	1973
Camel skins (thousands)	20.2	24.8	21.1	21.6
Sheep skins (thousands)	277.3	1839.4	2393.6	2353.9
Goat skins (thousands)	863.6	924.5	943.3	815.7
Sheep wool (thousands of tons)	14.8	18.6	18.6	20.1
Camel hair (thousands of tons)	3.8	3.4	3.2	3.0
Goat wool (thousands of tons)	1.1	1.1	0.9	0.9
Goat fluff (thousands of tons)	1.2	1.2	1.0	1.1

Fig. 10.11. Semi-desert pastures in Mongolian People's Republic (courtesy of TASS).

for pasturing any cattle (5.8%). Thus, most of the rangelands (over 83%) are good and medium pastures, even though their production is comparatively low.

Measures on the efficient use of semi-desert rangelands and on their improvement have been conducted in the MPR on a national scale, since the beginning of the Republic. Work has been carried out on the improvement of water supplies to pastures, on the creation of fodder reserves against a dearth of food in cases of dzuds [3], on making inventories of forage areas and taking measures for their improvement and preservation. Over-grazing is known to greatly decrease production of range vegetation, resulting in only a few species of resistant, but poorly palatable or unpalatable, plants remaining. Areas of highly productive pastures are degraded in this way into wastelands with very low yields of forage, usually around wells, cattle camps and human settlements.

Range condition and the nutritive value of pasture forage differ greatly between the summer—autumn and winter—spring seasons. The maximum yields of forage in semi-deserts occur late in summer, at the end of July or the beginning of August. In wintertime, depending on the type of range and weather conditions, from 20 to 50% of the maximum summer reserve remains.

[3] "Dzud" — lack of forage. There are several forms of dzud: white dzud — cattle cannot reach forage through deep snow (after a heavy snowfall); black dzud — absence of snow in winter in a waterless area or failure of grass crops in a summer drought; rime or ice crust which cattle cannot break to get through to forage; hoof dzud — when pastures are tramped down by too many animals. The greatest losses are caused by the white and black dzuds and natural calamities in spring when animals are particularly weak.

Plants of the semi-desert ranges have the highest nutritive value during the early stages of growth and at the time of ear formation. The value gradually decreases towards the end of the growing season, and is reduced catastrophically after the autumn frosts, when the plants perish. During the winter the nutritive value remains at a low level until the beginning of the new season. On the whole, the nutritive value of winter forage is between 20 and 50% of that of spring and early summer.

An advantageous feature of Mongolian rangelands is their year-round suitability for pasturing, and this is the reason why animal husbandry plays the most important role in the national economy, both at present and in the past. For many centuries the economy of Mongolia developed on the basis of nomadism, which in turn was based on animal husbandry.

Pasturing is done on a traditional system, with migration of animals to seasonal pastures, according to the time of the year and the condition of pastures. Migration territories are changed 2—4 times a year, and the radius of migration is not large, normally less than 100 km. In drought years and in times of heavy snowfall animals are sometimes driven over distances of several hundred kilometres. There are various kinds of migration that have been formed under the influence of natural features and historical causes.

Year-round pasturing is practised according to a free-range system of grazing within a broad, traditional system of pasture rotation which has been developed over the centuries. Depending on the weather conditions of each year and on yields of herbs, the pasture rotation systems are varied. No fencing of pastures has been practised yet.

Animal production from natural pastures has definite advantages over stall feeding. The vastness of the territory, the variety of forage suitable for different kinds of animals, the high nutritive value of herbs, their good quality in dry condition, the thin snow cover in normal winters — all make it possible to obtain valuable products of animal husbandry with a low input of finance and labour. Virtually all animal fodder is obtained from natural ranges.

In spite of all these advantages, the dependence of animal husbandry on natural conditions often results in considerable, at times catastrophic, losses of livestock. These losses occur as a result of the shortage, or even absence, of forage on pastures in winter and springtime, during times of abundant snowfalls, or in dry summers. This dependence still exists in the MPR at present and is a serious obstacle in the way of planned development of animal husbandry. Abundant snowfalls and rime or ice crust result in black and white dzuds which lead to severe losses of sheep. In 1944—1945 there was an especially severe dzud in the Mongolian rangelands, which resulted in the loss of about 5 million small-horned animals (Anonymous, 1971).

In order to eliminate the dependence of animal husbandry on unfavourable weather conditions, the government of the MPR after the victory of the People's Revolution in 1921 undertook, and still is undertaking, a number of measures for stabilizing the forage and fodder base of animal husbandry. They involve both the reasonable use of the range and improvement of grazing management (for example, through increasing water supplies, better use of pasturage) and the creation of insurance reserves, against lack of fodder, in the form of hay, agricultural products and sown herbs.

At present animal husbandry in the semi-arid region of Central Asia is still nomadic, but it is conducted on socialistic lines. The principal economic units are livestock-breeding associations. This allows a better organization of labour and gives a higher yield of agricultural products. Migrations of sheep are conducted in a more organized manner. For wintering, the best protected locations have been chosen to build over 50 000 walled-in, covered and closed shelters ("khashans") to protect animals against cold winds.

Provision of a sufficient number of watering places is an important condition for increasing the number of livestock and raising their productivity. Wells and artesian boreholes are the main sources of water supply. The government of the MPR devotes considerable attention to the construction of wells and boreholes, since it facilitates more efficient utilization of the ranges. In the arid regions of the MPR alone there are over 25 000 wells and over 1000 boreholes varying from 35 to 125 m in depth.

Additional feeding of cattle is practised in cases of the dearth of food, great droughts, ice crusting, or excessive snowfalls. In these cases insurance reserves of fodder are carried to the region of calamity by automobiles or aeroplanes, as shown in Fig. 10.12.

Fig. 10.12. Transportation of fodder reserves (courtesy of TASS).

The network of wells is expanded and improved every year. The most common wells are those lined with stone or logs. There are also wells lined with stone in their lower part and reinforced with log framework in the upper (Derevianko, 1959).

All types of wells are protected against decay with a seal of compact clay. The depth of wells varies according to geomorphological conditions — in the piedmont plains they are deeper than in the inter-montane valleys.

Water is raised to the surface by different methods; by wind turbines, pumps or with the aid of animals. In most regions the water is of sufficiently high quality to be potable for people and animals.

The south-eastern part of Gobi, Middle Gobi and partly the Southern Gobi and Gobi-Altai aimaks cannot be classed, hydrologically, as desert, since they have abundant ground waters, numerous wells and water of good quality, while the debit of water sources, their number, and mineralization of water do not correspond to desert hydrology. But along with such water-sufficient regions, there are areas in the Gobi where no wells or springs occur over several tens of kilometres. Such areas are not populated by Man since cattle-breeders cannot use pastures that have no wells or springs at distances of a few kilometres.

Much work has also been carried out on improving the retention of spring snow and the construction of water reservoirs.

A Soviet-Mongolian hydrological expedition has studied the ground water of the semi-arid zone for a number of years and has found rich underground reservoirs, some of which are already being used. The increase in the number of wells and artesian boreholes has enabled the use of previously unused pastures.

Agriculture

Agriculture is poorly developed in the semi-desert zone of Central Asia, due primarily to unfavourable climatic conditions. There are millions of hectares in the semi-desert zone which are potentially irrigable, if water could be found.

At present the developed lands are confined mainly to the rare settlements or river valleys, and agriculture there is mainly of an auxiliary nature, with vegetables and berry bushes being the main crops. On larger areas in the river valleys millet and forage herbs are grown. The principal obstacle to the introduction of forage plants is their low frost resistance. Indigenous forage herbs are better adapted to the climate and are also preferred by the animals. Experiments have shown that many of the indigenous plants of Central Asia can be used for the improvement of natural pastures (Kalinina, 1974).

In recent years, coinciding with a marked increase in the settled population in the semi-desert zone, there has developed a need for supplying these people with vegetables and other agricultural products. This has stimulated the development of small oasis-irrigated agriculture. For this purpose dams and water reservoirs are built on temporary streams. Water supplied by such small-scale schemes is used for the irrigation of fields, for the needs of animal husbandry and for supplying the nearby settlements.

Game and hunting

In the semi-deserts of Central Asia there are numerous species of wild animals which are hunted. The most important predators are wolves (*Canis lupus*), korsak fox (*Vulpes vulpes caragana*) and wild manul cat (*Felis manul*). Many rodents are hunted for their fur, e.g. the tarbagan marmot (*Marmota sibirica*), gophers (*Citellus eversmanni*) and tolai hare (*Lepus tolai*). Of the ungulates the antelopes "dzeren" (*Gazella gutturosa*) and "dzheiran" (*Gazella subgutturosa*) are hunted.

The species most prized by hunters is the tarbagan marmot, whose pelts are in great demand on the world market. Moreover, the specially treated bones and fat of the tarbagan are used for industrial purposes such as greasing. The area of its habitat extends from the semi-desert zone north into the steppe zone. Mongolian hunters annually return between one and two million tarbagan pelts, depending on such factors as the amount of precipitation, yield of herbs, diseases of animals, etc.

At present tarbagan hunting is regarded as a special branch of the hunting economy. Accordingly, there are particular programmes for pelt delivery and records are taken of the number of pelts collected from each region. At the same time measures are being developed for protecting these animals from epizootics, primarily plague. Hunting seasons and shooting quotas have been decreed by the government to prevent a decrease in the population. For the greatest part of the year shooting of tarbagans is prohibited, which ensures both their safety at breeding time and a high grade of pelts.

Besides tarbagan pelts, tens of thousands of korsak fox pelts and thousands of wolf pelts are annually procured (Table 10.5). Desert antelopes are hunted for meat and pelts.

TABLE 10.5

State purchases of furs in arid zones of the Mongolian People's Republic, in thousands of pelts (data from the National Economy of the MPR in 1973; Anonymous, 1974b)

	1960	1965	1970	1975
Tarbagan marmot	1304.2	1203.8	1222.6	967.7
Korsak fox	26.5	36.2	20.7	—
Wolf	3.5	3.6	4.5	4.3

Utilization of mineral resources, urbanization and transport facilities

In recent decades industrial development of natural resources has begun in the semi-deserts of Central Asia, particularly the mining of mineral resources. For instance, rich deposits of lead have been found in the South Gobi (Churun-Khordte). Coal has been found in the Tsogot-Tsetsei district of the same region. This coal is also used locally.

The exploitation of oil deposits in the Dzun-Bayan and Tsagan-Elisu in the East Gobi region is being started. The salt lakes of Eastern Gobi contain large deposits of Glauber and common salts which are partially worked. Greater development of these deposits is prevented by the absence of good roads.

The process of urbanization in the semi-desert zone of Central Asia is comparatively slow, and develops in the form of the expansion of existing settlements connected with railroads and oil-fields.

The Trans-Mongolia railroad, Ulan-Ude—Ulan-Bator—Dzahamin-Ude, built between 1947 and 1955, crosses the semi-desert zone in the eastern part of Mongolia and has a great influence on the development of some branches of the national economy. The railroad is approached by numerous earth roads from distant regions of the semi-desert, over which the products of animal husbandry can be more rapidly transported; also the population of animal-breeding settlements can be better supplied with different goods.

Internal transportation is done extensively by automobile and animal (camels, horses) transport, although the latter is becoming increasingly rare.

CONCLUSION

The scale and variety of exploitation of natural resources of the semi-desert zone of Central Asia has notably increased in recent years. The beneficial changes in animal husbandry are especially marked. The development of the semi-desert by industrial methods is presently at an early stage. However, there is a definite tendency toward their further development.

The future prospects of the utilization of natural resources of Central Asia are quite favourable. Man has not yet fully explored all the possibilities for their future utilization.

Animal husbandry on rangelands will remain the principal branch of economy in the next few years, and it will be developed still further. The number of sheep can be increased considerably by the organization of proper grazing management, water supply to distant, still unused, pastures, increase of crop yields (carrying capacity) of pastures by improving them and by giving additional fodder to animals in times of forage shortage and by mechanization of some labour-consuming operations (shearing of sheep, raising water to the surface, etc.).

Moreover, the production of animal husbandry can be increased by selective breeding of new, more productive animals, giving higher yields of meat and wool, as well as by the improvement of measures aimed at combatting animal diseases. The first results of selective breeding are quite hopeful. They show possibilities for the improvement of local, traditional breeds of sheep, which are well adapted to the drastic climatic conditions.

As animal husbandry in the semi-deserts of Central Asia becomes more settled every year, so its technical facilities increase. The power facilities will increase correspondingly, and this will in turn facilitate the development of all forms of transport, especially automobile transport.

Expansion of the scale of work on the utilization of natural wealth of the semi-arid zone of Central Asia will stimulate further growth of settled populations of these regions. At present the population density there is between 0.01 and 0.04 km^{-2} (see Fig. 10.7), but it will gradually increase. This will necessitate paying more attention to the development of plant-growing and increasing the area of irrigated lands. There are sufficient soils suitable for agriculture in the semi-arid regions of Central Asia, such as the brown and light-chestnut soils of the piedmont plains and inter-montane depressions. Climatic conditions are quite favourable for the cultivation of fruits, melons, water-melons, and for sheep-breeding. New small oases will appear in the semi-desert, with water tanks, artesian boreholes and wells. The considerable reserves of ground waters will permit an increase in irrigated lands and the population of the semi-desert settlements will consequently be supplied with vegetables, fruit, melons, etc.

REFERENCES

Anonymous, 1948. The Animal Kingdom of the U.S.S.R., Vol. II. Desert Zone. Akademia Nauk SSSR, Leningrad, 420 pp.

Anonymous, 1974a. Climatological Means for Foreign Asia, Vol. I. Gidrometeoizdat, Leningrad.

Anonymous, 1974b. National Economy MPR, Ulan-Bator.

Anonymous, 1971. Mongolian People's Republic, Nauka, Moscow.

Badamajab, D., Bazargur, D. and Budjab, J., 1966. Gazarzuyn Atlas (Geographical Atlas). Bugd Nayramdah Mongol Ard Uls, Ulan-Bator.

Bespalov, N.D., 1957. Soils of the Mongolian People's Republic. Academia Nauk SSSR. Mongolskaja Mommissia, Trudy, N 41, Moscow.

Derevianko, P.A., 1959. Agricultural Water Supply of the Mongolian People's Republic. Akademia Nauk SSSR. Mongolskaya Kommissia, Trudy, N 21, Moscow.

Grubov, V.I., 1955. Synopsis of the Flora of the Mongolian People's Republic. Akademia Nauk SSSR, Leningrad.

Kalinina, A.W., 1974. Main types of pastures of Mongolia (their structure and productivity). Biological Structure and Natural Condition of the Mongolian People's Republic, Vol. 2. Lengingrad.

Meigs, P., 1953. World distribution of arid and semi-arid homoclimates. In: Review of Research on Arid Zone Hydrology. Arid Zone Programme. Vol. I, UNESCO, Paris, pp. 203—210.

Murzayev, E.M., 1952. Mongolian People's Republic. Geografgiz, Moscow.

Yunatov, A.A., 1950. General outlines of the vegetation cover in Mongolian People's Republic. Akademia Nauk SSSR. Mongolskaya Kommissia, Trudy, N 39, Moscow.

Chapter 11

MANAGEMENT OF NATURAL VEGETATION IN THE SEMI-ARID ECOSYSTEMS OF MEXICO

F. MEDELLIN-LEAL and A. GOMEZ-GONZALEZ

Instituto de Investigación de Zonas Desérticas, Universidad Autónoma de San Luis Potosi, S.L.P. (Mexico)

INTRODUCTION

The Mexican Republic lies entirely within the Northern Hemisphere, but because of affinities and cultural characteristics, it has been classified by Europeans as part of Central America. The average inhabitant of the U.S.A. regards North America as ending at the southern frontier of his country, and Mexico is thought of as vaguely "South of the Border". It is along this border that the extensive Mexican arid zones are located as a natural prolongation of the United States deserts. Even the names used for these arid zones within the United States are derived from words of Mexican origin.

The Mexican arid and semi-arid zones stretch discontinuously over approximately 1 million km^2. There are eight arid zones, namely: (1) Sonoran, (2) Chihuahuan, (3) Queretan, (4) Hidalguense, (5) Poblan, (6) Guerrerense, (7) Oaxacan, and (8) Yucatec. These names are derived from names of the States they embrace. Although they almost reach central Mexico they do not really extend into Central America.

Biogeographically the first four zones have neartic affinities, despite the most characteristic plant in all of them being *Larrea divaricata*, whose nearest relatives are found in South American deserts, particularly in Argentina. The last four zones have been poorly studied but they show a more outstanding neotropical affinity.

Despite this first major difference, all of the Mexican arid zones are close to the Tropic of Cancer, and have consequently been classified as principally warm, although marked differences in diurnal and annual temperature variations occur. Diurnal variation can reach 22°C during the driest months, i.e. in winter. Rainfall generally occurs during summer.

The second most important factor is the orographic shadow projected by two mountain chains, the Sierra Madre Oriental and the Sierra Madre Occidental, which form a V-shape. They prevent the humid east winds reaching the central and western regions of the continent, thereby causing the Chihuahuan, Queretan, Hidalguense and Poblan arid zones and, in part, the Sonoran and Oaxacan arid zones. The last two zones also suffer as a result of their situa-

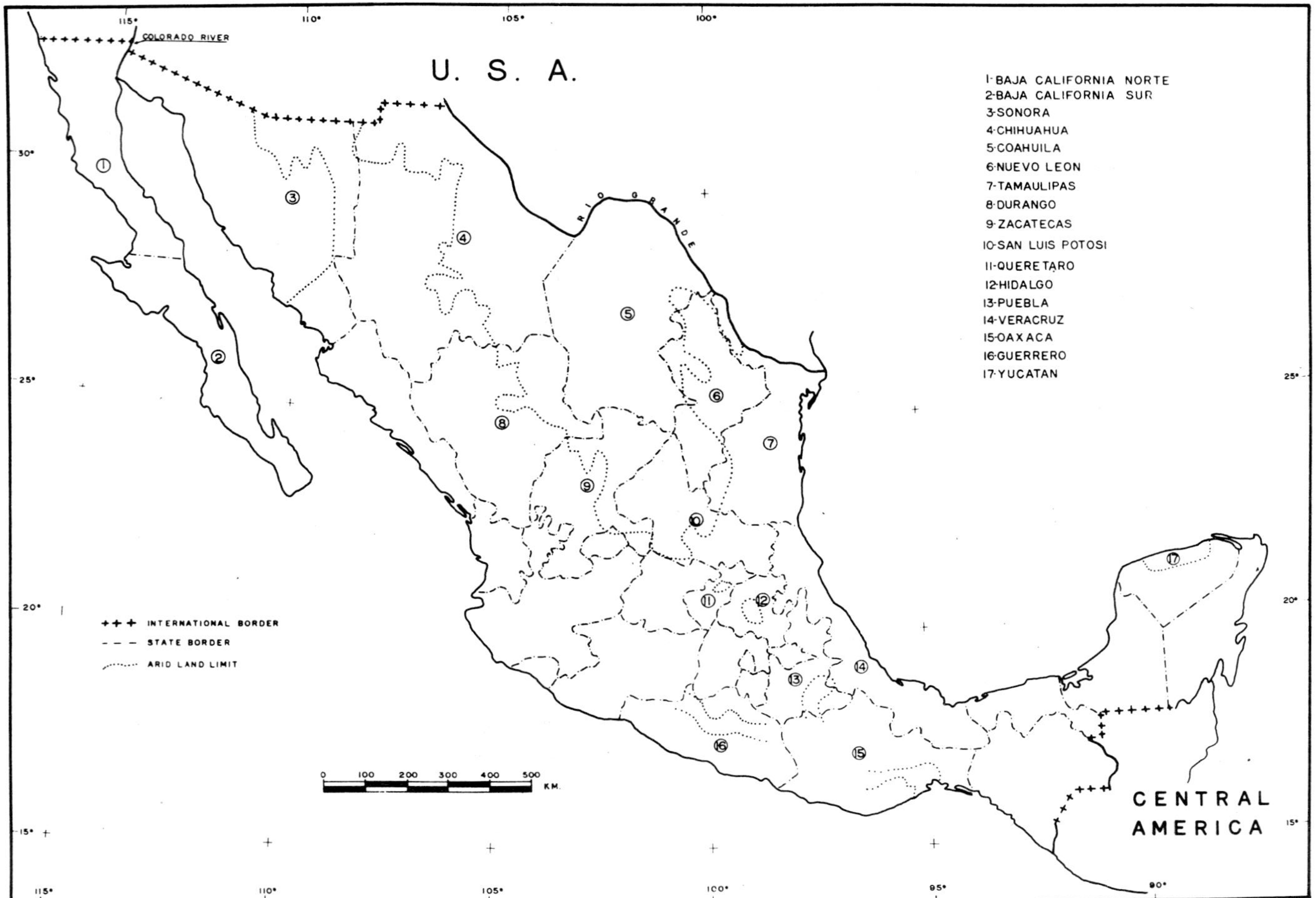

Fig. 11.1. Distribution of arid lands over the political divisions of the Mexican Republic.

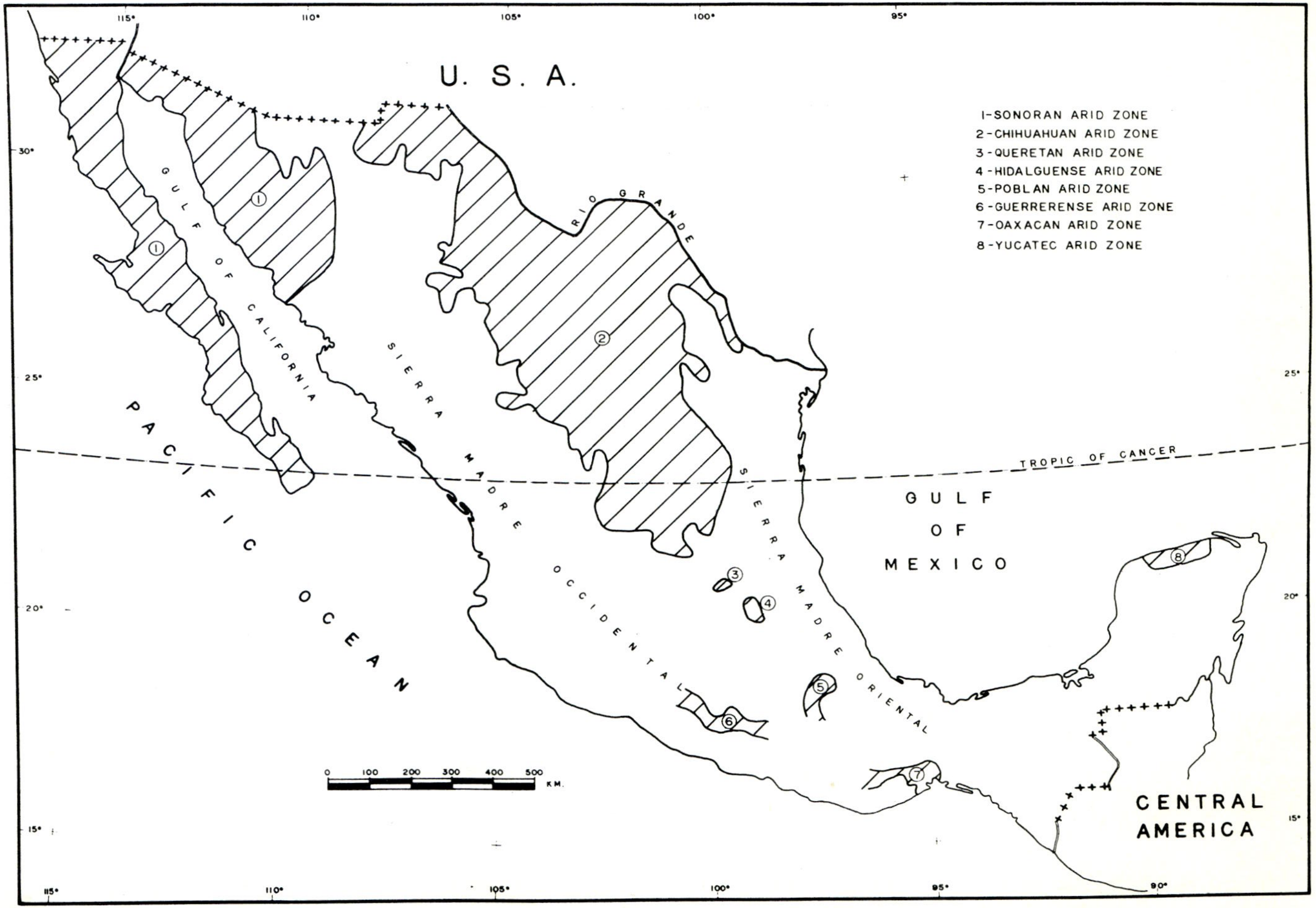

Fig. 11.2. Physiographic features of the arid lands in Mexico.

tion at the western end of the continent, adjacent to the cold sea current which descends in a N—SE direction along the entire western North American coast. The Guerrerense arid zone is almost exclusively the result of an orographic shadow since it is located where both Sierras Madres join in the south-central part of the country. The small Yucatec arid zone is found in a region whose winds are always humid, but the peninsula of Yucatan is extremely flat and favours a free aeolian circulation that discharges its humidity further south. Lastly, continentality increases in intensity in a SE—NW direction, reaching its highest point at the north-central portion of the Chihuahuan arid zone.

DESCRIPTION OF THE ECOSYSTEMS

Physiography and limits

The Mexican section of the Sonoran arid zone is divided into peninsula and continental regions by the Gulf of California. The Sonoran arid zone is separated from the Chihuahuan arid zone by the Sierra Madre Occidental which is only arid or semi-arid at the lower parts of its east and west flanks. The Sonoran and Chihuahuan zones extend into California, Arizona, New Mexico and Texas, where they merge to form a large, arc-shaped arid area. In Mexico they form more or less continuous undulating plains rising slowly from sea-level, in the Mexican Sonoran, to approximately 2400 m above sea-level on the south-east of the Chihuahuan zone.

The peninsula part of the Sonoran zone is bounded by the Pacific Ocean to the west and south, and the Gulf of California to the east. The continental part is limited by the western slopes of the Sierra Madre Occidental in the east, the Gulf of California in the west and the Mayo River in the south. The southern border though, is ill-defined owing to a gradual decrease in aridity, and to human interference in the form of irrigation which is lessening the effects of the arid climate. The Sonoran desert ranges over the States of Baja California Norte, Baja California Sur and Sonora.

The Chihuahuan arid zone is bounded on the west by the eastern spurs of the Sierra Madre Occidental, and on the east by the western flank of the Sierra Madre Oriental. The latter is perhaps the sharpest boundary of the whole Mexican desert, since in a transect of less than 30 km one progresses from desert, through oak forest to the low caducifolious forest of a truly humid climate. To the north, as previously stated, this zone continues into the U.S.A. The southern boundary is very difficult, if not impossible, to establish, given that in the course of more than 100 km one passes from the arid zone proper, through a series of regions, most of them mountainous and probably aridized by man over many centuries, down to the State of Guanajuato, where some isolated areas have seasonal and irrigated agriculture.

The Chihuahuan desert ranges over the eastern portion of Chihuahua and Coahuila, north, west and south of Nuevo León; the north-west and south-west of Tamaulipas; the north-east portion of Durango and Zacatecas; a small portion of Aguascalientes, and the western half of San Luis Potosi. For the purposes of this paper, we also include here the zone which Miranda (1955) named "Tamaulipec arid zone", and Medellin-Leal named "Jaumave arid zone" (quoted by Aranda-Gómez, 1973).

The outstanding topographical features of the Sonoran and Chihuahuan zones are a mountainous chain along the eastern portion of the peninsula of Baja California, both Sierras Madres, and many isolated hills, increasing in number towards the south. The principal rivers are the Colorado, which flows from the north into the Gulf of California and the Rio Bravo (Rio Grande or Rio del Norte) which forms part of the Mexico—U.S.A. boundary and is, potentially at least, the natural drainage of the Chihuahuan arid zone. There are few permanent streams in these two zones.

Scattered throughout the plains are many endorheic depressions. Ephemeral streams flow into these such that temporary ponds occasionally appear. These depressions, called "bolsones", constitute excellent places for evaporation and large amounts of salt are deposited. The most famous of these bolsones, known as Mapimi, occurs in Durango.

The small Queretan arid zone itself constitutes a non-endorheic depression belonging to the basin of the Pánuco River, between the villages of Vizarrón and Peñamiller. The Hidalguense arid zone, commonly known as the "Valle del Mezquital", is an inter-mountain valley, also of the Pánuco basin, populated from antiquity by Otomi natives. Biogeographically these two zones, both c. 1700 m above sea-level, can be considered as extensions of the Chihuahuan arid zone.

The Poblan arid zone has been much less studied. It ranges over the Tehuacán-Zapotitlán de las Salinas Valley at 1600 m above sea-level. This zone penetrates a little into the northern part of the State of Oaxaca and also into the State of Veracruz, for which reason it has been called the "Veracruzan arid zone" (Ramos-Alvarez and González-Medrano, 1972). The Guerrerense arid zone embraces a series of canyons on the upper part of the Balsas River basin, at an altitude of 400 to 1100 m. The Oaxacan arid zone, named by others the "Tehuantepecan arid zone" (Miranda, 1955), includes the Tehuantepec river basin from about 100 m to sea-level. Although they are small, these three zones are difficult to outline exactly owing both to sudden changes in the landscape and to prolonged human disturbance. It is likely that they were originally continuous.

Finally, the Yucatec arid zone, which on maps always appears exaggerated, constitutes a small E—W strip approximately 12 km wide by 150 km long, at the northern end of the Peninsula de Yucatán, almost at sea-level.

Climate

Fig. 11.3 is a climatic map of the arid zones in Mexico, drawn according to Meigs' System (quoted by McGinnies et al., 1968). It differs in detail from maps by other authors, who used the same criteria (McGinnies et al., 1968) but the differences reflect the specific aims of this paper. There are 13 climatic types in the Mexican arid zones, eight of which belong to the semi-arid regions. The climatic regions are defined according to Meigs (quoted by McGinnies et al., 1968) as follows: E, extremely arid; A, arid; S, semi-arid; a, without a designated rainy season; b, summer rains; c, winter rains. The first digit indicates the average temperature of the coldest month and the second the average temperature of the warmest one, according to the following scale: (1) 0—10°C; (2) 10—20°C; (3) 20—30°C; (4) more than 30°C.

In short, it can be stated that approximately 58% of the dry areas in Mexico are semi-arid; about 38% arid and only 4% belong to extremely arid lands.

Table 11.1 and Figs. 11.4 and 11.5 were drawn up from meteorological data from San Luis Potosi, this station being representative of the semi-arid zones of the central and northern part of the country. Meteorological data from other semi-arid regions are shown in Table 11.2 (data from García, 1973). The data show that, in general, a higher precipitation is offset by higher temperatures.

Vegetation

Vegetation studies of the Mexican arid zones have been conducted by both Mexican and foreign investigators. Three criteria have been used, namely climate, physiography and, more recently, vegetation physiognomy. Flores Mata et al., (1971a, b) drew up a vegetation map (Scale 1 : 2 000 000) using data from papers published up to that time. Their classification is similar to that by Ellenberg and Mueller-Dombois (see Table 11.3) in 1967 (Mueller-Dombois and Ellenberg, 1974).

More detailed maps are now being drawn up by the State as an aid to range management. The terminology used differs from that of Flores-Mata et al. (1971a, b) but their zones can also be classified by the criteria of Ellenberg and Mueller-Dombois (see Table 11.4). Their classification is shown in Table 11.5.

The vegetation of the Mexican semi-arid ecosystems is very difficult to map since it changes with both topography and soil type, both of which vary a great deal over short distances. However, in accordance with Flores-Mata et al. (1971a, b), the vegetation can be reduced to seven principal types: Chaparral, Matorral Submontano, Matorral Desertico Microfilo, Matorral

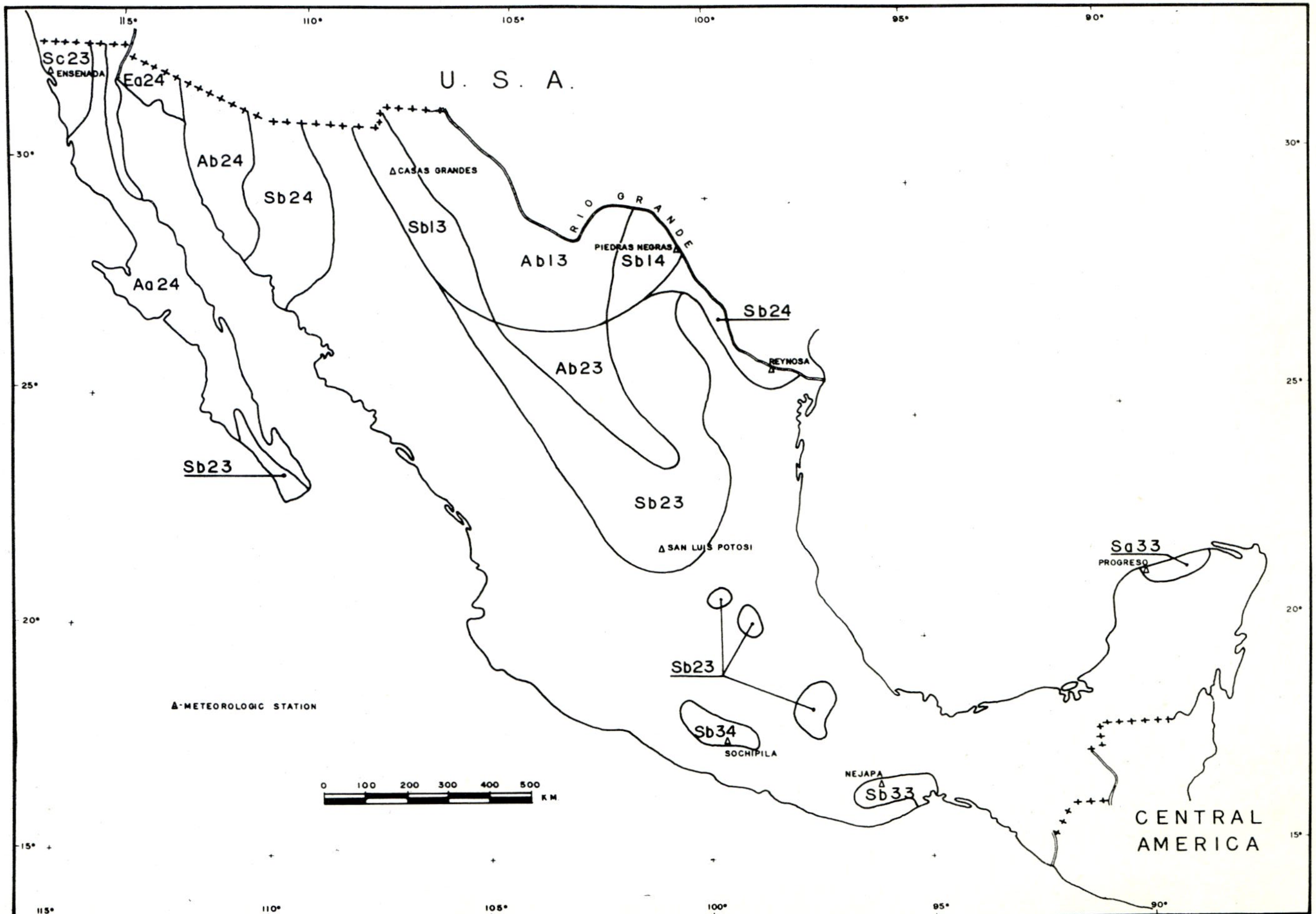

Fig. 11.3. Climates of the Mexican arid lands (according to Meigs' system, see text).

TABLE 11.1

Meteorological data from San Luis Potosi (Sb 23-Chihuahuan Desert)

Latitude	22°09′10″N
Longitude	100°58′38″W
Altitude (m above MSL)	1877
Mean barometric pressure (millibars)	817.3
Mean annual temperature (°C)	17.6
Mean maximum temperature (°C) (Usually registered during May)	37.3
Mean minimum temperature (°C) (Usually registered during January)	—3.7
Mean number of frosts per year	20
Normal time when frosts occur	October—March
Early frosts	September
Late frosts	May
Mean annual precipitation (mm)	361
Mean annual evaporation (mm)	1900
Mean annual relative humidity (%)	50
Mean annual number of clear days	243
Mean annual number of cloudy days	122
Mean annual number of rainy days	61

Crasicaule, Matorral Desertico Rosetofilo, Mezquital and Zacatal. They are defined as follows.

(a) Chaparral — very dense groups of shrubby oaks with small leathery leaves. Several species of *Quercus* are present, accompanied by other plants of similar growth form, e.g. *Arctostaphylos* spp. and *Amelanchier* spp.

(b) Matorral Submontano — small deciduous trees with small leaves, and

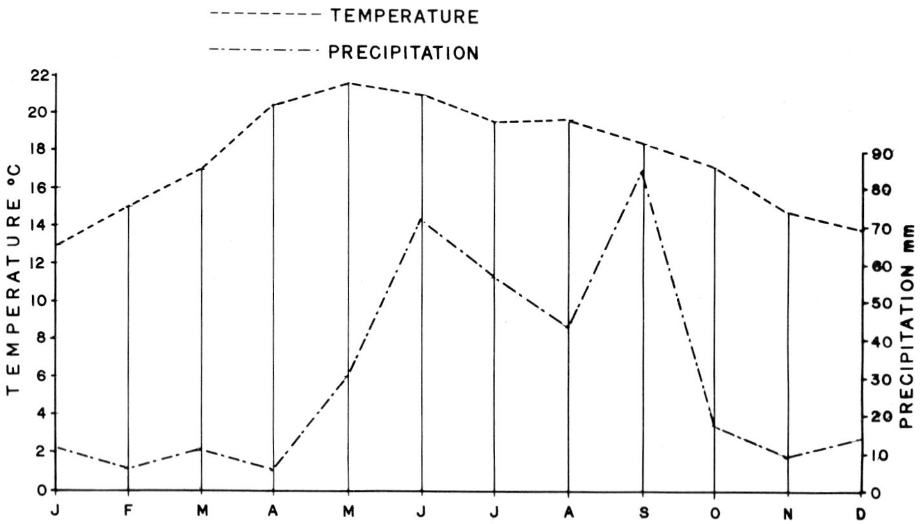

Fig. 11.4. Mean monthly temperatures and precipitation at San Luis Potosi.

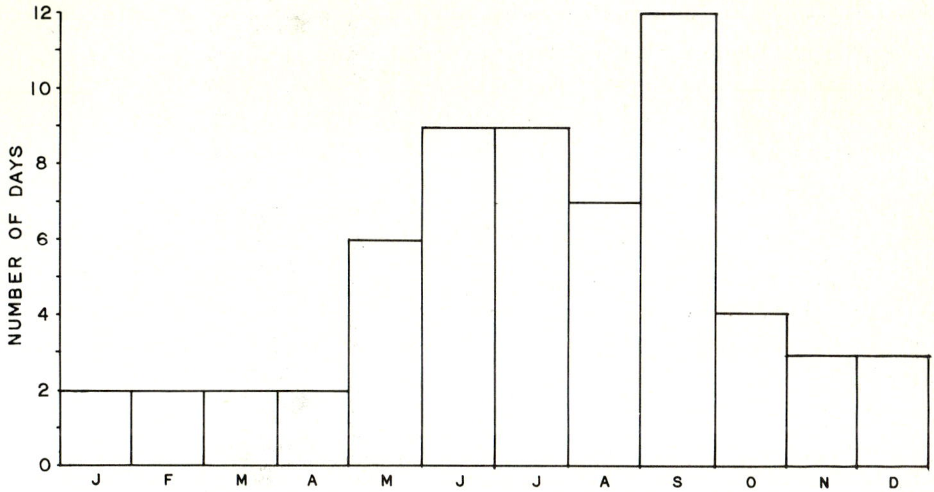

Fig. 11.5. Mean monthly number of rainy days at San Luis Potosi.

some members of the Cactaceae. This vegetation-type is floristically very rich, the most characteristic genera being *Leucaena*, *Cordia*, *Bumelia*, *Helietta*, *Gochnatia*, *Celtis*, *Flourensia*, *Mimosa*, *Colubrina*, *Lysiloma*, *Pseudosmodingium* and *Pistacia*.

(c) Matorral Desertico Microfilo — dominated by small-leaved, usually resinous, shrub species. The characteristic genera are *Larrea*, *Flourensia*, *Celtis* and *Rhus*.

(d) Matorral Crasicaule — the most characteristic type of vegetation of the Mexican arid zones and also the best known by the rest of the world. It is characterized by the abundance of Cactaceae of various sizes from a few cm up to more than 20 m, sometimes found in very dense dumps. The most characteristic genera are *Opuntia* (more than 100 species) *Carnegia*, *Lemaireocereus*, *Neobuxbaumia*, *Myrtillocactus* and *Cephalocereus*.

(e) Matorral Desertico Rosetofilo — as the name suggests this vegetation type is made up of large (sometimes enormous) plants with leaves joined into rosettes, with or without a conspicuous stem. The most common genera are *Agave* (numerous species), *Yucca*, *Hechtia*, *Dasylirion* and *Nolina*.

(f) Mezquital — small-leaved tree and shrub vegetation. *Prosopis* spp. are always dominant but the accompanying species vary with aridity and altitude.

(g) Zacatal — open grassland vegetation which stretches from the northern border to the centre of Mexico. Strictly speaking, there are two different vegetation types; one, represented by the Zacatal itself, is the climatic climax formed by grasses of the genera *Bouteloua*, *Muhlenbergia*, *Buchloe*, *Sporobolus*, *Lycurus*, *Heteropogon*, *Elyoneurus*, *Stipa*, *Andropogon*, *Trichachne*, *Hilaria*, etc., sometimes associated with *Larrea* and *Prosopis*; the other is an

TABLE 11.2

Meteorological data from some semi-arid regions in Mexico

Climate	Desert	Example	Latitude N	Longitude W	Altitude (m)	Average annual temp. (°C)	Average annual precipitation (mm)
Sa 33	Yucatec	Progreso	21°17′	89°40′	2	25.4	469
Sb 13	Chihuahuan	Casas Grandes	30°24′	107°56′	1473	18.0	259
Sb 14	Chihuahuan	Piedras Negras	28°42′	100°31′	320	21.6	545.3
Sb 23	Chihuahuan	San Luis Potosi	22°09′	100°58′	1877	17.6	361
Sb 24	Chihuahuan	Reynosa	26°04′	98°17′	38	23.0	423.8
Sb 33	Oaxacan	Nejapa	16°37′	95°59′	570	25.2	595.1
Sb 34	Guerrerense	Sochipala	17°50′	99°37′	1100	23.9	684
Sc 23	Sonoran	Ensenada	31°53′	116°38′	13	16.7	290

TABLE 11.3

Physiognomic-ecological classification of plant formations in the semi-arid ecosystems of Mexico

Class	Subclass	Group	Formation [a]	Key [b]
Matorral scrub	Mainly evergreen scrub	Evergreen broadleaved scrubland	Chaparral	III, A, 1, d
			Matorral Submontano	III, A, 1, e
	Extremely xeromorphic (subdesert) shrublands	Mainly evergreen subdesert shrublands	Matorral Desertico Microfilo	III, C, 1, a, 2
			Matorral Crasicaule	III, C, 1, a, 3
			Matorral Desertico Rosetofilo	III, C, 1, a, 4
			Mezquital	III, C, 2, a
Terrestrial herbaceous communities	Steppes and related grassland		Zacatal	V, B

[a] Flores-Mata et al. (1971a,b).
[b] Mueller-Dombois and Ellenberg (1974).

edaphic climax which grows on soils with high surface deposits of salts, resulting from evaporation. Examples are species of *Distichlis* and *Eragrostis*. The zacatal is probably a type of transition vegetation between the semi-arid zone and more humid regions. On the other hand, it seems to occupy the same ecological niche as the Matorral Crasicaule.

Although in many places extensive areas of each of the above-mentioned vegetation types can be found in a more or less pure condition, there are other places where they form mosaics, or merge to form very wide ecotones.

The genera *Opuntia*, *Agave*, *Yucca*, *Acacia* and *Fouquieria* deserve special attention. They are represented by several species which are found in all types of vegetation described and which are so typical of the Mexican landscape. Not as ubiquitous, but present in all the well-established arid areas, we find *Prosopis* and *Larrea*. In the areas abandoned by Man exotic weeds such as *Salsola kali* and *Nicotiana glauca* are becoming increasingly abundant.

If should be noted that some humid areas, with quite different types of vegetation, are found within the Mexican arid zones. These are caused solely by altitude and though they are present, are too small to have either economic or ecological influence over the areas surrounding them. In these humid regions the most frequent types of vegetation are pine and pine—oak forest.

TABLE 11.4

Physionomic-ecological classification of plant formations in the semi-arid ecosystems of Mexico used for ranching [a]

Class	Subclass	Group	Formation [a]	Key [b]
Matorral	Mainly evergreen scrub	Evergreen broadleaved-shrubland	Matorral mediano esclerifolio	III, A, 1, d
			Matorral alto subinerme	III, A, 1, e
	Extremely xeromorphic (subdesert)	Mainly evergreen subdesert shrublands	Matorral inerme parvifolio	III, C, 1, a, 2
			Matorral crasicaule	III, C, 1, a, 3
			Matorral crasi-rosulifolio espinoso	III, C, 1, a, 4
			Matorral mediano espinoso con espinas laterales	III, C, 2, a
Zacatal terrestrial herbaceous communities	Steppes and related grassland	Mid-grass steppes	Pastizal amacollado arbosufrutescente	V, B, 2, d
		Short-grass steppes	Pastizal mediano arbosufrutescente	V, B, 3, d
			Pastizal mediano abierto	V, B, 3, c
			Pastizal halofito arbosufrutescente	V, B, 5, d
			Pastizal halofito abierto	V, B, 5, c

[a] COTECOCA - Mexico (unpublished, 1974).
[b] Mueller-Dombois and Ellenberg, 1974.

Fauna [1]

Until recently the most conspicuous element of the semi-arid zacatales must have been the bison (*Bison bison*). This fact is supported by the first maps of these regions drawn by the Spanish conquerors. However, with the

[1] We are grateful to Dr. Nicolás Vázquez-Rosillo for the critical revision of this subject.

TABLE 11.5

Herbaceous production of the semi-arid ecosystems used for ranching [a]

Formation	Key [b]	ADM [c]
Matorral Mediano Esclerifolio	III, A, 1, d	210 ± 20
Matorral Alto Subinerme	III, A, 1, e	329 ± 55
Matorral Inerme Parvifolio	III, C, L, a, 2	172 ± 55
Matorral Crasicaule	III, C, 1, a, 3	368 ± 102
Matorral Crasirosulifolio Espinoso	III, C, 1, a, 4	223 ± 39
Matorral Mediano Espinoso Con Espinas Laterales	III, C, 2, a	200 ± 20
Pastizal Amacollado Arbosufrutescente	V, B, 2, d	382 ± 19
Pastizal Amacollado Abierto	V, B, 2, c	400 ± 30
Pastizal Mediano Arbosufrutescente	V, B, 3, d	380 ± 55
Pastizal Mediano Abierto	V, B, 3, c	395 ± 46
Pastizal Halofito Arbosufrutescente	V, B, 5, d	325 ± 20
Pastizal Halofito Abierto	V, B, 5, c	400 ± 25

[a] COTECOCA - Mexico (unpublished, 1974).
[b] Classification of Mueller-Dombois and Ellenberg, 1974.
[c] Annual above-ground dry matter production in kg ha^{-1}.

spread of Europeans through the continent this species rapidly became extinct. In Mexico today these animals are only found in zoos, and even they were all imported from the U.S.A.

Recently, owing to misuse, bad legislation, little concern by the authorities and poor public education, the major fauna from a great part of the Mexican semi-arid zones has been decimated. Nevertheless, the Mexican desert fauna is still rich, in both species and numbers. Jaeger (1957) gives an accurate account of the fauna, and we mention here only the most important animals.

Nearly all the naturally occurring arthropods are present in large numbers. Two moths are noteworthy: *Tegeticula yuccasella* and *T. mexicana* which effect pollination in plants of the genus *Yucca* (Bastida, 1964). The arachnids present include spiders, scorpions and solfugids. Centipedes and millipedes represent the myriapods.

Perhaps the most conspicuous animals are the vertebrates. Among the amphibians are small toads of the genus *Scaphiopus*; among the reptiles saurians of all sizes are found belonging to the genera *Sceloporus*, *Heloderma*, *Phrynosoma*, etc.; the remarkable desert turtles (*Gopherus*) and all types of snakes, e.g. *Crotalus* (cascabel), *Pituophis* (alicante).

There is a rich variety of birds belonging to the genera *Bubo* (buo), *Strix* (tecolote), *Falco* (halcón), *Polyborus* (quebrantahuesos) *Cathartes* (zopilote), *Callipepla* (codorniz), *Certonyx* (perdiz), *Lophortyx* (perdiz), *Dryobates* (carpintero), *Phalaenoptilus* (tapacaminos), *Corvus* (cuervo), *Zenaidura* (paloma) and numerous small birds of many species, including the best known bird of all American deserts, the "correcaminos" of the genus *Geococcyx*.

The large mammals are practically all extinct. The venado bura (*Odocoileus hemionus*), the berrendo (*Antilocapra americana*), the borrego cimarrón (*Ovis canadensis mexicanus*) and the lobo (*Canis lupus*) are all being decimated by hunting. Urgent protective measures are needed to prevent such species becoming so depleted that re-establishment would be difficult.

Also hunted, but still more or less abundant, are the venado cola blanca or salton (*Odocoileus virginianus*), the jabali mexicano (*Pecari tajacu*), the armadillo (*Dasypus novemcintus*), the perrito de las praderas (*Cynomys ludovicianus*) and the puma (*Felis concolor*).

Among the less remarkable, more abundant mammals are the tlacuache (*Didelphis marsupialis*), the mapache (*Procyon lotor*), the pintorrabo or cacomixtle (*Bassariscus astutus*), the coati or osito narigón (*Nasua narica*), the zorillos (*Mephitis mephitis*, *Spilogale putorius* and *Conepatus leuconotus*), the comadreja (*Mustela frenata*), the tejón (*Taxidea taxus*), the zorras (*Vulpes macrotis* and *Urocyon cinereoargenteus*) and of course bats, among which the genus *Leptonycteris* is noteworthy as a pollinating agent of the genus *Carnegia*, a cactus known as "sahuaro" in the Sonoran desert.

The coyote (*Canis latrans*) and the gato montez (*Lynx rufus*) are still very abundant. Despite official and private attempts to exterminate them, they are not uncommonly seen, even during the day. Later on, we will discuss the ecological imbalance which these intermittent campaigns have produced.

The rodents, rabbits and hares deserve a special mention since some of them are adapted to the various arid conditions; moreover, they sometimes occur in such numbers as to constitute a plague. Apart from the perrito de las praderas (mentioned earlier) the other more common species are the ardilla de tierra (*Citellus* spp.), the tuza (*Geomys* spp.), the raton (*Perognathus* spp., *Onychomys* spp., *Reithrodontomys* spp., *Peromyscus* spp. etc.), the rata (*Neotoma* spp., *Sigmodon* spp.) and the rata-canguro of the genus *Dipodomys*, which is possibly the best desert-adapted mammal. *Lepus* spp. and *Sylvilagus* spp. are representative genera of the hares and rabbits.

Of the forementioned plants and animals, we consider the following genera to be best adapted to the Mexican arid zones: among the plants, *Larrea* (Zygophyllaceae), *Fouquieria* (Fouquieriaceae), *Opuntia* (Cactaceae) and *Agave* (Agavaceae); among the animals, *Heloderma* (reptile), *Geococcyx* (bird) and *Dipodomys* (mammal).

The human population

According to a census (unfortunately unpublished) drawn up in 1974 by the Comisión Nacional de Zonas Aridas, a little more than 8 000 000 inhabitants live scattered over the arid and semi-arid zones. Most of them live in small communities of fewer than 200 inhabitants, in areas which are difficult to reach and which lack even the most basic public services. It is not surprising, therefore, that a high percentage of the population are illiterate and malnourished, and that diseases associated with unhygienic water supplies are rife.

The population represents a mixture of two cultures, as a result of miscegenation between Spaniards and the indigenous people. The peasants who inhabit the semi-arid zones are therefore heirs of a cultural mosaic which is reflected in their adoption of European practices, together with the ancient ones which evolved in the country.

This cultural background, coupled with the dispersed population, has precluded the establishment of modern techniques, and has had profound effects on the ecosystems concerned, since their vegetation and fauna are exploited (not managed) to a maximum. Thus, almost all non-toxic plants and animals are regularly consumed as food (insects, caterpillars, snakes, lizards, rats and all parts of many Cactaceae).

ECOSYSTEM MANAGEMENT

Historical background

With the arrival of the European in America, the continent began to experience a series of changes. For Mexico, the most important of these was the establishment of population centres in the central and northern parts of the country, stopping at the limits of the arid zones only until the discovery of gold and silver producing mines among the States of San Luis Potosi, Zacatecas, Durango and Chihuahua extended them into the desert. This human colonization was accompanied by horses, donkeys, cows, sheep, goats and pigs, all of which increased rapidly in numbers. The effect of livestock upon the surrounding vegetation must have been very pronounced. Mining activities required so much supporting timber and wood for fuel that timber had eventually to be brought by wheelbarrows from distances of up to 15 km.

At the same time, cultivation of land to feed the population was necessary, regardless of soil conditions and despite low production and high costs (Echeverría, 1960). This practice became so firmly established among the inhabitants that it continues even today, and is one of the principal causes of deterioration of the semi-arid ecosystems. Nevertheless, semi-arid conditions are so unfavourable for arable agriculture that ranching soon prevailed as the major type of land-use.

When villages and cities were founded they were settled according to the Spanish customs of land ownership, which dictated that each town should have an area of individual properties, together with a community property, the "ejido", meaning "place at the town's edge". All the town's inhabitants had the right to exploit the resources of the communal ejidal lands, e.g. construction materials, water, woods and grasslands. This system still prevails in isolated, ancient towns. The pre-Cortesian people had a community system represented by the "Calpullalli" and the "Altepetalli" (Echeverría, 1960).

Private properties were assigned to conquerors as a reward for distinguished service; they were called "Peonias" (about 500 ha) and were large enough to cultivate wheat, corn, fruit trees, to obtain firewood and sufficient grass for 20 cows, 100 goats and 10 pigs.

The properties increased in size according to military rank, and in some cases they even exceeded 100 000 ha, producing large ranches with little or no cultivation. Each of these large properties, called "Haciendas", constituted a complete rural unit.

From early colonial times, those in power are almost the only ones that have exploited the communal lands; and various laws have been passed in an attempt to limit these abuses; for example the many "Leyes de Indias", the "Ley de desamortizaciones of 1856", the "Constitución Política of 1917" and the "Ley de asentamientos humanos of 1976".

Up until 1877 there was very little, if any, management of the natural resources. Man merely made use of forests and grasslands without doing anything more than developing essential watering points. No concern was given to management, since the land apparently gave everything in abundance. This concept is reflected in Humboldt's statements (1822) comparing the country to an inexhaustible horn of plenty; this dangerous idea has prevailed among some social and political classes up to the present day.

Undoubtedly, the first pasture land management practices in America were a result of the introduction of the "Toro de Lidia", the Spanish fighting bull. Full use was made of Spanish tradition and experience in the raising of these animals.

The 1877—1910 government encouraged and strengthened the establishment of cattle ranges. This was probably the period when the rangelands were best used (Gentry, 1957). As evidence of that period one can still find numerous long stone-walls which served to control grazing. Usually a rotation of two camps was used, in which livestock were kept in one camp all year, while the other rested.

Vegetation changes

The early management systems were based on intuition and were basically inefficient. Nevertheless, there are still some regions in which climax or near-climax vegetation occurs, especially where human population densities are low. There is, however, much evidence of vegetation changes. Early explorers such as Cabeza de Vaca (in Durango during Colonial times — Gentry, 1957), Beltrami (in 1830 around Peotillos in the valley of San Luis

Potosi, Villa de Arriaga, S.L.P., Ojuelos and Cienaga de Mata up to Lagos de Moreno, Jal. — Glantz, 1964) and Tomás de Cuellar (in 1869, quoted by Tavera-Alfaro, 1964), commented on the vegetation. If one compares their travel records with the present vegetation, extensive changes have taken place. The studies of Hasting and Turner (1965) provide clear evidence of the changes which have occurred.

The factors influencing vegetation change in Mexico can be grouped under the headings geology, hydrology, climate, fire, overgrazing and man.

(1) Geology. There is evidence that large lake basins existed during the Tertiary from Chihuahua to Oaxaca (Trujillo, 1975). These basins were interrupted by a series of volcanic eruptions. This primitive connection would explain the possible relation of the Queretan and Hidalguense arid zones to the Chihuahua arid zones, since many genera are common to both areas (Miranda, 1955; Rzedowski and Medellin-Leal, 1958).

(2) Hydrology. Many hydrological changes have occurred and are still occurring at the present time. Road and highway constructions have had an important influence on the shallow waterways which they intercept, e.g. the highways from San Luis Potosi to Zacatecas and from Zacatecas to Concepción del Oro. There has been little technical regulation of these road building schemes, probably owing to the lack of short-term evidence of damage.

(3) Climate. Records of climatological stations in the arid zone are relatively recent. Only one station has more than 100 years of observations (San Luis Potosi), and for that reason we have insufficient information to hypothesize a change of climate. However, normal variations within the rainfall patterns characteristic of these regions may induce cyclic changes in vegetation physiognomy. Two years of drought can seriously damage the grass cover. The reverse condition, very humid years, can have the same effect. For example, in 1955 when many hurricanes occurred in the Caribbean Sea and the Gulf of Mexico, heavy rain fell on the NW portion of the Rioverde plain in the State of San Luis Potosi. This area had supported a "mezquital", but after this several-month inundation the *Prosopis* died, leaving in its place a grassland community.

(4) Fire. There is no evidence of spontaneous fires, but fire has been used by man since pre-colonial times. In Durango, Gentry (1957) tells of the way natives made a circle of fire to enclose animals. At present, fire preserves the grass cover in some areas, and in the "mezcaleras" of San Luis Potosi it is the method used to eliminate *Agave*'s dry leaves. Fire is also used to eliminate ruderal vegetation along the highways traversing the arid zones. Unfortunately, these fires often get out of control and invade surrounding grazing lands.

(5) Over-grazing. After the revolutionary period of 1910, patterns of land ownership changed. The number of proprietors increased markedly and the Hacienda as a unit of production came to an end. At the same time the rangelands were supporting the maximum possible number of livestock, and their forage production declined (Gentry, 1957).

According to the Centro de Investigación del Desarrollo-COPARMEX (1965) only 12.5% of the ranches from the arid zone were not eroding, and over-grazing was found in all but a few ranches in the State of Coahuila. This has made the delimitation of the "matorral desertico microfilo" difficult, and for this reason also only one type of zacatal can be recognized, "grassland" and "shrub-grassland" having been included within the "matorral" group; see the revisions of Rzedowski (1966, p. 117) and Flores-Mata et al. (1971b, p. 50).

(6) Man. During the Colonial period the Spanish could not suppress the native Indian tribes of the arid zone and so were forced to bring natives from the centre of Mexico (Tlaxcaltecans) to work in the royal mines of the arid zone. As previously stated, the present rural population in the arid zones is the result of this hybridization, and their activities consist largely of subsistence agriculture close to or around the mines, regardless of soil and costs.

The practice of ploughing the land, in regions with such a low probability of obtaining a corn or bean harvest, has resulted in widespread destruction of the climax vegetation. There is a popular belief that a rich cover of grasses indicated that corn (which is Mexico's staple crop) can be cultivated there. This is incorrect and results in the field being abandoned after 2 or 3 years. The increasing population, and the fact that most inhabitants cultivate corn but do not raise livestock is causing a rapid decline in the situation.

During Colonial times the sources of carbohydrate were limited and there were few methods of cultivating food in desert areas. For this reason, fruits of wild Cactaceae and *Prosopis* were used widely as the only food supply, for both animals and humans. This is how the large areas of "nopaleras", "magueyeras" and "mezquiteras" (types of secondary vegetation almost entirely composed of *Opuntia*, *Agave* and *Prosopis*, respectively), in San Luis Potosi and Zacatecas must have developed.

The most exploited species were, and still are, *Agave atrovirens*, *A. asperrima*, *Prosopis laevigata* and *Opuntia streptacantha.* A candy named "queso de tuna", which lasts for a long time without deteriorating, and an alcoholic drink "colonche", are made from the "tunas" (fruits of the last species mentioned). During the dry season (8 months) livestock on the range are usually fed with the stems of *Opuntia* spp. after burning-off the spines. Frequently, pieces which have not been ingested survive and act as propagules. Terminal leaf buds of *Yucca* spp. are often given to feed range livestock. These practices undoubtedly increased the natural extension of the "matorral crasicaule" to such a degree that it is now impossible to distinguish natural areas from induced ones.

The high density of *Agave* in grassland areas is also caused by artificial propagation, since another sweet drink named "aguamiel" is obtained from *A. atrovirens*. This drink is non-alcoholic but can be fermented to form "pulque". A liquor named "mezcal" is obtained from *A. asperrima*. This is similar to "tequila" except for a "sui generis" flavour and smell. The "pencas" or *Agave* leaves are collected to feed dairy cattle. The presence of *Opuntiae* and *Agavaceae* outside their normal range is a strong indication that man has lived there.

The "mezquite" (*Prosopis* spp.) legumes contain sugars from which "atole" (a gruel) is made for human consumption. The pods are also used to feed dairy cattle and goats, and as a result of this seeds are dispersed in the dung under ideal conditions for germination. The spread of *Prosopis* in this way is similar to that described by Glendening and Paulsen (1955) in the south of the U.S.A. It has led to difficulties in identifying its original distribution, and classification of the "mezquital" formation is accordingly confused (c.f. Rzedowski's revision, 1966, and Flores-Mata et al., 1971b).

Few people have studied plant succession in the semi-arid zones of Mexico. Hernández-Xolocotzi (1957) classified the grasses according to their forage quality and value as indicators of salinity, over-grazing and toxic substances, but perhaps the person who has contributed most to our knowledge of secondary vegetation is Rzedowski (1966). Despite the lack of basic successional studies, the Dyksterhuis system (1949) has been adopted for practical reasons. This classifies successional stages into four groups: excellent, good, regular and poor. The method is used by the Mexican state service (COTECOCA-SAG). Details of its application are not possible because of lack of regional studies and because the technicians who apply the system have an agricultural but not ecological training and find it difficult to identify plants. The method itself assumes that succession is an ordered, linear series but it is difficult to recognize the various stages. The method also considers the climax as specific, but Heady (1973) finds there may be convergence of succession and that it is not always of specific character, but only in life forms.

Range management

An investigation by the Centro de Investigación del Desarrollo-COPARMEX (1965) of 605 ranches in the north-central arid lands of the country found that 3.8% of the ranches were not fenced and 27% had only a peripheral fence, giving a single paddock. In these two types of property, which range from 300 ha to 10 000 ha, there was no rotational grazing. 38% of the ranches used rotational grazing and, among these, ranches having from 2 to 5 paddocks were most common. The most common ranch size was between 5000 and 10 000 ha.

With regard to water supply, 8.6% did not have permanent water and 21% depended on one supply point only, thus limiting possible management. 17% had two, 10% three, 8% four and 8.6% five water points.

Improvement practices of one kind or another were being used in only 58% of the ranches. Improved practices were shrub control, removal of toxic plants, rodent control, prescribed burning, reseeding, soil conservation and fertilization, but these were used only on areas smaller than 100 ha.

The problem of shrub invasion, as mentioned in "causes of vegetation change" is very important in the formation of arid zones. It occurs in practically all types of grassland. Studies of shrub control have demonstrated that forage production can be increased up to ten times, e.g. in Nuevo Leon (Huss and Aguirre, 1973), depending on the condition of the grass cover prior to shrub removal (Echavarría, 1972).

According to the Céntro de Investigación del Desarrollo-COPARMEX (1965), only 12% of the ranches studied had any programme of shrub control, and most of these were in areas not exceeding 100 ha. This is mainly due to the cost of shrub control methods and the lack of defined boundaries of rural properties in the arid zones.

A few studies have indicated toxic plant species in various regions, e.g. González and Martínez (1958), Zalazar-Estrada (1959) and Gómez-González (1966). According to Gómez-González, 47% of these species have a neartic distribution, 25% are cosmopolitan, 22% are endemic, 4% American and 2% paleartic. Classified by their life-form, 31% are hemicryptophytes, 26% therophytes, 26% phanerophytes, 15% chamaephytes and 2% geophytes. 46% of these species are increased by over-grazing.

Livestock poisoning by toxic plants is complex and depends upon various coinciding factors, since animals frequently eat these plants without showing signs of debilitation or death. The toxicity also varies according to regional ecotypes. For example, *Drymaria arenarioides* is very toxic in Chihuahua (González and Martínez, 1958) but recent studies show that it is not toxic from Zacatecas to Hidalgo; *Acacia berlandieri* and *Prosopis juliflora* are toxic in Texas (Sperry et al., 1964) while in Mexico, they constitue an important part of the range livestock diet. In arid zones many plants accumulate toxic substances but do not represent a danger to livestock either because they are not preferentially eaten or because they cause subclinical damage which is difficult to detect. Toxic plants are most dangerous when the stock are weakened by drought and lack of forage.

Some small herbivores vary in relation to vegetation conditions. "Ratacanguro" (*Dipodomys* spp.) prefer the most degraded range conditions, increasing in number under these circumstances but decreasing when succession nears the climax (Fitch, 1948). Other species prefer to live in grasslands in good condition (Reynolds and Haskell, 1949). Studies of vegetation in Mexico do not normally mention the importance of rodents or other wild animals in vegetation succession and perhaps the first mention of this problem was by Muñoz (quoted by Carrera, 1973a) in the State of Chihuahua. Later, Valadez (1969) attempted to control the "rata-canguro" (*Dipodomys* spp.) with poisoned bait. Carrera (1973a) also used toxic baits to control the "topo" (*Thomomys* spp.) and tried to evaluate the amount of forage consumed by hares (Carrera, 1973b). The unrestricted control of wild fauna through poisoned baits without an understanding of their ecology is not generally beneficial (Fogden and Fogden, 1975). For example, there is evidence that many herbivorous species help control the number of shrub seedlings (Fogden, P., 1975). The density of hares is a function of the number of predators, but as the latter — "coyote" (*Canis latrans*) "puma" (*Felis concolor*) and "gato montez" (*Lynx rufus*) — are being systematically reduced they are very scarce in some places, allowing an increase in the number of hares, especially, at present, in Chihuahua. For this reason, a systematic campaign against hares has been suggested using ecologically undesirable means, such as poisons and fire arms. M. Fogden (1975) however, has demonstrated that it is possible to cut down the very high density of hares by 80% by controlling the shrubs, and by an additional 10% by increasing the grass cover. However, he also remarked that control of small "zacatal" herbivores by poisoning and hunting is not worthwhile owing to the possibility of new individuals moving into the treated areas.

Grass species have been reseeded into areas where mean annual precipitation fluctuates between 300 and 400 mm in order to restore the "zacatal". Examples of native species are *Bouteloua curtipendula*, *Leptochloa dubia*, *Trichachne californica* and the introduced *Eragrostis lehmanniana*, *E. superba*, *Panicum antidotale* and *Sorghum almum*. In regions with a mean annual rainfall of 400 to 500 mm *Cenchrus ciliaris*, *Chloris gayana*, *Eragrostis curvula*, *E. chloromelas*, *Panicum maximum* and *P. coloratum* have been used.

Although there is enough information on how to seed the "zacatal" (Echavarría, 1973) efforts by individual ranchers are uncommon and have only been used in areas of less than 100 ha. During 1972—1973 there was a serious effort by the Federal government to restore the "zacatal" and 3000 ha of "zacatal" were planted in the State of Zacatecas. This proved to be most productive (see Table 11.6). However, there was a lack of governmental understanding, vision and persistence and after the first year of production, following the establishment of the "zacatal", the programme was suspended, despite good prospects.

Livestock

The most abundant type of livestock in the semi-arid ecosystems is cattle. The dominant breed is Hereford (40%), followed by native cattle (27%) descended from the various Spanish stocks, and then the Zebu (*Bos indicus*) (25%). Aberdeen Angus, Santa Gertrudis and Charolais are found in smaller

TABLE 11.6

Increase in forage production through reseeding within the "Pastizal Mediano Arbosufru-tescente" in Zacatecas on the brush control exclosure

Locality	Treatment	ADM [a] (kg ha^{-1})	Reseeding/Exclosure
Boquilla de Abajo	Reseeding	941	2.17
	Exclosure	434	
Emiliano Zapata	Reseeding	602	2.50
	Exclosure	240	
La Encantada	Reseeding	1430	2.22
	Exclosure	645	
Sotoles	Reseeding	1024	2.03
	Exclosure	505	

[a] Annual dry matter production.

numbers. Sheep are second in importance but their population will tend to decrease because of very high demand for these animals, unless the Federal government enforces intelligent policies to encourage their successful utilization. 69% belong to the native breed, descended from the Spanish-Churra breed which is in turn descended from Rambouillet: 20% are considered pure Rambouillet. Goats are very important in the "matorral" areas among the States of Coahuila, Nuevo Leon and San Luis Potosi and the most common breeds are Nubian, native Granadine and their crossbreed. The ratio of sheep to goats is 1.2 : 1.

Utilization of wild plants for industry and human consumption

Within the "matorral desertico rosetofilo", the two dominant species *Agave lechequilla* ("lechuguilla") and *Yucca carnerosana* ("palma samandoca") are exploited in a primitive way. A hard fibre named "ixtle" is obtained from them and manufactured into cords and sacks for export (Marroquín et al., 1964). Their harvest constitutes one of the most important activities of the inhabitants within these areas. Another species of this same vegetation type is *Euphorbia antisyphilitica* ("candelilla") from which wax is obtained after boiling.

During World War II, large amounts of "guayule" (*Parthenium argentatum*), a component of the "matorral desertico microfilo", were exploited to produce low-quality rubber. The government is presently trying to revive this activity, but the results are still uncertain.

Ripe fruits and inflorescences of *Yucca* spp. are consumed as human food, and in times of emergency they are also given to animals.

Large quantities of fruits "tunas" from almost all species of *Opuntia* ("nopal") *Myrtillocactus geometrizans* ("garambullo") and *Lemaireocareus* spp. ("pitayo") are used for human consumption. Young *Opuntia* stems are eaten as vegetables, and the ripe plants, after the removal of thorns with portable burners, are given to the livestock in dry seasons to offset the lack of water.

RECOMMENDED MANAGEMENT PRACTICES

The semi-arid region of Mexico which borders on the southern United States has long been an exporter of calves to the United States for fattening. In 1973, this region exported 435 000 head of cattle to the U.S.A., which amounted to 76% of all cattle imported by the United States.

There are a number of management techniques used to improve the semi-arid ecosystems. One of these is the division of pastures with barbed wire fences. While three to five strands of barbed wire are normally sufficient for cattle and horses, it is necessary to use two extra strands for sheep. While the wells are scarce and of low production ($2\ l\ s^{-1}$), they are sufficient for the drinking requirements of livestock. The majority of these wells are powered by windmills or diesel energy. The use of plastic pipeline that can support pressures up to $7\ kg\ cm^{-2}$ has made possible the installation of watering places at distances as far as 10 km from the wells, in order to improve the forage utilization. An even distribution of watering places, about 5 km apart, is considered to result in the best utilization of the forage, and at the same time overgrazing is reduced to a minimum. An even distribution of salt licks also helps to prevent over-grazing.

The techniques used to promote the conservation of soil water are: contour forrows, chiselling and pitting. The latter is the most advantageous. In the absence of wells, small earthen dams are constructed in the low portions of a basin which in some cases can retain water for up to 6 months. This last technique, however, is questionable as the dams act as evaporation surfaces and in some cases the water is lost more rapidly.

To increase the vegetation cover when it is not necessary to seed, it is preferable to manipulate the ecosystem to promote the valuable forage species by means of increased retention of moisture, removal of undesirable species by mechanical or chemical methods, reducing the number of animals, rotating pastures and, finally, by leaving 40 to 50% of the pasture ungrazed.

Biological methods are used to reduce the undesirable plants. Dicotyledenous weeds are controlled by forcing sheep to over-graze the area until they are eliminated, thereby using the preference of sheep for this class of plants, to the benefit of the grass and cattle grazing. Shrubs are controlled by increasing the number of goats.

With the exception of the border zone, where normally only cattle are bred, it is usual to find cattle, goats, sheep and horses grazing together without any apparent damage. This condition agrees with the mosaic vegetation mentioned in other chapters and with the grass habits of these domestic animals.

As a final note we feel that there are some general recommendations regarding the semi-arid zones, which should be considered at this point. Studies of the semi-arid ecosystems have been thorough in the areas of Sonora and Chihuahua, particularly in regard to species inventories, but unfortunately this does not apply to the other semi-arid zones of the country. The advanced studies in the zones located to the north are largely the result of their proximity to the U.S.A. Because they are less accessible, the other zones are poorly known, though this has not and does not seem to present an obstacle to their residents who intensively exploit them.

There has also been a tendency to study the same few species, particularly those with demonstrated utility. For example, in Mexico, the *Opuntia* and *Agave* have been heavily but superficially studied. Much emphasis has been placed on the studies of plants which were important economically, in some cases critically so, at certain points in history, e.g. *Parthenium*, which was used as a source of rubber during the Second World War. We do not want to imply that the classic lines of investigation should be given up, but rather that the great number of plants about which we know practically nothing should not be ignored or disregarded.

The specific recommendations are as follows.

(1) To initiate a systematic, intensive study of the most important plants in the arid zones, including the taxonomic, agronomic and genetic identification of both species and varieties for the different ecological regions. No basic studies relating to the practices of cultivation, pruning and fertilization in the arid areas exist, so it is difficult to recommend management practices.

(2) To develop mechanized and/or industrialized processing for plant products. The harvest occurs in a 2-month period and this creates considerable problems relating to both its processing and conservation.

(3) To develop experimental programmes in applied ecology in each ecological region in order to learn the techniques necessary for sustained maximum production.

(4) To develop a system of seed exchange between countries with arid zones to allow evaluation for the possible introduction of useful plants from one country to another where it appears feasible.

(5) To develop the important potential of meat, wool and fur production from wild animals, and their possible expansion if the ecological conditions permit.

(6) To study the system of runoff for each catchment area in order to make maximum use of the periodic rain water for cattle or natural vegetation and to avoid flood hazards.

(7) Through an extension service, make the public aware of the modern methods of range management for their respective ecological zones, by the

use of visual demonstrations on experimental plots.

(8) In the semi-arid environments arable agriculture is only possible in limited areas where runoff from the catchment area is large. However, when the ranches are broken up indiscriminately with small agricultural areas, interfering with management of the grazing lands, the country remains impoverished.

(9) Finally, it is desirable to establish several "sanctuary zones" in the semi-arid regions of Mexico to ensure the preservation of the natural resources and to provide an adequate source of semi-arid plant and animal germplasm for the future.

REFERENCES

Aranda-Gómez, J.J., 1973. Reporte del servicio social efectuado en el Instituto de Investigación de Zonas Desérticas sobre documentación geológica de las zonas áridas de México. Tesis Ingeniero Géologo. San Luis Potosí, Universidad Autónoma, 289 pp.

Bastida, L., 1964. Algunas observaciones sobre la biología de *Tegeticula mexicana* Bat. (Lep: Prodoxidae). An. Inst. Biol., Univ. Nac. Auton. Mex., Ser. Zool., 35: 95—104.

Carrera, L.J., 1973a. Evaluación de tres tipos de control de topo *Thomomys* spp. Pastizales, 4(4): 7.

Carrera, L.J., 1973b. Consumo de forraje por la liebre *Lepus* spp. Pastizales, 4(4): 10.

Centro de Investigación del Desarrollo-COPARMEX, 1965. Estudio integral preliminar sobre la ganadería de la zona norte de la Républica Mexicana. Inventario de recusos ganaderos del norte de México IV. México, El Centro, 235 pp.

Dyksterhuis, E.J., 1949. Condition and management of rangeland based on quantitive ecology. J. Range Manage., 2: 104—115.

Echavarría, S., 1972. Incremento de la producción forrajera de Pastiza-les semiáridos invadidos por arbustos indeseables. Pastizales, 3(1): 11—14.

Echavarría, S., 1973. Evaluación del comportamiento de 10 especies de zacates nativos introducidos sembrados de temporal en la región central de Chihuahua. Pastizales, 4(2): 2—8.

Echeverría, L.M., 1960. La ganadería mexicana. México, Dpto. Inv. Ind. Banco de México, 188 pp.

Fitch, H.S., 1948. Ecology of the California ground squirrel on grazing land. Am. Midl. Nat, 39: 513—597.

Flores-Mata, G., Jiménez-López, J., Madrigal-Sánchez, X., Moncayo Ruiz, F. and Takaki-Takaki, F., 1971a. Mapa de Tipos de Vegetación de la República Mexicana. Esc. 1 : 2 000 000. México. Dirección de Agrología de la Dirección General de Estudios de la Subsecretaría de Planeación de la Secretaría de Recursos Hidráulicos.

Flores-Mata, G., Jiménez-López, J., Madrigal-Sánchez, X., Moncayo Ruiz, F. and Takaki-Takaki, F., 1971b. Memoria del Mapa de Tipos de Vegetación de la República Mexicana. México. Dirección de Agrología de la Dirección General de Estudios de la Subsecretaría de Planeación de la Secretaría de Recursos Hidráulicos, 59 pp.

Fogden, M., 1975. Control de las liebres por manipulación de su habitat. Pastizales, 6(4): 13.

Fogden, P., 1975. Efecto económico de liebres y roedores sobre los pastizales. Pastizales, 6(4): 12.

Fogden, M. and Fogden, P., 1975. Efecto de la Fauna silvestre sobre la dispersión y germinación, supervivencia de semillas y plántulas de mezquite. Pastizales, 6(1): 2—6.

García, E., 1973. Modificaciones al sistema de clasificación climática de Köeppen. UNAM. México, 245 pp.

Gentry, H.S., 1957. Los Pastizales de Durango. Mex., Inst. Mex. Rec. Nat. Renov., 361 pp.

Glantz, M. (Ed.), 1964. Viajes en México. Crónicas extranjeras. México. Secretaría de Obras Públicas, 502 pp.

Glendening, G.E. and Paulsen, H.A., 1955. Reproduction and establishment of Velvet Mezquite as related to invasion of semi-desert grassland. U.S.D.A. Tech. Bull. 1127, 50 pp.

Gómez-González., A., 1966. Plantas tóxicas para el ganado en la zona semiárida de San Luis Potosí. Tesis Biólogo. México. Esc. Nal. Cien. Bíol. Inst. Pol. Nal., 75 pp.

González, M.H. and Martínez, F., 1958. Las Plantas tóxicas al ganado en los pastizales de Chihuahua. México, Oficina de Estudios Especiales de la Secretaría de Agricultura, Folleto de Divulgación N. 28, 27 pp.

Hasting, J.R. and Turner, R.M., 1965. The Changing Mile. University of Arizona Press, Tucson, 317 pp.

Heady, H.F., 1973. Structure and function of climax. In: D.N. Hyder (Editor), Arid Shrublands; Proceedings of the Third Workshop of the United States/Australia rangelands Panel, Tucson,

Arizona, 26 March—5 April 1973. Denver. Society for Range Management, pp. 73—80.
Hernández-Xolocotzi, E., 1957. Los Pastizales mexicanos. In: Instituto Mexicano de Recursos Naturales Renovables. Mesas Redondas sobre problemas en la industria agropecuaria en México. México, El Instituto, pp. 3—78.
Huss, D.L. and Aguirre, E.L., 1973. Fundamentos de Manejo de Pastizales. Monterrey I.T.E.S.M., 227 pp.
Jaeger, E.C., 1957. The North American Deserts. Standford University Press, pp. 164—224.
Marroquín, S.J., Borja, G., Velázquez, R. and de la Cruz, J.A., 1964. Estudio ecológico, dasonómico de las zonas áridas del Norte de México. México. INIF. Secretaría de Agricultura and Ganadería Publicación Especial No. 2, 166 pp.
McGinnies, W.G., Goldman, G.J. and Paylore, P. (Editors), 1968. Deserts of the World. University of Arizona Press, Tucson, pp. 21, 26.
Miranda, F., 1955. Formas de vida vegetales y el problema de la delimitación de las zonas áridas de México. In: Instituto Mexicano de Recursos Naturales Renovables. Mesas Redondas Sobre Problemas de las Zonas Áridas de México. México, El Instituto, pp. 85—109.
Mueller-Dombois, D. and Ellenberg, H., 1974. Aims and Methods of Vegetation Ecology. Wiley, New York, 547 pp.
Ramos-Alvarez, C.H. and González-Medrano, F., 1972. La Vegetación de la zona árida veracruzana. An. Inst. Biol. Univ. Nal. Auton. México, Ser. Bot., 43: 77—100.
Reynolds, H.G. and Haskell, H.S., 1949. Life history notes on price Bailey pocket mice of southern Arizona. J. Mammal., 30: 150—156.
Rzedowski, J., 1966. Vegetación del Estado de San Luis Potosí. Acta Cient. Potosina, 5: 5—291 (1 map).
Rzedowski, J. and Medellin-Leal, F., 1958. El límite sur de distribución geográfica de *Larrea tridentata*. Acta Cient. Potosina, 2: 133—147.
Sperry, O.E., Dollahite, J.W., Hoffman, G.O. and Camp, B.J., 1964. Texas plants poisonous to livestock. Texas Agric. Exp. Stn, B-1028, 59 pp.
Tavera-Alfaro, X. (Editor), 1964. Viajes en México. Crónicas mexicanas. México. Secretaría de Obras Públicas, 450 pp.
Trujillo, J.A., 1975. Las cuencas lacustres terciarias del Altiplano Mexicano.Edición del Autor.s.d.t.
Valadez, E.G., 1969. Combate de la rata-canguro (*Dipodomys* sp.) en los pastizales de Chihuahua. Circ. Campana, 10: 1—8.
Zalazar-Estrada, P., 1959. Investigación preliminar de las plantas potencialmente capaces de causar daño al ganado en el Norte de Nuevo Léon; Tesis Ingeniero Agrónomo, Monterrey Inst. Tec. Sup., 89 pp.

Section III

DISCUSSION: MANAGEMENT PRINCIPLES FOR SEMI-ARID ECOSYSTEMS

B.H. WALKER

MANAGEMENT PRINCIPLES FOR SEMI-ARID ECOSYSTEMS

B.H. WALKER

INTRODUCTION

The accounts in the preceding chapters of human interference in semi-arid ecosystems cover very different time-scales. Significant changes induced by Man were evident more than 5000 years ago in the Middle East, and probably in India and Asia, whereas in some parts of Africa they began only in this century.

In the areas of ancient habitation the changes in ecosystem structure and function are difficult to trace, but where significant impact is recent and recorded, an interesting pattern emerges. East and Southern Africa, the south-western United States, Australia and, to some extent, Mexico, show an initial boom period with a dramatic decrease in indigenous herbivores coupled with a rapid increase in domestic livestock — mainly cattle, sheep and goats (although the unintended explosion of rabbits in Australia can be linked to this phase). Severe ecological degradation, usually triggered by a period of drought, then resulted in an economic collapse. Animal numbers declined markedly (often as a result of death), and this in turn has been followed by a gradual improvement through the process of secondary succession. The management strategies which were subsequently developed seem to have benefitted from the earlier mistakes.

In Mexico, conditions reached an optimum state around the turn of the century, and have since declined. In the Sahel, and in parts of India and Southern and East Africa, the situation is likewise declining. Elsewhere, although conditions can by no means be described as good, some sort of stability has been regained, albeit with a different vegetation structure. There is, however, widespread concern that a new and potentially more serious decline is imminent (and has in fact already started) owing to the human population explosion in these areas. Desertification, desertization, or however else it might be described, is proceeding in the Sahel and in the Middle East, where population pressures are most severe and management options most limited. It remains to be seen whether management can adapt sufficiently in the presently stable areas, or whether they, too, will follow the pattern of Mexico, where a previously stable system gave way under population pressure to result in a less desirable form of land-use.

The nature of the "new" stable systems just referred to warrants further examination. The best-known examples are the striking change from grassland to succulent, spiny and other scrub in the south-western United States, the Chenopodiaceous shrub land to mixed grassland change in Australia, and the grassland to shrubland changes in Southern Africa (notably the development of the False Karoo). In the sense of Noy-Meir (1975) and May (1977) these changes may be viewed as locally stable points in systems which are characterized by having multiple stable points. Unlike the example given by Noy-Meir (1975), though, these new states of the system are characterized by having a different species composition. They do not merely represent different amounts of the same vegetation and herbivores, but reflect a change in the structure of the system itself. The change from one form to the other took place in a continuous fashion, but intermediate combinations of grasses and shrubs were unstable. So, once started, the change to the shrub-dominated composition was quite rapid, and although the structure of the vegetation is now quite different it is still valid to consider the two forms as different phases of the same system.

Provided the variables of the system remain sufficiently close to the equilibrium point (i.e., the changes in the relative amounts of the components of the system are within the domain of attraction of that equilibrium point), then, following any departure from the equilibrium configuration of the system, the tendency will be to converge back to that structure. When (through excessive use by Man) the system is so altered that it is pushed beyond the domain of attraction, it does not entirely collapse, but flips into a new and different region of the system's state space, and is attracted to a new stable state. So far, there is little evidence that a system so changed can be manoeuvered back into its original domain.

What is the significance of this? In their account of Israel's semi-arid areas, Noy-Meir and Seligman refer to the apparent resilience (cf. Holling, 1973) of these ecosystems. Globally, the systems do appear to be remarkably resilient, but the changes just described imply a much lower local resilience. Put in another way, these ecosystems are subjected to marked inter-seasonal variation in driving forces, and they react by exhibiting a very unstable pattern of inter-seasonal herbaceous production, but with a more stable non-herbaceous component. Their resilience, however, is stretched to cope with the natural variation in growth conditions, and when Man's efforts are superimposed the limits may be exceeded, and the change to a new equilibrium point takes place. The degrees of resilience of the systems in their new states are unknown, but two points are noteworthy: the shrubland or succulent phases (whether the original or derived) are apparently more stable (in that inter-seasonal variation is less), but they are also less productive. It is difficult to find evidence of a natural ecosystem that has exhibited more than two stable equilibria. One is left with the impression that these systems have naturally evolved to a state of reasonably well-balanced stability and productivity, and that under the pressure of excessive use they generally change to an alternative, more resilient but less productive, structure.

The above speculation is an attempt to explain some of the observed dynamics of semi-arid ecosystems. It is, however, purely speculative and the question to be answered is: are these new states really stable equilibria, or is it merely a case of Man's continued interference keeping the systems away from their original and perhaps only, equilibrium points?

The answer to this and other theoretical questions concerning the dynamics of semi-arid ecosystems has important practical consequences, since a thorough understanding of their dynamics is a pre-requisite for ecologically sound management and optimal use of these resources. But management cannot be suspended until the research has been done, and there is a need for guidelines to semi-arid ecosystem management. This was the motivation for this book, and the following section is a brief, integrated summary of the main points emerging from the preceding chapters. They are treated under two headings: Determinants, and Management Principles, and represent firstly an attempt to pull together those common factors which seem to operate in semi-arid areas, and then to assemble a set of principles which form a sound ecological basis for management and should be used as guidelines in the formulation of particular management strategies. The various schemes and activities employed will differ from region to region, according to differences in many other factors, such as available capital, amount and quality of available labour, the system of land tenure, culture and tradition, and so on.

DETERMINANTS OF SEMI-ARID ECOSYSTEMS

Each semi-arid region of the world is unique in a number of ways, exhibiting its own peculiar features as a result of particular factors which exert a significant influence on the system. In other respects, and usually in their more important characteristics, these regions show a high degree of similarity. It is towards these latter, unifying characteristics that this section is aimed. It is difficult to rank these common determinants in any order of importance, since the quantitative data for such an analysis are lacking. The list of points has been kept purposefully short and includes only those factors which were regarded as significant in at least two of the preceding chapters.

(1) A low and highly variable rainfall. All of the authors have stressed the over-riding importance of the high variability in rainfall. The variability is both temporal and spatial.

Any one site experiences marked inter-seasonal variation, and it is common for intra-seasonal droughts to occur as well. In any one season rainfall received at sites just a few km apart may differ by more than 100%.

(2) Largely as a result of the rainfall pattern, primary production is highly variable, and in at least two areas (Southern Africa and Central Asia) inter-seasonal variations in excess of 500% have been recorded. The variability increases with increasing aridity.

(3) Animal, and subsequently human, populations responded to the variability in rainfall and production through the evolution of migration patterns. The regular migrations of the large game herds in East Africa reflect the more regular seasonal rainfall and production pattern there, as opposed to the irregular migrations and movements of large herbivores in Southern Africa. The traditional nomadic behaviour of the people in the Sahel, Middle East and Asia reflects their adaptation to the variation in primary production and water supplies.

It is of considerable importance to note that many of the worst problems associated with the management of the semi-arid world stem from the fact that increasing human populations have stopped the movements of both wild animals, and of people and their domestic herds. We shall return to this point in the next section.

(4) Soils are characteristically low in organic matter (which enhances fluctuations in soil moisture), but are generally quite fertile (with the notable exception of South America), since they are subjected to little, if any, leaching.

(5) A factor generally regarded as being of considerable importance (in Australia, the Sahel, Southern and East Africa, India and the U.S.A.) is the development of a surface cap or crust on exposed soils. It results from the explosive action of unimpeded raindrops, and forms an effective seal, significantly reducing the infiltration of water into the soil, and leading to run-off losses and soil erosion. The caps may be compounded by the development of a micro-floral mat, largely of blue-green algae and/or lichens. This mat is regarded in some areas (Central Asia) as favourable, enhancing soil water properties due to their high water retention, and in others (Sahel, Southern Africa) as unfavourable, owing to their hydrophobic nature when dry.

(6) Lack of leaching, and high evaporation rates where water flows into an area, lead to a fairly common occurrence of high salinity soils. The vegetation on these soils generally has a high salt content and leads to very marked increases in the water needs of animals feeding on it. In Asia it results in a seasonal pattern of only spring and autumn use of such areas, when the salt content is sufficiently low.

(7) Surface water is universally in short supply. Most animals are restricted to an effective grazing zone of about 5 km around water supplies, although in Israel the distance is apparently greater. Under natural conditions it results in a very uneven pattern of utilization, and where Man has introduced domestic livestock the zones around water points have degraded to the

extent where many of them are now effectively desert.

(8) The vegetation is generally a mixture of woody and herbaceous species. The former play an important role because, unlike the higher rainfall regions, browse in the form of tree and shrub leaves and shoots, is both palatable and highly nutritious. The time of the year when it is most evident, and the fact that browse production varies less from one year to another than does herbaceous production, are considered to be of great importance.

It is also noteworthy that semi-arid woody and succulent plants are of considerable direct value to man — as described in Chap. 11.

The relationships between the woody and herbaceous components are as yet not well understood. Evidence of increased herbaceous production from clearing trials needs to be considered in the light of the ecosystems' long-term stability, and the value of the browse component, as discussed in virtually all of the chapters.

(9) Fire is generally not an important factor. This point will be expanded in the next section.

(10) Temperature extremes are a feature of these areas. Low humidities cause very high diurnal variation, and seasonal variation is also marked. The tropical and sub-tropical zones do not suffer from low temperatures, but high temperatures are a major obstacle for plants and animals. In temperate zones, winter conditions are a major determinant of animal distribution and survival.

(11) Periodic outbreaks of insect populations are a feature of these environments. In India and Africa, for example, harvester termites are regarded as a serious problem, competing with domestic livestock for available grazing.

MANAGEMENT PRINCIPLES FOR SEMI-ARID ECOSYSTEMS

The principles which follow are again general principles that apply to the semi-arid world. Specific details of management which are concerned with particular schemes are not considered. Because priorities change from area to area it is not meaningful to put these principles into a definite order of importance. They are therefore considered to be more or less equivalent, although it would be true to say that those given greatest emphasis by most authors are presented first. The management principles are followed by a few general recommendations.

Management principles

Arable crops

(1) Dry-land cultivation should be opportunistic and limited to run-on areas and other favoured sites as in Mexico and Central Asia. Continued

research may allow for development of reliable, annual crops (cf. Noy-Meir's and Seligman's account of Israel's achievements), but in the absence of a proven system it should remain a subsidiary form of land-use within a subsistence framework. Singh and Joshi go so far as to recommend that in India there should be no expansion of arable areas beyond those already present. The degradation resulting from cultivation and crop failure is sufficient justification for such a recommendation.

Livestock production

(1) It is of the utmost importance that stocking rate should remain within the carrying capacity of the vegetation. In Australia, the U.S.A., Southern and East Africa and Asia this point is made very strongly. It is difficult to achieve because the carrying capacity itself may vary between years by 500% or more. The temptation on the part of the manager is to stock to the level of the better years, or at least the "average" year (which may never occur), with the result that severe over-grazing occurs in the lean years. The recovery achieved by the ecosystem during the best years is less than the degradation occurring in the poor years, and the all too familiar pattern of a downward spiral becomes evident.

The problem stems from the fact that only recently have most of these areas been subjected to permanent settlement and regular use with fixed numbers of herbivores (as discussed in the first point under Determinants of these ecosystems). The next four principles are aimed at alleviating the problem.

(2) The development of regional or national fodder banks. Barnes (Africa) and Petrov (Asia) have both suggested this approach, and it may be achieved by (i) harvesting excess material (at the correct stage for necessary food value) during years of high rainfall, and (ii) the use of irrigated pastures. The fodder banks would enable a stocking rate at or above the carrying capacity of the "average" years since it would fill in the troughs during the drier years when natural fodder is in short supply.

(3) Seeding natural range with improved grasses. Herbel (U.S.A.), Medellin-Leal and Gomez-Gonzalez (Mexico) and Petrov (Asia) have recommended this technique, though the economics of it may sometimes be doubtful, particularly where management expertise is not of a very high standard.

(4) Related to the previous point is the importance of maintaining as high a proportion of perennial grasses as possible. This is stressed for Southern and East Africa and the Sahel, where there is clear evidence that inter-seasonal variation is markedly higher where the vegetation is comprised of mostly annual species. One of the first stages in retrogressive succession, resulting from over-grazing, is the death of perennial grasses and their replacement by annuals.

(5) The development of flexible ranching systems, which allow for variable herd sizes and numbers of cattle occupying an area at any one time.

This objective is satisfied by the old nomadic grazing patterns in the Sahel (although they are losing their ability to cope with increased pressures), and the modern production and marketing systems employed in the south-western U.S.A.

(6) Animals should be allowed freedom of movement within defined areas. Traditional systems of penning animals at night, and herding them close to dwelling sites, lead to a markedly uneven utilization of vegetation, and animal performance deteriorates. Cattle, in particular, may do a considerable amount of their grazing at night during hot periods.

(7) Water points should be distributed evenly at intervals of approximately 10 km.

(8) Rotational grazing, particularly where the ecosystem has been somwhat degraded, may aid optimum utilization of the vegetation and can increase secondary productivity. Research into this aspect of grazing management has been extensive, all over the world, and the results are remarkably conflicting. In Australia the overall conclusion is that deferred grazing, of any kind, has not proved to be worthwhile, and light, continuous stocking is preferred. In the south-western U.S.A. results have not been consistent. Short-duration grazing has proved to be beneficial in some areas but not in others. It is noteworthy that it did not succeed where, within a season, the monthly forage production varied between 25 and 700 kg ha^{-1} as a result of variable rainfall. It may well be that the results reported have depended on the seasonal pattern of rainfall during the course of the experiments, and the rigidity with which the rotation was enforced.

In Southern Africa Barnes reports a general lack of experimental support for short-duration grazing in semi-arid areas. In India the evidence has been both for and against but the general conclusion is that "continuous controlled" grazing is preferred to deferred grazing. Early season deferment is, however, very beneficial.

The conflicting evidence may be partly due to the rigidity which is imposed on grazing experiments by the requirements for statistical analysis. From all that has been said earlier about these ecosystems, it becomes clear that management must be flexible and opportunistic. Frequent and irregular adjustment in time and intensity of grazing are called for, and their denial within a statistically designed experiment may well result in a pattern of use which is in fact quite different to that which it was intended to test. The conflicting evidence is not entirely due to experimental design, however, since the empirical evidence in Australia and Africa is also conflicting.

As a final point, it is worth noting the conclusions of Heady (1961), that the one generalization which might be made about this problem is that rotational grazing is often beneficial in restoring deteriorated ranges, but it is seldom superior to continuous grazing in range which is in good condition.

(9) Where a number of paddocks per herd are used there is a general consensus that, wherever possible, the fences be aligned along natural vegetation

boundaries, to avoid area selective grazing.

(10) There is a general need to reduce the density of woody vegetation in semi-arid savannas. Complete clearing is probably undesirable (cf. Walker, Chap. 2, and Wilson and Graetz, Chap. 3), but thinning to an open grassland with a few woody plants undoubtedly increases fodder production.

(11) Fire is undesirable in semi-arid ecosystems. It reduces the amount of litter cover on the soil, as well as the soil organic matter, both of which are in short supply. It is, however, seldom a serious problem, owing to a lack of sufficient fuel. Under particular circumstances it may be a useful tool to control the thickening-up of scrub, or to remove old, unpalatable moribund grass ("top-hamper"). As such, it should be used only on an opportunistic basis, and never in a regular burning schedule.

(12) Herbivore composition should be compatible with vegetation structure and the animals should be selected for their performance in the environment concerned. Almost all of the authors have noted the importance of using improved strains of livestock.

Mixed herds, consisting of various combinations of cattle, goats and sheep have been shown to make better use of vegetation and to result in increased secondary production in Mexico, Australia, Asia, the Sahel and in Southern Africa. The better use of, and development of new breeds from, indigenous adapted herbivores deserves particular attention.

(13) In order to avoid unnecessary loss of secondary production, the minimum requirement for animal management is disease control and mineral supplementation. The former usually includes drenching for internal parasites and spraying or plunge dipping in acaricide solutions for control of ticks.

Other forms of land-use

Three other forms of land-use require attention within this section.

(1) *Forestry.* The shortage of firewood is either becoming or is already a serious problem in most semi-arid areas. It was particularly noted in the chapters on Israel and India, and in the latter farm forestry is stressed as an important aspect of land-use.

In Central Asia it is important both as fodder and, more particularly, as a control mechanism for wind erosion and to prevent snow from being blown away during winter.

(2) *The use of indigenous plants by Man.* This aspect has received most attention in Mexico and the recommendations in the chapter on that country, regarding extended research into the use and cultivation of indigenous species, should be heeded in other regions.

(3) *Tourism.* The characteristics of the semi-arid ecosystems which limit their usefulness to man as food-producing areas, combine to make them

highly attractive in to-day's over-crowded world as refuges for vacationers. The climate, extensiveness, comparative quiet and solitude, and undeniable harsh beauty are attracting more and more people. In the U.S.A. and even Israel, this form of land-use is already causing problems.

From the point of view of management principles we know as yet too little to be definitive about people management. There is, however, one important principle which must be adhered to: the number of people must be kept within the carrying capacity of the system. Uncontrolled movements of people in these ecosystems can have long-lasting, deleterious effects. From personal observations, the vehicle tracks from a single vehicle crossing an inter-dune grassland in the Namib desert have so far persisted for 5 years. The semi-arid areas are perhaps less dramatic in their response than are deserts, but they are nevertheless easily damaged. The planning of tourism in these areas must include ecologists — and this implies proper ecologists with a sound biological training.

General recommendations

It would not be difficult at this point to present a long list of recommendations, few if any of which would have much chance of being implemented. The list has therefore been reduced to four main areas which seem to be of greatest importance in all semi-arid regions, and all of which should be attended to as soon as possible.

(1) The greatest stumbling block to improved management is the lack of effective extension services. In all but a few areas there is already sufficient knowledge on what to do to make enormous improvements in particular land-use practices. Research is far ahead of practice and the most important need is to improve and increase the extension services. This emerges as a major factor in semi-arid areas over all parts of the globe, in both backward and sophisticated societies.

(2) The various semi-arid regions need to be planned holistically, on an integrated basis. The Australian and Russian experiences deserve note, where land-system mapping, carrying capacity assessments and management plans are called for. Israel, India and the Sahel all recommend integrated planning of crops and livestock, at various levels. Regional land capability analysis followed by integrated land-use planning is an urgent requirement.

(3) Where large-scale development is proposed in a semi-arid ecosystem (usually in the form of some International Aid Programme), there should be a built-in requirement for an environmental or ecological impact assessment, as part of the proposals.

(4) In the less-developed, or peasant-occupied semi-arid areas, probably the greatest long-term danger is the continued growth of the human population. It is obviously a problem requiring solution outside these areas as well, since the growth is both by influx and natural increase. It is undoubt-

edly the most difficult problem to solve, and comtemplation of all the factors involved gives cause for depression in even the most optimistic of scientists. Nevertheless, it is necessary to mention it, as a final note to this book, because the future of the semi-arid areas undoubtedly depends on it's timely solution.

REFERENCES

Heady, H.F., 1961. Continuous vs. specialized grazing systems. A review and application to the California annual type. J. Range Manage., 14: 182—193.
Holling, C.S., 1973. Resilience and stability of ecological systems. Annu. Rev. Ecol. Systematics, 4: 1—24.
May, R.M., 1977. Thresholds and breakpoints in ecosystems with a multiplicity of stable states. Nature (London), 269: 471—477.
Noy-Meir, I., 1975. Stability of grazing systems: an application of predator—prey graphs. J. Ecol., 63: 459—481.

AUTHOR INDEX

The number(s) directly after the initial(s) of the author is (are) the page number(s) on which the author (or his work) is mentioned in the text. Numbers in italics give the page on which the complete reference is listed.

